Mathematics for Algorithm and Systems Analysis

Edward A. Bender
S. Gill Williamson
University of California, San Diego

DOVER PUBLICATIONS, INC.
Mineola, New York

Copyright

Copyright © 2005 by Edward A. Bender and S. Gill Williamson
All rights reserved.

Bibliographical Note

Mathematics for Algorithm and Systems Analysis is a new work, first published in book form by Dover Publications, Inc., in 2005.

Library of Congress Cataloging-in-Publication Data

Bender, Edward A., 1942–
  Mathematics for algorithm and systems analysis / Edward A. Bender, S. Gill Williamson.
    p. cm.
  Includes indexes.
  ISBN-13: 978-0-486-44250-1 (pbk.)
  ISBN-10: 0-486-44250-0 (pbk.)
    1. Computer science—Mathematics—Textbooks. 2. Algorithms—Textbooks. 3. Systems analysis—Textbooks. 4. Functions—Textbooks. 5. Decision trees—Textbooks. 6. Graph theory—Textbooks. I. Williamson, S. Gill (Stanley Gill) II. Title.

QA76.9.M35B46 2005
004'.01'51—dc22

2005045552

Manufactured in the United States by Courier Corporation
44250004   2013
www.doverpublications.com

Preface

Discrete mathematics is an essential tool in almost all subareas of computer science. Interesting and challenging problems in discrete mathematics arise in programming languages, computer architecture, networking, distributed systems, database systems, AI, theoretical computer science, and other areas.

The course. The University of California, San Diego, has a lower-division two-quarter course sequence in discrete mathematics that includes Boolean arithmetic, combinatorics, elementary logic, induction, graph theory and finite probability. These courses are core undergraduate requirements for majors in Computer Science, Computer Engineering, and Mathematics-Computer Science. This text, *Mathematics for Algorithm and System Analysis*, was developed for the second quarter and *A Short Course in Discrete Mathematics* was developed for the first quarter. Because some students transfer into the second quarter of the course without having taken the first quarter, there is some overlap between the two texts and, with appropriate students, this text could be used without the first.

This book consists of four units of study (Counting and Listing—CL; Functions—Fn; Decision Trees and Recursion—DT; and Basic Concepts of Graph Theory—GT), each divided into four sections. Each section contains a representative selection of problems. These vary from basic to more difficult, including proofs for study by mathematics students or honors students. The first three sections in units CL and Fn are primarily a review of material in *A Short Course in Discrete Mathematics* needed for this course.

The review questions. "Multiple Choice Questions for Review" appear at the end of each unit. The explanatory material in this book is directed towards giving students the mathematical language and sophistication to recognize and articulate the ideas behind these questions and to answer questions that are similar in concept and difficulty. Many variations of these questions have been successfully worked on exams by most beginning students using this book at UCSD.

Students who master the ideas and mathematical language needed to understand these review questions gain the ability to formulate, in the neutral language of mathematics, problems that arise in various applications of computer science. This skill greatly facilitates their ability to discuss problems in discrete mathematics with other computer scientists and with mathematicians.

Table of Contents

> Asterisks (stars) are used in the text to mark more difficult material that is not needed in later sections.

Unit CL: Basic Counting and Listing

Section 1: Lists with Repetitions 1
set, list, multiset, sequence, word, permutation, k-set, k-list, k-multiset, k-lists with repetition, rule of product, Cartesian product, lexicographic order (lex order), dictionary order, rule of sum, composition of a positive integer

Section 2: Lists Without Repetition 9
k-lists without repetition, Stirling's formula for approximating $n!$, circular arrangements, words from a collection of letters

Section 3: Sets 13
set intersection, set union, set difference, set complement, symmetric difference, set product (Cartesian product), binomial coefficients, generating functions, binomial theorem, full house (card hand), two pairs (card hand), rearranging words, multinomial coefficients, card hands and multinomial coefficients, recursions, set partitions, Stirling numbers of the second kind ($S(n, k)$), straight (card hand), Bell numbers B_n

Section 4: Probability and Basic Counting 28
sample space, selections done uniformly at random, event, probability function, combining events, Venn diagrams, odds, hypergeometric probabilities, fair dice, geometric probability, principle of inclusion exclusion, birthday problem

Multiple Choice Questions for Review 41

Unit Fn: Functions

Section 1: Some Basic Terminology 45
direct product, intersection, union, symmetric difference, domain, range, codomain, one-line notation, surjection, onto, injection, one-to-one, bijection, permutation, relation, functional relation, two-line notation

Section 2: Permutations 51
composition, cycle, cycle form of permutation, involution, permutation matrices, derangements

Section 3: Other Combinatorial Aspects of Functions............. 58
image, inverse image, coimage, image size and Stirling numbers, strictly increasing, strictly decreasing, weakly increasing, weakly decreasing, monotone, multisets, lists without repetition, restricted growth functions and partitions

Section 4: Functions and Probability............................. 65
random variable, probability function, event, probability distribution function, expectation, covariance, variance, standard deviation, correlation, independent events, independent random variables, product spaces, generating random permutations, joint distribution function, marginal distributions, binomial distribution, Poisson distribution, normal distribution, standard normal distribution, cumulative distribution, central limit theorem, normal approximation to binomial, Poisson approximation to binomial, Tchebycheff's inequality

Multiple Choice Questions for Review............................. 85

Unit DT: Decision Trees and Recursion

Section 1: Basic Concepts of Decision Trees....................... 89
decision trees, vertices, root, edges, degree of vertex, down degree, child, parent, leaves, internal vertex, height of leaf, path to vertex, traversals of decision tree, depth first vertices, depth first edges, breadth first, preorder, postorder, length-first lex order, dictionary order, permutations in lex order, partial permutation, rank of leaf, direct insertion order for permutations, backtracking, Latin squares, domino coverings, strictly decreasing functions, unlabeled balls into boxes, isomorph rejection

Section 2: Recursive Algorithms 103
recursive algorithm, simplest case reduction, recursive algorithm for 0-1 sequences, sorting by recursive merging, recursive approach, recursive solutions, local description for permutations in lex order, recursive description of Towers of Hanoi, decision tree for Towers of Hanoi, recursion and stacks, configuration analysis of Towers of Hanoi, abandoned leaves and RANK, characteristic functions and subsets, Gray code for subsets, decision tree for Gray code for subsets, local description of Gray code, Towers of Hanoi with four poles

Section 3: Decision Trees and Conditional Probability............ 115
conditional probability, independent events, Venn diagrams, probabilities of leaves, probabilities of edges, probabilistic decision trees, decision trees and Bayesian methods, Bayes' theorem, multiplication theorem for conditional probabilities, sequential sampling, the SAT problem, first moment method, tournaments, gambler's ruin problem

Section 4: Inductive Proofs and Recursive Equations 128
induction, recursive equations, induction hypothesis, inductive step, base case, prime factorization, sum of first n integers, local description, recurrence relation,

binomial coefficients $C(n,k)$, Stirling numbers $S(n,k)$, guessing solutions to recurrences, linear two term recurrence, constant coefficients, characteristic equation, two real roots, one real root, complex roots, recursion for derangements, Fibonacci recurrence relation, recurrence relation for derangements

Multiple Choice Questions for Review 140

Unit GT: Basic Concepts in Graph Theory

Section 1: What is a Graph? 145
computer network example, simple graph, graph, vertices, edges, set theoretic description of graph, pictorial description of a graph, incidence function, vertices joined by an edge, adjacent vertices, edge incident on a vertex, simple graphs are graphs, form of a graph, equivalence relations, equivalence classes, blocks, binary relations, reflexive, symmetric, transitive, equivalent forms, isomorphism of graphs, graph isomorphism as an equivalence relation, degree of a vertex, loops, parallel edges, isolated vertices, degree sequences and isomorphism, random graphs

Section 2: Digraphs, Paths, and Subgraphs 157
flow of commodities, directed graph, digraph, simple digraph, simple graphs as simple digraphs, directed loops, digraphs and binary relations, symmetric binary relations and simple graphs with loops, complete simple graphs, path, trail, walk, vertex sequence, walk implies path, restrictions of incidence functions, subgraphs, subgraph induced by edges, subgraph induced by vertices, cycles, connected graphs, connected components and equivalence classes, connectivity in digraphs, Eulerian trail, Eulerian circuit, Hamiltonian cycle, Hamiltonian graph, bicomponents of graphs, bipartite graphs, oriented simple graphs, antisymmetry, order relations, Hasse diagrams, covering relations, counting trees

Section 3: Trees 168
tree, alternative definitions of a tree, rooted graph, rooted tree, parent, child, sibling, leaf, internal vertices, unique paths in trees, rooted plane tree, RP-tree, traversing RP-trees, depth first sequences, breadth first sequences, spanning trees, minimum spanning trees, greedy algorithms, Prim's algorithm, Kruskal's algorithm, lineal or depth-first spanning trees, algorithm for depth-first spanning trees, bipartite graphs and depth first spanning trees, degree sequences of trees, binary trees, full binary trees, height and leaf restrictions in binary trees

Section 4: Rates of Growth and Analysis of Algorithms 181
comparing algorithms, machine independence, example of finding the maximum, Θ notation, O notation, properties of Θ and O, Θ as an equivalence relation, sufficiently large, eventually positive, asymptotic, "little oh" notation, using Θ to compare polynomial evaluation algorithms, average running time, tractable, intractable, graph coloring problem, traveling salesman problem, clique problem, NP-complete problems, NP-hard, NP-easy, chromatic number of a graph, almost good algorithms, almost correct algorithms, close algorithms, polynomial time, exponential time, Θ and series, Θ and logs

[vii]

Multiple Choice Questions for Review 195

Solutions to Exercises 199

Notation Index 239

Subject Index 241

Unit CL

Basic Counting and Listing

Section 1: Lists with Repetitions

We begin with some matters of terminology and notation. Two words that we shall often use are *set* and *list*. (Lists are also called *strings*.) Both words refer to collections of objects. There is no standard notation for lists. Some of those in use are

apple banana pear peach

apple, banana, pear, peach

and   (apple, banana, pear, peach).

The notation for sets is standard: the items are separated by commas and surrounded by curly brackets as in

{apple, banana, pear, peach}.

The curly bracket notation for sets is so well established that you can normally assume it means a set — but beware, some mathematical software systems use { } (curly brackets) for lists.

What is the difference between a set and a list? "Set" means a collection of distinct objects in which the order doesn't matter. Thus

{apple, peach, pear}   and   {peach, apple, pear}

are the same sets, and the set {apple, peach, apple} is the same as the set {apple, peach}. In other words, repeated elements are treated as if they occurred only once. Thus two sets are the same if and only if each element that is in one set is in both. In a list, order is important and repeated objects are usually allowed. Thus

(apple, peach)   (peach, apple)   and   (apple, peach, apple)

are three different lists. Two lists are the same if and only if they have exactly the same items in exactly the same positions. Thus, "sets" and "lists" represent different concepts: A list is *always ordered* and a set has *no repeated elements*.

Example 1 (Using the terminology) People, in their everyday lives, deal with the issues of "order is important" and "order is not important." Imagine that Tim, Jane, and Linda are going to go shopping for groceries. Tim makes a note to remind himself to get apples and bananas. Tim's note might be written out in an orderly manner, or might just be words randomly placed on a sheet of paper. In any case, the purpose of the note is to remind him to buy some apples and bananas and, we assume, the order in which these items are noted is not important. The number of apples and bananas is not specified in the note. That will be determined at the store after inspecting the quality of the apples and bananas. The best model for this note is a set. Tim might have written

Basic Counting and Listing

{apples, bananas}. We have added the braces to emphasize that we are talking about sets. Suppose Jane wrote {bananas, apples} and Linda wrote {apples, bananas, apples}. Linda was a bit forgetful and wrote apples twice. It doesn't matter. All three sets are the same and all call for the purchase of some apples and some bananas. If Linda's friend Mary had made the note {peaches, bananas, oranges} and Linda and Mary had decided to combine their notes and go shopping together, they would have gone to the store to get {apples, peaches, bananas, oranges}.

There are times when order is important for notes regarding shopping trips or daily activities. For example, suppose Tim makes out the list (dentist, bookstore, groceries). It may be that he regards it as important to do these chores in the order specified. The dentist appointment may be at eight in the morning. The bookstore may not be open until nine in the morning. He may be planning to purchase milk at the grocery store and does not want the milk to be sitting in the car while he goes to the bookstore. In a list where order matters, the list (dentist, bookstore, groceries, dentist) would be different than (dentist, bookstore, groceries). The first list directs Tim to return to the dentist after the groceries, perhaps for a quick check that the cement on his dental work is curing properly.

In addition to the sets and lists described above, there is another concept that occurs in both everyday life and in mathematics. Suppose Tim, Jane, and Linda happen to go the grocery store and are all standing in line at the checkout counter with bags in hand containing their purchases. They compare purchases. Tim says "I purchased 3 bananas and 2 apples." Jane says, "I purchased 2 bananas and 3 apples." Linda says, "I purchased 3 apples and 2 bananas." Jane and Linda now say in unison "Our purchases are the same!" Notice that repetition (how many bananas and apples) now matters, but as with sets, order doesn't matter (Jane and Linda announced their purchases in different order but concluded their purchases were the same). We might use the following notation: Tim purchased {2 apples, 3 bananas}, Jane purchased {3 apples, 2 bananas}, Linda purchased {2 bananas, 3 apples}. Another alternative is to write {apple, apple, banana, banana, banana} for Tim's purchase. All that matters is the number of apples and bananas, so we could have written

{apple, banana, apple, banana, banana}

for Tim's purchase. Such collections, where order doesn't matter, but repetition does matter are called *multisets* in mathematics. Notice that if Tim and Jane dumped their purchases into the same bag they would have the combined purchase {5 apples, 5 bananas}. Combining multisets requires that we keep track of repetitions of objects. In this chapter, we deal with sets and lists. We will have some brief encounters with multisets later in our studies. ☐

To summarize the concepts in the previous example:

List: an ordered collection. Whenever we refer to a list, we will indicate whether the elements must be distinct.[1]

Set: a collection of distinct objects where order does not matter.

---

[1] A list is sometimes called a *string*, a *sequence* or a *word*. Lists are sometimes called vectors and the elements components.

Section 1: Lists with Repetitions

Multiset: a collection of objects (repeats allowed) where order does not matter.[2]

The terminology "k-list" is frequently used in place of the more cumbersome "k-long list." Similarly, we use k-set and k-multiset. Vertical bars (also used for absolute value) are used to denote the number of elements in a set or in a list. We call $|A|$ "the number of elements in A" or, alternatively, "the cardinality of A." For example, if A is an n-set, then $|A| = n$.

We want to know how many ways we can do various things with a set. Here are some examples, which we illustrate by using the set $S = \{x, y, z\}$.

1. How many ways can we *list*, without repetition, all the elements of S? This means, how many ways can we arrange the elements of S in a list so that each element of S appears exactly once in each of the lists. For $S = \{x, y, z\}$, there are six ways: xyz, xzy, yxz, yzx, zxy and zyx. Notice that we have written the list (x, y, z) simply as xyz since there is no possibility of confusion. (These six lists are all called permutations of S. People often use Greek letters like π and σ to indicate a permutation of a set.)

2. How many ways can we construct a k-list of distinct elements from a set? When $k = |S|$, this is the previous question. If $k = 2$ and $S = \{x, y, z\}$, there are six ways: xy, xz, yx, yz, zx and zy.

3. If the list in the previous question is allowed to contain repetitions, what is the answer? There are nine ways for $S = \{x, y, z\}$: xx, xy, xz, yx, yy, yz, zx, zy and zz.

4. If, in Questions 2 and 3, the order in which the elements appear doesn't matter, what are the answers? For $S = \{x, y, z\}$ and $k = 2$, the answers are three and six, respectively. We are forming 2-sets and 2-multisets from the elements of S. The 2-sets are $\{x, y\}$, $\{x, z\}$ and $\{y, z\}$. The 2-multisets are the three 2-sets plus $\{x, x\}$, $\{y, y\}$ and $\{z, z\}$.

5. How many ways can the set S be partitioned into a collection of k pairwise disjoint nonempty smaller sets?[3] With $k = 2$, the set $S = \{x, y, z\}$ has three such: $\{\{x\}, \{y, z\}\}$, $\{\{x, y\}, \{z\}\}$ and $\{\{x, z\}, \{y\}\}$.

We will learn how to answer these questions without going through the time-consuming process of listing all the items in question as we did for our illustration.

How many ways can we construct a k-list (repeats allowed) using an n-set? Look at our illustration in Question 3 above. The first entry in the list could be x, y or z. After any of these there were three choices (x, y or z) for the second entry. Thus there are $3 \times 3 = 9$ ways to construct such a list. The general pattern should be clear: There are n ways to choose each list entry. Thus

Theorem 1 (k-lists with repetitions) *There are n^k ways to construct a k-list from an n-set.*

This calculation illustrates an important principle:

Theorem 2 (Rule of Product) *Suppose structures are to be constructed by making a sequence of k choices such that, (1) the ith choice can be made in c_i ways, a number*

---

[2] *Sample* and *selection* are often used in probability and statistics, where it may mean a list or a multiset, depending on whether or not it is ordered.

[3] In other words, each element of S appears in *exactly one* of the smaller sets.

Basic Counting and Listing

independent of what choices were made previously, and (2) each structure arises in exactly one way in this process. Then, the number of structures is $c_1 \times \cdots \times c_k$.

"Structures" as used above can be thought of simply as elements of a set. We prefer the term structures because it emphasizes that the elements are built up in some way; in this case, by making a sequence of choices. In the previous calculation, the structures are k-lists, which are built up by adding one element at a time. Each element is chosen from a given n-set and $c_1 = c_2 = \ldots = c_k = n$.

Definition 1 (Cartesian Product) *If $C_1, \ldots, C_k$ are sets, the Cartesian product of the sets is written $C_1 \times \cdots \times C_k$ and consists of all k-lists $(x_1, \ldots, x_k)$ with $x_i \in C_i$ for $1 \le i \le k$.*

For example, $\{1,2\} \times \{x\} \times \{a,b,c\}$ is a set containing the six lists $1xa$, $1xb$, $1xc$, $2xa$, $2xb$ and $2xc$.

A special case of the Rule of Product is the fact that the number of elements in $C_1 \times \cdots \times C_k$ is the product $|C_1| \cdots |C_k|$. Here C_i is the collection of ith choices and $c_i = |C_i|$. This is only a special case because the Rule of Product would allow the *collection* C_i to depend on the previous choices $x_1, \ldots, x_{i-1}$ as long as the *number* c_i of possible choices does not depend on $x_1, \ldots, x_{i-1}$.

Here is a property associated with Cartesian products that we will find useful in our later discussions.

Definition 2 (Lexicographic order) *If $C_1, \ldots, C_k$ are lists of distinct elements, we may think of them as sets and form the Cartesian product $P = C_1 \times \cdots \times C_k$. The lexicographic order on P is defined by saying that $(a_1, \ldots, a_k) <_L (b_1, \ldots, b_k)$ if and only if there is some $t \le k$ such that $a_i = b_i$ for $i < t$ and $a_t < b_t$. Usually we write $(a_1, \ldots, a_k) < (b_1, \ldots, b_k)$ instead of $(a_1, \ldots, a_k) <_L (b_1, \ldots, b_k)$, because it is clear from the context that we are talking about lexicographic order.*

Often we say *lex order* instead of lexicographic order. If all the C_i's equal

$$(0,1,2,3,4,5,6,7,8,9)$$

then lex order is simply numerical order of k digit integers with leading zeroes allowed. Suppose that all the C_i's equal (<space>, A, B, ..., Z). If we throw out those elements of P that have a letter following a space, the result is dictionary order. For example, BAT, BATTERS and BATTLE are in lex order. Why? All agree in the first three positions. The fourth position of BAT is <space>, which precedes all letters in our order. Similarly, BATTERS comes before BATTLE because they first differ in the fifth position and E < L. Unlike these two simple examples, the C_i's usually vary with i.

Section 1: Lists with Repetitions

Example 2 (A simple count) The North-South streets in Rectangle City are named using the numbers 1 through 12 and the East-West streets are named using the letters A through H. The most southwesterly intersection occurs where 1 and A streets meet. How many blocks are within the city?

Each block can be labeled by the streets at its southwesterly corner. These labels have the form (x, y) where x is between 1 and 11 inclusive and y is between A and G. (If you don't see why 12 and H are missing, draw a picture and look at southwesterly corners.) By the Rule of Product there are $11 \times 7 = 77$ blocks. In this case the structures can be taken to be the descriptions of the blocks. Each description has two parts: the names of the north-south and East-West streets at the block's southwest corner. □

Example 3 (Counting galactic names) In a certain land on a planet in a galaxy far away the alphabet contains only 5 letters which we will transliterate as A, I, L, S and T in that order. All names are 6 letters long, begin and end with consonants and contain two vowels which are not adjacent to each other. Adjacent consonants must be different. The list begins with LALALS, LALALT, LALASL, LALAST, LALATL, LALATS, LALILS and ends with TSITAT, TSITIL, TSITIS, TSITIT. How many possible names are there?

The possible positions for the two vowels are $(2,4)$, $(2,5)$ and $(3,5)$. Each of these results in two isolated consonants and two adjacent consonants. Thus the answer is the product of the following factors:

1. choose the vowel locations (3 ways);
2. choose the vowels ($2 \times 2 = 4$ ways);
3. choose the isolated consonants ($3 \times 3 = 9$ ways);
4. choose the adjacent consonants ($3 \times 2 = 6$ ways).

The answer is $3 \times 4 \times 9 \times 6 = 648$. This construction can be interpreted as a Cartesian product as follows. C_1 is the set of lists of possible positions for the vowels, C_2 is the set of lists of vowels in those positions, and C_3 and C_4 are sets of lists of consonants. Thus

$$C_1 = \{(2,4), (2,5), (3,5)\} \quad C_2 = \{AA, AI, IA, II\}$$

$$C_3 = \{LL, LS, LT, SL, SS, ST, TL, TS, TT\} \quad C_4 = \{LS, LT, SL, ST, TL, TS\}.$$

For example, $((2,5), IA, SS, ST)$ in the Cartesian product corresponds to the word SISTAS. □

Here's another important principle, the proof of which is self evident:

Theorem 3 (Rule of Sum) *Suppose a set T of structures can be partitioned into sets $T_1, \ldots, T_j$ so that each structure in T appears in exactly one T_i, then*

$$|T| = |T_1| + \cdots + |T_j|.$$

Basic Counting and Listing

Example 4 (Counting galactic names again) We redo the previous example using the Rule of Sums. The possible vowel (V) and consonant (C) patterns for names are CCVCVC, CVCCVC and CVCVCC. Since these patterns are disjoint and cover all cases, we may compute the number of names of each type and add the results together. For the first pattern we have a product of six factors, one for each choice of a letter: $3 \times 2 \times 2 \times 3 \times 2 \times 3 = 216$. The other two patterns also give 216, for a total of $216 + 216 + 216 = 648$ names.

This approach has a wider range of applicability than the method we used in the previous example. We were only able to avoid the Rule of Sum in the first method because each pattern contained the same number of vowels, isolated consonants and adjacent consonants. Here's an example that requires the Rule of Sum. Suppose a name consists of only four letters, namely two vowels and two consonants, constructed so that the vowels are not adjacent and, if the consonants are adjacent, then they are different. There are three patterns: CVCV, VCVC, VCCV. By the Rule of Product, the first two are each associated with 36 names, but VCCV is associated with only 24 names because of the adjacent consonants. Hence, we cannot choose a pattern and then proceed to choose vowels and consonants. On the other hand, we can apply the Rule of Sum to get a total of 96 names. □

Example 5 (Smorgasbord College committees) Smorgasbord College has four departments which have 6, 35, 12 and 7 faculty members. The president wishes to form a faculty judicial committee to hear cases of student misbehavior. To avoid the possibility of ties, the committee will have three members. To avoid favoritism the committee members will be from different departments and the committee will change daily. If the committee only sits during the normal academic year (165 days), how many years can pass before a committee must be repeated?

If T is the set of all possible committees, the answer is $|T|/165$. Let T_i be the set of committees with no members from the ith department. By the Rule of Sum $|T| = |T_1| + |T_2| + |T_3| + |T_4|$. By the Rule of Product

$|T_1| = 35 \times 12 \times 7 = 2940$ $|T_3| = 35 \times 6 \times 7 = 1470$

$|T_2| = 6 \times 12 \times 7 = 504$ $|T_4| = 35 \times 12 \times 6 = 2520$.

Thus the number of years is $7434/165 = 45^+$. Due to faculty turnover, a committee need never repeat — if the president's policy lasts that long. □

Whenever we encounter a new technique, there are two questions that arise:

- *When* is it used?   • *How* is it used?

For the Rules of Sum and Product, the answers are intertwined:

> Suppose you wish to count the number of structures in a set and that you can describe how to construct the structures in terms of subconstructions that are connected by "ands" and "ors." *If this leads to the construction of each structure in a unique way*, then the Rules of Sum and Product apply. To use them, replace "ands" by products and "ors" by sums. Whenever you write something like "Do A AND do B," it should mean "Do A AND then do B" because the Rule of Product requires that the choices be made sequentially. Remember that the number of ways to do B *must not depend on* the choice for A.

Section 1: Lists with Repetitions

Example 6 (Applying the sum–product rules) To see how this technique is applied, let's look back at Example 5. A committee consists of either

1. One person from Dept. 1 AND one person from Dept. 2 AND one person from Dept. 3, OR

2. One person from Dept. 1 AND one person from Dept. 2 AND one person from Dept. 4, OR

3. One person from Dept. 1 AND one person from Dept. 3 AND one person from Dept. 4, OR

4. One person from Dept. 2 AND one person from Dept. 3 AND one person from Dept. 4.

The number of ways to choose a person from a department equals the number of people in the department. ◻

Until you become comfortable using the Rules of Sum and Product, look for "and" and "or" in what you do. This is an example of the useful tactic:

> **Step 1**: Break the problem into parts.
> **Step 2**: Work on each piece separately.
> **Step 3**: Put the pieces together.

Here Step 1 is getting a phrasing with "ands" and "ors;" Step 2 is calculating each of the individual pieces; and Step 3 is applying the Rules of Sum and Product.

Exercises for Section 1

The following exercises will give you additional practice on lists with repetition and the Rules of Sum and Product.

In each exercise, indicate how you are using the Rules of Sum and Product.

1.1. Suppose a bookshelf contains five discrete math texts, two data structures texts, six calculus texts, and three Java texts. (All the texts are different.)

   (a) How many ways can you choose one of the texts?

   (b) How many ways can you choose one of each type of text?

1.2. How many different three digit positive integers are there? (No leading zeroes are allowed.) How many positive integers with at most three digits? What are the answers when "three" is replaced by "n?"

1.3. Prove that the number of subsets of a set S, including the empty set and S itself, is $2^{|S|}$.

1.4. Suppose $n > 1$. An n-digit number is a list of n digits where the first digit in the list is not zero.

Basic Counting and Listing

   (a) How many n-digit numbers are there?

   (b) How many n-digit numbers contain no zeroes?

   (c) How many n-digit numbers contain at least one zero?
   Hint: Use (a) and (b).

1.5. For this exercise, we work with the ordinary alphabet of 26-letters.

   (a) Define a "4-letter word" to be any list of 4 letters that contains *at least* one of the vowels A, E, I, O and U. How many 4-letter words are there?

   (b) Suppose, instead, we define a "4-letter word" to be any list of 4 letters that contains *exactly* one of the vowels A, E, I, O and U. How many 4-letter words are there?

1.6. In a certain land on a planet in a galaxy far away the alphabet contains only 5 letters which we will transliterate as A, I, L, S and T in that order. All names are 5 letters long, begin and end with consonants and contain two vowels which are not adjacent to each other. Adjacent consonants must be different. How many names are there?

1.7. A *composition* of a positive integer n is a list of positive integers (called *parts*) that sum to n. The four compositions of 3 are 3; 2,1; 1,2 and 1,1,1.

   (a) By considering ways to insert plus signs and commas in a list of n ones, obtain the formula 2^{n-1} for the number of compositions of n. To avoid confusion with the Rule of Sum, we'll write this plus sign as $\oplus$. (The four compositions 3; 2,1; 1,2 and 1,1,1 correspond to $1 \oplus 1 \oplus 1$; $1 \oplus 1,1$; $1,1 \oplus 1$ and 1,1,1, respectively.)

   (b) List all compositions of 4.

   (c) List all compositions of 5 with 3 parts.

1.8. In Example 3 we found that there were 648 possible names. Suppose that these are listed in the usual dictionary order. The last word in the first third of the dictionary is LTITIT (the 216th word). The first word in the middle third is SALALS. Explain.

1.9. There is another possible lexicographic order on the names in Example 3 (Counting galactic names) that gives rise to a "nonstandard" lex order on this list of names. Using the interpretation of the list of names as the Cartesian product of the lists $C_1 \times C_2 \times C_3 \times C_4$, we can lexicographically order the entire list of names based on the following linear orderings of the C_i, $i = 1, 2, 3, 4$:

$$C_1 = ((2,4), (2,5), (3,5)) \quad C_2 = (\text{AA}, \text{AI}, \text{IA}, \text{II})$$

$$C_3 = (\text{LL}, \text{LS}, \text{LT}, \text{SL}, \text{SS}, \text{ST}, \text{TL}, \text{TS}, \text{TT}) \quad C_4 = (\text{LS}, \text{LT}, \text{SL}, \text{ST}, \text{TL}, \text{TS}).$$

What are the first seven and last seven entries in this lex ordering?
Hint: The lex ordering can be done entirely in terms of the sets C_i and then translated to the names as needed. Thus the first two entries in the list $C_1 \times$

$C_2 \times C_3 \times C_4$ in lex order are (2,4)(AA)(LL)(LS) and (2,4)(AA)(LL)(LT). The last two are (3,5)(II)(TT)(TL) and (3,5)(II)(TT)(TS). These translate to LALALS and LALALT for the first two and TLITIT and TSITIT for the last two.

1.10. Recall that the size of a multiset is the number of elements it contains. For example, the size of $\{a, a, b\}$ is three.

(a) How many 4-element multisets are there whose elements are taken from the set $\{a, b, c\}$? (An element may be taken more than once; for example, the multiset $\{c, c, c, c\}$.)

(b) How many multisets are there whose elements are taken from the set $\{a, b, c\}$?

Section 2: Lists Without Repetition

What happens if we do not allow repeats in our list? Suppose we have n elements to choose from and wish to form a k-list with no repeats. How many lists are there? We can choose the first entry in the list AND choose the second entry AND $\cdots$ AND choose the kth entry. There are $n - (i - 1) = n - i + 1$ ways to choose the ith entry since $i - 1$ elements have been removed from the set to make the first part of the list. By the Rule of Product, the number of lists is $n(n-1)\cdots(n-k+1)$. Using the notation $n!$ for the product of the first n integers and writing $0! = 1$, you should be able to see that this answer can be written as $n!/(n-k)!$, which is often designated by $(n)_k$ and called the *falling factorial*. Some authors write the falling factorial as $n^{\underline{k}}$. We have proven

Theorem 4 (k-lists without repetition) *When repeats are not allowed, there are $n!/(n-k)! = (n)_k$ k-lists that can be constructed from an n-set. (When $k > n$ the answer is zero.)*

When $k = n$, a list without repeats is simply a *linear ordering* of the set. We frequently say "ordering" instead of "linear ordering." An ordering is sometimes called a "permutation" of S. Thus, we have proven that a set S can be (linearly) ordered in $|S|!$ ways.

Example 7 (Lists without repeats) How many lists without repeats can be formed from a 5-set? There are $5! = 120$ 5-lists without repeats, $5!/1! = 120$ 4-lists without repeats, $5!/2! = 60$ 3-lists, $5!/3! = 20$ 2-lists and $5!/4! = 5$ 1-lists. By the Rule of Sum, this gives a total of 325 lists, or 326 if we count the empty list. ☐

Basic Counting and Listing

Example 8 (Linear arrangements) How many different ways can 100 people be arranged in the seats in a classroom that has exactly 100 seats? Each seating is simply an ordering of the people. Thus the answer is 100!. Simply writing 100! probably gives you little idea of the size of the number of seatings. A useful approximation for factorials is given by *Stirling's formula*.

Theorem 5 (Stirling's formula) $\sqrt{2\pi n}\,(n/e)^n$ approximates $n!$ *with a relative error less than* $1/10n$.

We say that $f(x)$ approximates $g(x)$ with a *relative error* of $|f(x)/g(x) - 1|$. Thus, the theorem states that $\sqrt{2\pi n}\,(n/e)^n/n!$ differs from 1 by less than $1/10n$. When relative error is multiplied by 100, we obtain "percentage error." By Stirling's formula, we find that 100! is nearly 9.32×10^{157}, which is much larger than estimates of the number of atoms in the universe. □

We can extend the ideas of the previous example. Suppose we still have 100 seats but have only 95 people. We need to think a bit more carefully than before. One approach is to put the people in some order, select a list of 95 seats, and then pair up people and seats so that the first person gets the first seat, the second person the second seat, and so on. By the general formula for lists without repetition, the answer is $100!/(100-95)! = 100!/120$. We can also solve this problem by thinking of the people as positions in a list and the seats as entries! Thus we want to form a 95-list using the 100 seats. According to Theorem 4, this can be done in $100!/(100-95)!$ ways.

Lists can appear in many guises. As seen in the previous paragraph, the people could be thought of as the positions in a list and the seats the things in the list. Sometimes it helps to find a reinterpretation like this for a problem. At other times it is easier to tackle the problem starting over again from scratch. These methods can lead to several approaches to a problem. That can make the difference between a solution and no solution or between a simple solution and a complicated one. You should practice using both methods, even on the same problem.

Example 9 (Circular arrangements) How many ways can n people be seated on a Ferris wheel with exactly one person in each seat? Equivalently, we can think of this as seating the people at a circular table with n chairs. Two seatings are defined to be "the same" if one can be obtained from the other by rotating the Ferris wheel (or rotating the seats around the table).

If the people were seated in a straight line instead of in a circle, the answer would be $n!$. Can we convert the circular seating into a linear seating (i.e., a list)? In other words, *can we convert the unsolved problem to a solved one*? Certainly — simply cut the circular arrangement between two people and unroll it. Thus, to arrange n people in a linear ordering,

>first arrange them in a circle   AND   then cut the circle.

According to our AND/OR technique, we must prove that each linear arrangement arises in *exactly one way* with this process.

Section 2: Lists Without Repetition

- Since a linear seating can be rolled up into a circular seating, it can also be obtained by unrolling that circular seating. Hence each linear seating arises *at least once*.

- Since the people at the circular table are all different, the place we cut the circle determines who the first person in the linear seating is, so each cutting of a circular seating gives a different linear seating. Obviously two different circular seatings cannot give the same linear seating. Hence each linear seating arises *at most once*.

Putting these two observations together, we see that each linear seating arises *exactly once*. By the Rule of Product,

$$n! = (\text{number of circular arrangements}) \times n.$$

Hence the number of circular arrangements is $n!/n = (n-1)!$.

Our argument was somewhat indirect. We can derive the result by a more direct argument. For convenience, let the people be called 1 through n. We can read off the people in the circular list starting with person 1. This gives a linear ordering of $\{1, \ldots, n\}$ that starts with 1. Conversely, each such linear ordering gives rise to a circular ordering. Thus the number of circular orderings equals the number of such linear orderings. Having listed person 1, there are $(n-1)!$ ways to list the remaining $n-1$ people.

If we are making circular necklaces using n distinct beads, then the arguments we have just given prove that there are $(n-1)!$ possible necklaces provided we are not allowed to flip necklaces over.

What happens if the beads are not distinct? For example, suppose there are three blue beads and three yellow beads. There are just two linear arrangements associated with the circular arrangement BYBYBY, namely (B,Y,B,Y,B,Y) and (Y,B,Y,B,Y,B). But there are six linear arrangements associated with the circular arrangement BBBYYY. Thus, the approach we used for distinct beads fails, because the number of lists associated with a necklace depends on the necklace. For now, you only need to be aware of this complication. □

We need not insist on "no repetitions at all" in lists. There are natural situations in which some repetitions are allowed and others are not allowed. The following example illustrates one such way that this can happen.

Example 10 (Words from a collection of letters — first try) How many "words" of length k can be formed from the letters in ERROR when no letter may be used more often than it appears in ERROR? (A "word" is any list of letters, pronounceable or not.) You can imagine that you have 5 tiles, namely one E, one O, and three R's. The answer is not 3^k even though we are using 3 different letters. Why is this? Unlimited repetition is not allowed so, for example, we cannot have EEE. On the other hand, the answer is not $(3)_k$ since R can be repeated some. Also, the answer is not $(5)_k$ even though we have 5 tiles. Why is this? The formula $(5)_k$ arises if we have 5 *distinct* objects; however, our 3 tiles with R are identical. At present, all we can do is carefully list the possibilities. Here

Basic Counting and Listing

they are in alphabetical order.

$k = 1$:   E, O, R

$k = 2$:   EO, ER, OE, OR, RE, RO, RR

$k = 3$:   EOR, ERO, ERR, OER, ORE, ORR, REO, RER, ROE, ROR, RRE, RRO, RRR

$k = 4$:   EORR, EROR, ERRO, ERRR, OERR, ORER, ORRE, ORRR, REOR, RERO,

RERR, ROER, RORE, RORR, RREO, RRER, RROE, RROR, RRRE, RRRO

$k = 5$:   EORRR, ERORR, ERROR, ERRRO, OERRR, ORERR, ORRER, ORRRE,

REORR, REROR, RERRO, ROERR, RORER, RORRE, RREOR, RRERO,

RROER, RRORE, RRREO, RRROE

This is obviously a tedious process. We shall return to this type of problem in the next section. □

Exercises for Section 2

The following exercises will give you additional practice with lists with restricted repetitions.

In each exercise, indicate how you are using the Rules of Sum and Product.

It is instructive to first do these exercises using only the techniques introduced so far and then, after reading the next section, to return to these exercises and look for other ways of doing them.

2.1. We want to know how many ways 3 boys and 4 girls can sit in a row.

(a) How many ways can this be done if there are no restrictions?

(b) How many ways can this be done if the boys sit together and the girls sit together?

(c) How many ways can this be done if the boys and girls must alternate?

2.2. Repeat the previous exercise when there are 3 boys and 3 girls.

2.3. What are the answers to the previous two exercises if the table is circular?

2.4. How many ways are there to form a list of two distinct letters from the set of letters in the word COMBINATORICS? three distinct letters? four distinct letters?

2.5. How many ways are there to form a list of two letters from the set of letters in the word COMBINATORICS if the letters cannot be used more often than they appear in COMBINATORICS? three letters?

2.6. We are interested in forming 3 letter words ("3-words") using the letters in LITTLEST. For the purposes of the problem, a "word" is any list of letters.

(a) How many words can be made with no repeated letters?

(b) How many words can be made with unlimited repetition allowed?

(c) How many words can be made if repeats are allowed but no letter can be used more often than it appears in LITTLEST?

2.7. By 2050 spelling has deteriorated considerably. The dictionary defines the spelling of "relief" to be any combination (with repetition allowed) of the letters R, L, F, I and E subject to certain constraints:

- The number of letters must not exceed 6.
- The word must contain at least one L.
- The word must begin with an R and end with an F.
- There is just one R and one F.

(a) How many spellings are possible?

(b) The most popular spelling is the one that, in dictionary order, is five before the spelling RELIEF. What is it?

*2.8. By the year 2075, further deterioration in spelling has occurred. The dictionary now defines the spelling of "relief" to be any combination (with repetition allowed) of the letters R, L, F, I and E subject to these constraints:

- The number of letters must not exceed 6.
- The word must contain at least one L.
- The word must begin with a nonempty string of R's and end with a nonempty string of F's, and there are no other R's and F's.

(a) How many spellings are possible?

(b) The most popular spelling is the one that, in dictionary order, is five before the spelling RELIEF. What is it?

*2.9. Prove that the number of lists without repeats that can be constructed from an n-set is very nearly $n!e$. Your count should include lists of *all* lengths from 0 to n. *Hint*: Recall that from Taylor's Theorem in calculus $e^x = 1+x+x^2/2!+x^3/3!+\cdots$.

Section 3: Sets

We first review some standard terminology and notation associated with sets. When we discuss sets, we usually have a "universal set" U in mind, and the sets we discuss are subsets of U. For example, $U = \mathbb{Z}$ might be the integers. We then speak of the natural numbers

Basic Counting and Listing

$\mathbb{N} = \{0, 1, 2, \ldots\}$, the positive integers $\mathbb{N}^+$, the odd integers $\mathbb{N}_o$, etc., thinking of these sets as subsets of the "universal set" $\mathbb{Z}$.

Definition 3 (Set notation) *A set is an unordered collection of distinct objects. We use the notation $x \in S$ to mean "x is an element of S" and $x \notin S$ to mean "x is not an element of S." Given two subsets (subcollections) of U, X and Y, we say "X is a subset of Y," written $X \subseteq Y$, if $x \in X$ implies that $x \in Y$. Alternatively, we may say that "Y is a superset of X." $X \subseteq Y$ and $Y \supseteq X$ mean the same thing. We say that two subsets X and Y of U are equal if $X \subseteq Y$ and $Y \subseteq X$. We use braces to designate sets when we wish to specify or describe them in terms of their elements: $A = \{a, b, c\}$, $B = \{2, 4, 6, \ldots\}$. A set with k elements is called a k-set or set with cardinality k. The cardinality of a set A is denoted by $|A|$.*

Since a set is an unordered collection of distinct objects, the following all describe the same 3-element set

$$\{a, b, c\} = \{b, a, c\} = \{c, b, a\} = \{a, b, b, c, b\}.$$

The first three are simply listing the elements in a different order. The last happens to mention some elements more than once. But, since a set consists of distinct objects, the elements of the set are still just a, b, c. Another way to think of this is:

> Two sets A and B are equal if and only if every element of A is an element of B and every element of B is an element of A.

Thus, with $A = \{a, b, c\}$ and $B = \{a, b, b, c, b\}$, we can see that everything in A is in B and everything in B is in A. You might think "When we write a set, the elements are in the order written, so why do you say a set is not ordered?" When we write something down we're stuck — we have to list them in some order. You can think of a set differently: Write each element on a separate slip of paper and put the slips in a paper bag. No matter how you shake the bag, it's still the same set.

For the most part, we shall be dealing with finite sets. Let U be a set and let A and B be subsets of U.

- The sets $A \cap B$ and $A \cup B$ are the *intersection* and *union* of A and B.

- The set $A \setminus B = \{x : x \in A, x \notin B\}$ is the *set difference* of A and B. It is also written $A - B$.

- The set $U \setminus A$ or A^c is the *complement* of A (relative to U). The complement of A is also written A' and $\sim A$.

- The set $A \oplus B = (A \setminus B) \cup (B \setminus A)$ is the *symmetric difference* of A and B.

- The set $A \times B = \{(x, y) : x \in A, y \in B\}$ is the *product* or *Cartesian product* of A and B.

Section 3: Sets

Example 11 (Cardinality of various sets) Recall that $|S|$, the cardinality of the set S is its size; that is, the number of elements in the set.

By the Rule of Product, $|A \times B| = |A| \times |B|$. (The first multiplication is Cartesian product; the second is multiplication of numbers.) Also, by the Rule of Product, the number of subsets of A is $2^{|A|}$. To see this, notice that for each element of A we have two choices — include the element in the subset or not include it.

What about things like $|A \cup B|$ and $|A \oplus B|$? They can't be expressed just in terms of $|A|$ and $|B|$. To see this, note that if $A = B$, then $|A \cup B| = |A|$ and $|A \oplus B| = |\emptyset| = 0$. On the other hand, if A and B have no common elements, $|A \cup B| = |A| + |B|$ and $|A \oplus B| = |A| + |B|$ as well. Can we say anything in general? Yes. We'll return to this later. □

The algebraic rules for operating with sets are also familiar to most beginning university students. Here is such a list of the basic rules. In each case the standard name of the rule is given first, followed by the rule as applied first to ∩ and then to ∪.

Theorem 6 (Algebraic rules for sets) *The universal set U is not mentioned explicitly but is implicit when we use the notation $\sim X = U - X$ for the complement of X. An alternative notation is $X^c = \sim X$.*

Associative:	$(P \cap Q) \cap R = P \cap (Q \cap R)$	$(P \cup Q) \cup R = P \cup (Q \cup R)$
Distributive:	$P \cap (Q \cup R) = (P \cap Q) \cup (P \cap R)$	$P \cup (Q \cap R) = (P \cup Q) \cap (P \cup R)$
Idempotent:	$P \cap P = P$	$P \cup P = P$
Double Negation:	$\sim \sim P = P$	
DeMorgan:	$\sim(P \cap Q) = \sim P \cup \sim Q$	$\sim(P \cup Q) = \sim P \cap \sim Q$
Absorption:	$P \cup (P \cap Q) = P$	$P \cap (P \cup Q) = P$
Commutative:	$P \cap Q = Q \cap P$	$P \cup Q = Q \cup P$

These rules are "algebraic" rules for working with ∩, ∪, and $\sim$. You should memorize them as you use them. They are used just like rules in ordinary algebra: whenever you see an expression on one side of the equal sign, you can replace it by the expression on the other side.

We use the notation $\mathcal{P}(A)$ to denote the set of all subsets of A and $\mathcal{P}_k(A)$ the set of all subsets of A of size (or cardinality) k. (In the previous example, we saw that $|\mathcal{P}| = 2^{|A|}$.) Let $C(n, k) = |\mathcal{P}_k(A)|$ denote the number of different k-subsets that can be formed from an n-set. The notation $\binom{n}{k}$ is also frequently used. These are called *binomial coefficients* and are read "n choose k." How do we compute $C(n, k)$?

Can we rephrase the problem in a way that converts it to a list problem, since we know how to solve those? In other words, can we relate this problem, where order does not matter, to a problem where order matters?

Let's consider all possible orderings of each of our k-sets. This gives us a way to construct all lists with distinct elements in two steps: First construct a k-set, then order it.[4] We can order a k-set by forming a k-list without repeats from the k-set. By Theorem 4

---

[4] We used an idea like this in Example 9 when we counted circular lists with distinct elements.

Basic Counting and Listing

of Section 2, we know that this can be done in $k!$ ways. By the Rule of Product, there are $C(n,k)\,k!$ distinct k-lists with no repeats. By Theorem 4 again, this number is $n(n-1)\cdots(n-k+1) = n!/(n-k)!$. Dividing by $k!$, we have

Theorem 7 (Binomial coefficient formula) *The value of the binomial coefficient is*

$$\binom{n}{k} = C(n,k) = \frac{n(n-1)\cdots(n-k+1)}{k!} = \frac{n!}{k!\,(n-k)!}.$$

Furthermore $\binom{n}{k} = \binom{n}{n-k}$.

Example 12 (Computing binomial coefficients) Let's compute some binomial coefficients for practice.

$$\binom{7}{3} = \frac{7 \times 6 \times 5}{3!} = 35,$$

because $n=7$, $k=3$ and so $n-k+1 = 5$. Alternatively,

$$\binom{7}{3} = \frac{7!}{3!\,4!} = \frac{1 \times 2 \times 3 \times 4 \times 5 \times 6 \times 7}{(1 \times 2 \times 3)(1 \times 2 \times 3 \times 4)},$$

which again gives 35 after some work.

How about computing $\binom{12}{10}$? Using the formula $\frac{12(11)\cdots(3)}{10!}$ involves a lot of writing and then a lot of cancellation (there are common factors in the numerator and denominator). There is a quicker way. By the last sentence in the theorem, $\binom{12}{10} = \binom{12}{2}$. Now we have $\binom{12}{2} = \frac{12 \times 11}{2!} = 66$. ☐

*Example 13 (A generating function for binomial coefficients)** We'll now approach the problem of evaluating $C(n,k)$ in another way. In other words, we'll "forget" the formula we just derived and start over with a new approach. You may ask "Why waste time using another approach when we've already gotten what we want?" We gave a partial answer to this earlier. Here is a more complete response.

- By looking at a problem from different viewpoints, we may come to understand it better and so be more comfortable working similar problems in the future.

- By looking at a problem from different viewpoints, we may discover that things we previously thought were unrelated have interesting connections. These connections might open up easier ways to solve some types of problems and may make it possible for us to solve problems we couldn't do before.

- A different point of view may lead us to a whole new approach to problems, putting powerful new tools at our disposal.

In the approach we are about to take, we'll begin to see a powerful tool for solving counting problems. It's called "generating functions" and it lets us put calculus and related subjects to work in combinatorics.

Suppose that $S = \{x_1, \ldots, x_n\}$ where $x_1, x_2, \ldots$ and x_n are variables as in high school algebra. Let $P(S) = (1+x_1)\cdots(1+x_n)$. The first three values of $P(S)$ are

Section 3: Sets

$n = 1:$ $1 + x_1$

$n = 2:$ $1 + x_1 + x_2 + x_1 x_2$

$n = 3:$ $1 + x_1 + x_2 + x_3 + x_1 x_2 + x_1 x_3 + x_2 x_3 + x_1 x_2 x_3.$

From this you should be able to convince yourself that $P(S)$ consists of a sum of terms where each term represents one of the subsets of S as a product of its elements. Can we reach some understanding of why this is so? Yes, but we'll only explore it briefly now. The understanding relates to the Rules of Sum and Product. Interpret plus as OR, times as AND and 1 as "nothing." Then $(1 + x_1)(1 + x_2)(1 + x_3)$ can be read as

- include the factor 1 in the term OR include the factor x_1 AND
- include the factor 1 in the term OR include the factor x_2 AND
- include the factor 1 in the term OR include the factor x_3.

In other words

- omit x_1 OR include x_1 AND
- omit x_2 OR include x_2 AND
- omit x_3 OR include x_3.

This is simply a description of how to form an arbitrary subset of $\{x_1, x_2, x_3\}$. On the other hand we can form an arbitrary subset by the rule

- include nothing in the subset OR
- include x_1 in the subset OR
- include x_2 in the subset OR
- include x_3 in the subset OR
- include x_1 AND x_2 in the subset OR
- include x_1 AND x_3 in the subset OR
- include x_2 AND x_3 in the subset OR
- include x_1 AND x_2 AND x_3 in the subset.

If we drop the subscripts on the x_i's, then a product representing a k-subset becomes x^k. We get one such term for each subset and so it follows that the coefficient of x^k in the polynomial $f(x) = (1 + x)^n$ is $C(n, k)$; that is,

$$(1 + x)^n = \sum_{k=0}^{n} C(n, k) x^k.$$

This expression is called a *generating function* for the binomial coefficients $C(n, k)$.

Can this help us evaluate $C(n, k)$? Calculus comes to the rescue through Taylor's Theorem! Taylor's Theorem tells us that the coefficient of x^k in $f(x)$ is $f^{(k)}(0)/k!$. Let $f(x) = (1 + x)^n$. Taking the k-th derivative of f gives

$$f^{(k)}(x) = n(n-1) \cdots (n - k + 1)(1 + x)^{n-k}.$$

[17]

CL-17

Basic Counting and Listing

Thus $C(n,k)$, the coefficient of x^k in $(1+x)^n$, is

$$C(n,k) = \frac{f^{(k)}(0)}{k!} = \frac{n(n-1)\cdots(n-k+1)}{k!}.$$

We conclude this example with

Theorem 8 (Binomial Theorem)

$$(x+y)^n = \sum_{k=0}^{n} \binom{n}{k} x^{n-k} y^k.$$

This follows from the identity $(1+x)^n = \sum_{k=0}^{n} C(n,k) x^k$: Since $(x+y)^n = x^n(1+(y/x))^n$, the coefficient of $x^n(y/x)^k$ in $(x+y)^n$ is $C(n,k)$. $\square$

To illustrate, $(x+y)^3 = \binom{3}{3}x^3 y^0 + \binom{3}{2}x^2 y^1 + \binom{3}{1}x^1 y^2 + \binom{3}{0}x^0 y^3$, which equals $x^3 + 3x^2y + 3xy^2 + y^3$.

Example 14 (Smorgasbord College programs) Smorgasbord College allows students to study in three principal areas: (a) Swiss naval history, (b) elementary theory and (c) computer science. The number of upper division courses offered in these fields are 2, 92, and 15 respectively. To graduate, a student must choose a major and take 6 upper division courses in it, and also choose a minor and take 2 upper division courses in it. Swiss naval history cannot be a major because only 2 upper division courses are offered in it. How many programs are possible?

The possible major-minor pairs are b-a, b-c, c-a, and c-b. By the Rule of Sum we can simply add up the number of programs in each combination. Those programs can be found by the Rule of Product. The number of major programs in (b) is $C(92,6)$ and in (c) is $C(15,6)$. For minor programs: (a) is $C(2,2) = 1$, (b) is $C(92,2) = 4186$ and (c) is $C(15,2) = 105$. Since the possible programs are constructed by

$$\Big(\text{major (b) AND } \big(\text{minor (a) OR minor (c)}\big)\Big)$$
$$\text{OR } \Big(\text{major (c) AND } \big(\text{minor (a) OR minor (b)}\big)\Big),$$

the number of possible programs is

$$\binom{92}{6}(1+105) + \binom{15}{6}(1+4186) = 75{,}606{,}201{,}671,$$

a rather large number. $\square$

Section 3: Sets

Example 15 (Card hands: Full house) Card hands provide a source of some simple sounding but tricky set counting problems. A standard deck of cards contains 52 cards, each of which is marked with two labels. The first label, called the "suit," belongs to the set
$$\text{suits} = \{\clubsuit, \heartsuit, \diamondsuit, \spadesuit\},$$
called club, heart, diamond and spade, respectively. (On the blackboard, we will use C, H, D and S rather than drawing the symbols.) The second label, called the "value" belongs to the set
$$\text{values} = \{2, 3, 4, 5, 6, 7, 8, 9, 10, J, Q, K, A\},$$
where J, Q, K and A are jack, queen, king and ace, respectively. Each pair of labels occurs exactly once in the deck. A hand is a subset of a deck. Two cards are a pair if they have the same values.

How many 5 card hands consist of a pair and a triple? (In poker, such a hand is called a "full house.")

To calculate this we describe how to construct such a hand:

- Choose the value for the pair AND
- Choose the value for the triple different from the pair AND
- Choose the 2 suits for the pair AND
- Choose the 3 suits for the triple.

This produces each full house exactly once, so the number is the product of the answers for the four steps, namely

$$13 \times 12 \times C(4,2) \times C(4,3) = 3{,}744. \qquad \square$$

Example 16 (Card hands: Two pairs) We'll continue with our poker hands. How many 5 card hands consist of two pairs? A description of a hand always means that there is nothing better in the hand, so "two pairs" means we don't have a full house or four of a kind.

The obvious thing to do is replace "triple" by "second pair" in the description for constructing a full house and add a choice for the card that belongs to no pair. This is not correct! Each hand is constructed twice, depending on which pair is the "second pair." Try it! What happened? Before choosing the cards for a pair and a triple, we can distinguish a pair from a triple because a pair contains 2 cards and a triple 3. We can't distinguish the two pairs, though, until the values are specified. This is an example of a situation where we can easily make mistakes if we forget that "AND" means "AND then." Here's a correct description, with "then" put in for emphasis.

- Choose the values for the two pairs AND then
- Choose the 2 suits for the pair with the larger value AND then
- Choose the 2 suits for the pair with the smaller value AND then
- Choose the remaining card from the 4×11 cards that have different values from the pairs.

[19]

CL-19

Basic Counting and Listing

The answer is

$$\binom{13}{2} \times \binom{4}{2} \times \binom{4}{2} \times 44 = 123{,}552. \qquad \Box$$

Example 17 (Rearranging MISSISSIPPI) We are going to count the ways to "rearrange" the letters in the word MISSISSIPPI. Before "rearranging" them, we should be precise about what we mean by "arranging" them. The distinct letters in the word MISSISSIPPI are I, M, S, and P. There are eleven letter positions in the word MISSISSIPPI which we can explicitly label as follows:

```
 1  2  3  4  5  6  7  8  9  10 11
 M  I  S  S  I  S  S  I  P  P  I
```

We can describe this placement of letters by a rule such as

$$\text{I} \leftarrow \{2,5,8,11\}, \quad \text{M} \leftarrow \{1\}, \quad \text{P} \leftarrow \{9,10\}, \quad \text{and} \quad \text{S} \leftarrow \{3,4,6,7\}.$$

If we remember the ordering (alphabetic in this case), I, M, P, S, then we can specify this arrangement by the *ordered partition*

$$(\{2,5,8,11\}, \{1\}, \{9,10\}, \{3,4,6,7\})$$

of the set $\{1, 2, \ldots, 11\}$.[5] We say that this ordered partition is of *type* $(4,1,2,4)$, referring to the sizes of the sets, in order, that make up the ordered partition. Each of these sets is called a *block* or, in statistics, a *cell*. In general, an ordered partition of a set T of type $(m_1, m_2, \ldots, m_k)$ is a sequence of *disjoint* sets $(B_1, B_2, \ldots, B_k)$ such that $|B_i| = m_i$, $i = 1, 2, \ldots, k$, and $\cup_{i=1}^{k} B_i = T$. Empty sets are allowed in ordered partitions. The set of all *rearrangements* of the letters in the word MISSISSIPPI corresponds to the set of all ordered partitions (B_1, B_2, B_3, B_4) of $\{1, 2, \ldots, 11\}$ of type $(4, 1, 2, 4)$. For example, the ordered partition $(\{1, 5, 7, 10\}, \{2\}, \{9, 11\}, \{3, 4, 6, 8\})$ corresponds to the placement

$$\text{I} \leftarrow \{1,5,7,10\}, \text{ M} \leftarrow \{2\}, \text{ P} \leftarrow \{9,11\}, \text{ and S} \leftarrow \{3,4,6,8\}$$

and leads to the "word"

```
 1  2  3  4  5  6  7  8  9  10 11
 I  M  S  S  I  S  I  S  P  I  P
```

Another, somewhat picturesque, way of describing ordered partitions of a set T is to think of ordered (i.e., labeled) boxes $(B_1, B_2, \ldots, B_k)$ into which we distribute the elements of T, m_i elements to box B_i, $i = 1, \ldots, k$. The next example takes that point of view and concludes that the number of such distributions of elements into boxes (i.e., the number of ordered partitions) is the *multinomial coefficient*

$$\binom{n}{m_1, m_2, \ldots, m_k} = \frac{n!}{m_1!\, m_2! \cdots m_k!}.$$

As a result, the number of rearrangements of the word MISSISSIPPI is the multinomial coefficient

$$\binom{11}{4,1,2,4} = \frac{11!}{4!\, 1!\, 2!\, 4!} = 34{,}650. \qquad \Box$$

---

[5] Note the use of $(\ldots)$ and $\{\ldots\}$ here: We have a list, indicated by $(\ldots)$. Each element of the list is a set, indicated by $\{\ldots\}$.

Example 18 (Multinomial coefficients) Suppose we are given k boxes labeled 1 through k and an n-set S and we are told to distribute the elements of S among the boxes so that the ith box contains exactly m_i elements. How many ways can this be done?

Let $n = |S|$. Unless $m_1 + \ldots + m_k = n$, the answer is zero because we don't have the right number of objects. Therefore, we assume from now on that

$$m_1 + \ldots + m_k = n.$$

Here's a way to describe filling the boxes.

- Fill the first box (There are $C(n, m_1)$ ways.[6]) AND
- Fill the second box (There are $C(n - m_1, m_2)$ ways.) AND
- • • •
- Fill the kth box. (There are $C(n - (m_1 + \ldots + m_{k-1}), m_k) = C(m_k, m_k) = 1$ ways.)

Now apply the Rule of Product, use the formula $C(p,q) = p!/q!\,(p-q)!$ everywhere, and cancel common factors in numerator and denominator to obtain $n!/m_1!\,m_2!\cdots m_k!$. To illustrate

$$\binom{12}{4}\binom{12-4}{3}\binom{12-4-3}{3} = \frac{12!}{4!\,8!}\,\frac{8!}{3!\,5!}\,\frac{5!}{3!\,2!} = \frac{12!}{4!\,3!\,3!\,2!},$$

which we write $\binom{12}{4,3,3,2}$. In general, this expression is written

$$\binom{n}{m_1, m_2, \ldots, m_k} = \frac{n!}{m_1!\,m_2!\cdots m_k!}$$

where $n = m_1 + m_2 + \ldots + m_k$ and is called a *multinomial coefficient*. In multinomial notation, the binomial coefficient $\binom{n}{k}$ would be written $\binom{n}{k,(n-k)}$. You can think of the first box as the k things that are chosen and the second box as the $n - k$ things that are not chosen.

As in the previous example (Example 17), we can think of the correspondence

objects being distributed $\Longleftrightarrow$ positions in a word

boxes $\Longleftrightarrow$ letters.

If the object "position 3" is placed in the box "D," then the letter D appears as the third letter in the word. The multinomial coefficient is then the number of words that can be made so that letter i appears exactly m_i times. A word can be thought of as a list of its letters. □

---

[6] Since m_1 things went into the first box, we have only $n - m_1$ left, from which we must choose m_2 for the second box.

Basic Counting and Listing

Example 19 (Distributing toys) Eleven toys are to be distributed among 4 children. How many ways can this be done if the oldest child is to receive only 2 toys and each of the other children is to receive 3 toys?

We can do this directly if we are used to thinking in terms of multinomial coefficients. We could also do it by converting the problem into one of our previous interpretations.

Here is the first: We want an ordered partition of 11 toys into 4 piles ("blocks") such that the first pile (for the oldest child) contains 2 and each of the 3 remaining piles contain 3 toys. This is an ordered partition of type (2,3,3,3). The number of them is $\binom{11}{2,3,3,3} = 92,400$.

Here is the second: Think of each child as a box into which we place toys. The number of ways to fill the boxes is, again, $\binom{11}{2,3,3,3}$. □

Example 20 (Words from a collection of letters — second try) Using the idea at the end of the previous example, we can more easily count the words that can be made from ERROR, a problem discussed in Example 10. Suppose we want to make words of length k. Let m_1 be the number of E's, m_2 the number of O's and m_3 the number of R's. By considering all possible cases for the number of each letter, you should be able to see that the answer is the sum of $\binom{k}{m_1,m_2,m_3}$ over all m_1, m_2, m_3 such that

$$m_1 + m_2 + m_3 = k, \quad 0 \leq m_1 \leq 1, \quad 0 \leq m_2 \leq 1, \quad 0 \leq m_3 \leq 3.$$

Thus we obtain

$$k = 1: \binom{1}{0,0,1} + \binom{1}{0,1,0} + \binom{1}{1,0,0} = 3$$

$$k = 2: \binom{2}{0,0,2} + \binom{2}{0,1,1} + \binom{2}{1,0,1} + \binom{2}{1,1,0} = 7$$

$$k = 3: \binom{3}{0,0,3} + \binom{3}{0,1,2} + \binom{3}{1,0,2} + \binom{3}{1,1,1} = 13$$

$$k = 4: \binom{4}{0,1,3} + \binom{4}{1,0,3} + \binom{4}{1,1,2} = 20$$

$$k = 5: \binom{5}{1,1,3} = 20.$$

This is better than in Example 10. Instead of having to list words, we have to list triples of numbers and each triple generally corresponds to more than one word. Here are the lists of triples for the preceding computations

$k = 1:$ (0,0,1) (0,1,0) (1,0,0)
$k = 2:$ (0,0,2) (0,1,1) (1,0,1) (1,1,0)
$k = 3:$ (0,0,3) (0,1,2) (1,0,2) (1,1,1)
$k = 4:$ (0,1,3) (1,0,3) (1,1,2)
$k = 5:$ (1,1,3)

□

Section 3: Sets

Example 21 (Forming teams) How many ways can we form 4 teams from 12 people so that each team has 3 members? This is another multinomial coefficient (ordered set partition) problem and the answer is $\binom{12}{3,3,3,3} = 554,400$.

Wait! We forgot to tell you that the teams don't have names or any other distinguishing features except who the team members are. The solution that gave 554,400 created a list of teams, so there was a Team 1, Team 2, Team 3 and Team 4. We can deal with this the same way we got the formula for counting subsets: To form a list of 4 teams, first form a set and then order it. Since 4 distinct things can be ordered in $4! = 24$ ways, we have $554,400 = 24x$ where x is our answer. We obtain 23,100.

If we told you in the first place that the teams were not ordered, you may not have thought of multinomial coefficients. This leads to two points.

- It may be helpful to impose order and then divide it out.
- We have found a way to count unordered partitions when all the blocks are the same size. This can be extended to the general case of blocks of various sizes but we will not do so.

Wait! We forgot to tell you that we are going to form 4 teams, pair them up to play each other in a contest, say the team with Alice plays the team with Bob, and the other two teams play each other. The winners then play each other. Now we have to form the teams and divide them into pairs that play each other. Let's do that. Suppose we have formed 4 unordered teams. Now we must pair them off. This is another unordered partition: The four teams must be partitioned into two blocks each of size 2. From what we learned in the previous paragraph, we compute $\binom{4}{2,2}$ and divide by $2!$, obtaining 3. Thus the answer is $23,100 \times 3 = 69,300$. □

Example 22 (Card hands and multinomial coefficients) To form a full house, we must choose a face value for the triple, choose a face value for the pair, and leave eleven face values unused. This can be done in $\binom{13}{1,1,11}$ ways. We then choose the suits for the triple in $\binom{4}{3}$ ways and the suits for the pair in $\binom{4}{2}$ ways. Note that we choose suits only for the cards in the hand, not for the "unused face values."

To form two pair, we must choose two face values for the pairs, choose a face value for the single card, and leave ten face values unused. This can be done in $\binom{13}{2,1,10}$ ways. We then choose suits for each of the face values in turn, so we must multiply by $\binom{4}{2}\binom{4}{2}\binom{4}{1}$.

Let's imagine an eleven card hand containing two triples, a pair and three single cards. You should be able to see that the number of ways to do this is

$$\binom{13}{2,1,3,7}\binom{4}{3}\binom{4}{3}\binom{4}{2}\binom{4}{1}\binom{4}{1}\binom{4}{1}.$$ □

We conclude this section with an introduction to recursions. Let's explore yet another approach to evaluating the binomial coefficient $C(n,k) = \binom{n}{k}$. Let $S = \{x_1, \ldots, x_n\}$. We'll think of $C(n,k)$ as counting k-subsets of S. Either the element x_n is in our subset or it is not. The cases where it is in the subset are all formed by taking the various $(k-1)$-subsets of $S - \{x_n\}$ and adding x_n to them. The cases where it is not in the subset are all formed

Basic Counting and Listing

by taking the various k-subsets of $S - \{x_n\}$. What we've done is describe how to build k-subsets of S from certain subsets of $S - \{x_n\}$. Since this gives each subset exactly once,

$$\binom{n}{k} = \binom{n-1}{k-1} + \binom{n-1}{k}$$

by the Rule of Sum.

The equation $C(n,k) = C(n-1, k-1) + C(n-1, k)$ is called a *recursion* because it tells how to compute $C(n, k)$ from values of the function with smaller arguments. This is a common approach which we can state in general form as follows.

Example 23 (Deriving recursions) To count things, you might ask and answer the question

> How can I construct the things I want to count of a given size by using the same type of things of a smaller size?

This process usually gives rise to a recursion.

Actually, we've cheated a bit in all of this because the recursion only works when we have some values to start with. The correct statement of the recursion is either

$$C(0,0) = 1,$$
$$C(0,k) = 0 \quad \text{for } k \neq 0 \quad \text{and}$$
$$C(n,k) = C(n-1, k-1) + C(n-1, k) \quad \text{for } n > 0;$$

or

$$C(1,0) = C(1,1) = 1,$$
$$C(1,k) = 0 \quad \text{for } k \neq 0, 1 \quad \text{and}$$
$$C(n,k) = C(n-1, k-1) + C(n-1, k) \quad \text{for } n > 1;$$

depending on how we want to start the computations based on this recursion. Below we have made a table of values for $C(n, k)$. Sometimes this tabular representation of $C(n, k)$ is called "Pascal's Triangle."

n \ k	0	1	2	3	4	5	6
0	1						
1	1	1					
2	1	2	1				
3	1	3	3	1			
4	1	4	6	4	1		
5	1	5	10	10	5	1	
6	1	6	15	20	15	6	1

$C(n,k)$

Sometimes it is easier to think in terms of "breaking down" rather than "constructing." That is, ask the question

Section 3: Sets

How can I break down the things I want to count into smaller things of the same type?

Let's look at the binomial coefficients again. What happens to the k-subsets of the set $S = \{x_1, \ldots, x_n\}$ if we throw away x_n? We then have subsets of $S \setminus \{x_n\} = \{x_1, \ldots, x_{n-1}\}$. The k-subsets of S that did not contain x_n are still k-subsets, but those that contained x_n have become $(k-1)$-subsets. We get all k-subsets and all $(k-1)$-subsets of $S \setminus \{x_n\}$ exactly once when we do this. Thus $C(n,k) = C(n-1,k) + C(n-1,k-1)$ by the Rule of Sum. $\square$

Example 24 (Set partitions) A *partition of a set* B is a collection of nonempty subsets of B such that each element of B appears in exactly one subset. Each subset is called a *block* of the partition. The 15 partitions of $\{1, 2, 3, 4\}$ by number of blocks are

1 block: $\{\{1,2,3,4\}\}$
2 blocks: $\{\{1,2,3\},\{4\}\}$ $\{\{1,2,4\},\{3\}\}$ $\{\{1,2\},\{3,4\}\}$ $\{\{1,3,4\},\{2\}\}$
           $\{\{1,3\},\{2,4\}\}$ $\{\{1,4\},\{2,3\}\}$ $\{\{1\},\{2,3,4\}\}$
3 blocks: $\{\{1,2\},\{3\},\{4\}\}$ $\{\{1,3\},\{2\},\{4\}\}$ $\{\{1,4\},\{2\},\{3\}\}$ $\{\{1\},\{2,3\},\{4\}\}$
           $\{\{1\},\{2,4\},\{3\}\}$ $\{\{1\},\{2\},\{3,4\}\}$
4 blocks: $\{\{1\},\{2\},\{3\},\{4\}\}$

Let $S(n,k)$ be the number of partitions of an n-set having exactly k blocks. These are called *Stirling numbers of the second kind*. Do not confuse $S(n,k)$ with $C(n,k) = \binom{n}{k}$. In both cases we have an n-set. For $C(n,k)$ we want to *choose a subset* containing k elements and for $S(n,k)$ we want to *partition the set* into k blocks.

What is the value of $S(n,k)$? Let's try to get a recursion. How can we build partitions of $\{1, 2, \ldots, n\}$ with k blocks out of smaller cases? If we take partitions of $\{1, 2, \ldots, n-1\}$ with $k-1$ blocks, we can simply add the block $\{n\}$. If we take partitions of $\{1, 2, \ldots, n-1\}$ with k blocks, we can add the element n to one of the k blocks. You should convince yourself that all k block partitions of $\{1, 2, \ldots, n\}$ arise in exactly one way when we do this. This gives us a recursion for $S(n,k)$. Putting n in a block by itself contributes $S(n-1,k-1)$. Putting n in a block with other elements contributes $S(n-1,k) \times k$ by the Rule of Product. By the Rule of Sum
$$S(n,k) = S(n-1, k-1) + k\, S(n-1, k).$$

Let's take a tearing down view. If we remove n from the set $\{1, \ldots, n\}$ and from the block of the partition in which it occurs:

- We get a partition counted by $S(n-1,k-1)$ if n was in a block by itself because that block disappears.

- We get a partition counted by $S(n-1,k)$ if n was in a block with other things. In fact, we get each of these partitions k times since n could have been in any of the k blocks.

This gives us our recursion $S(n,k) = S(n-1,k-1) + kS(n-1,k)$ again.

To illustrate, let's look at what happens when we remove 4 from our earlier list of 3-block partitions:

3 blocks: $\{\{1,2\},\{3\},\{4\}\}$ $\{\{1,3\},\{2\},\{4\}\}$ $\{\{1,4\},\{2\},\{3\}\}$ $\{\{1\},\{2,3\},\{4\}\}$
           $\{\{1\},\{2,4\},\{3\}\}$ $\{\{1\},\{2\},\{3,4\}\}$

Basic Counting and Listing

The partitions with singleton blocks {4} removed give us the partitions

$$\{\{1,2\},\{3\}\} \quad \{\{1,3\},\{2\}\} \quad \{\{1\},\{2,3\}\}.$$

Thus the partitions counted by $S(3,2)$ each occur once. The partitions in which 4 is not in a singleton block, with 4 removed, give us the partitions

$$\{\{1\},\{2\},\{3\}\} \quad \{\{1\},\{2\},\{3\}\} \quad \{\{1\},\{2\},\{3\}\}.$$

Thus the partitions counted by $S(3,3)$ (there's only one) each occur 3 times. Hence $S(4,3) = S(3,2) + 3S(3,3)$.

Below is the tabular form for $S(n,k)$ analogous to the tabular form for $C(n,k)$.

n \ k	1	2	3	4	5	6	7
1	1						
2	1	1					
3	1	3	1		$S(n,k)$		
4	1	7	6	1			
5	1	15	25	10	1		
6	1	31	90	65	15	1	
7	1	--	--	--	--	--	1

Notice that the starting conditions for this table are that $S(n,1) = 1$ for all $n \geq 1$ and $S(n,n) = 1$ for all $n \geq 1$. The values for $n = 7$ are omitted from the table. You should fill them in to test your understanding of this computational process. For each n, the total number of partitions of a set of size n is equal to the sum $S(n,1) + S(n,2) + \ldots S(n,n)$. These numbers, gotten by summing the entries in the rows of the above table, are called the *Bell numbers*, B_n. For example, $B_4 = 1 + 7 + 6 + 1 = 15$. □

Exercises for Section 3

3.1. How many 6 card hands contain 3 pairs?

3.2. How many 5 card hands contain a straight? A straight is 5 consecutive cards from the sequence A,2,3,4,5,6,7,8,9,10,J,Q,K,A without regard to suit.

3.3. How many compositions of n (sequences of positive integers called "parts" that add to n) are there that have exactly k parts? A composition of 5, for example, corresponds to a placement of either a "+" or a "," in the four spaces between a

Section 3: Sets

sequence of five ones: 1 1 1 1 1. Thus, the placement $1,1+1,1+1$ corresponds to the composition $(1,2,2)$ of 5 which has 3 parts.

3.4. How many rearrangements of the letters in EXERCISES are there?

3.5. In some card games only the values of the cards matter and their suits are irrelevant. Thus there are effectively only 13 distinct cards among 52 total. How many different ways can a deck of 52 cards be arranged in this case? The answer is a multinomial coefficient.

3.6. In a distant land, their names are spelled using the letters A, I, L, S, and T. Each name consists of seven letters. Each name begins and ends with a consonant, contains no adjacent vowels and never contains three adjacent consonants. If two consonants are adjacent, they cannot be the same. An example of a name is LASLASS.

(a) List the first 4 names in dictionary order.

(b) List the last 4 names in dictionary order.

(c) How many names are possible?

3.7. Prove $\binom{n}{k} = \binom{n}{n-k}$ and $\binom{n}{0} + \binom{n}{1} + \cdots + \binom{n}{n} = 2^n$.

3.8. For $n > 0$, prove the following formulas for $S(n,k)$:

$$S(n,n) = 1, \quad S(n,n-1) = \binom{n}{2}, \quad S(n,1) = 1, \quad S(n,2) = (2^n-2)/2 = 2^{n-1}-1.$$

3.9. Let B_n be the total number of partitions of an n element set. Thus

$$B_n = S(n,0) + S(n,1) + \cdots + S(n,n).$$

These numbers are called the *Bell numbers*.

(a) Prove that

$$B_{n+1} = \sum_{i=0}^{n} \binom{n}{i} B_{n-i},$$

where B_0 is defined to be 1.

Hint: Construct the block containing $n+1$ and then construct the rest of the partition. If you prefer tearing down instead of building up, remove the block containing $n+1$.

(b) Calculate B_n for $n \leq 5$.

3.10. We consider permutations $a_1, \ldots, a_9$ of 1,2,3,4,5,6,7,8,9.

(a) How many have the property that $a_i < a_{i+1}$ for all $i \leq 8$?

(b) How many have the property that $a_i < a_{i+1}$ for all $i \leq 8$ except $i = 5$?

Basic Counting and Listing

Section 4: Probability and Basic Counting

Techniques of counting are very important in probability theory. In this section, we take a look at some of the basic ideas in probability theory and relate these ideas to our counting techniques. This requires, for the most part, a minor change of viewpoint and of terminology.

Let U be a set and suppose for now that U is finite. We think of U as a "universal set" in the sense that we are going to be concerned with various subsets of U and their relationship with each other. In probability theory, the term "universal set" is replaced by *sample space*. Thus, let U be a sample space. We say that we "choose an element of U uniformly at random" if we have a method of selecting an element of U such that all elements of U have the same chance of being selected. This definition is, of course, self referential and pretty sloppy, but it has intuitive appeal to anyone who has selected people for a sports team, or for a favored task at camp, and attempted to be fair about it. We leave it at this intuitive level.

The quantitative way that we say that we are selecting uniformly at random from a sample space U is to say that each element of U has *probability* $1/|U|$ of being selected.

A subset $E \subseteq U$ is called an *event* in probability theory. If we are selecting uniformly at random from U, the probability that our selection belongs to the set E is $|E|/|U|$. At this point, basic probability theory involves nothing more than counting (i.e., we need to count to get $|E|$ and $|U|$).

A more general situation arises when the method of choosing is not "fair" or "uniform." Suppose $U = \{H, T\}$ is a set of two letters, H and T. We select either H or T by taking a coin and flipping it. If "heads" comes up, we choose H, otherwise we choose T. The coin, typically, will be dirty, have scratches in it, etc., so the "chance" of H being chosen might be different from the chance of T being chosen. If we wanted to do a bit of work, we could flip the coin 1000 times and keep some records. Interpreting these records might be a bit tricky in general, but if we came out with 400 heads and 600 tails, we might suspect that tails was more likely. It is possible to be very precise about these sort of experiments (the subject of statistics is all about this sort of thing). But for now, let's just suppose that the "probability" of choosing H is 0.4 and the probability of choosing T is 0.6. Intuitively, we mean by this that if you toss the coin a large number N of times, about $0.4N$ will be heads and $0.6N$ will be tails. The function P with domain $U = \{H, T\}$ and values $P(H) = 0.4$ and $P(T) = 0.6$ is an example of a "probability function" on a sample space U.

The more general definition is as follows:

Definition 4 (Probability function and probability space) *Let U be a finite sample space and let P be a function from U to $\mathbb{R}$ (the real numbers) such that $P(t) \geq 0$ for all t and $\sum_{t \in U} P(t) = 1$.*

- *P is called a probability function on U.*

- *The pair (U, P) is called a probability space.*

Section 4: Probability and Basic Counting

- We extend P to events $E \subseteq U$ by defining $P(E) = \sum_{t \in E} P(t)$. $P(E)$ is called the *probability of the event* E. (If $t \in U$, we write $P(t)$ and $P(\{t\})$ interchangeably.)

An element $t \in U$ is called an *elementary event* or a *simple event*.

Note that since $P(t) \geq 0$ for all t, it follows from $\sum P(t) = 1$ that $P(t) \leq 1$.

Think of U as a set of elementary events that can occur. Each time we do an experiment or observe something, *exactly one* of the elementary events in U occurs. Imagine repeating this many times. Think of $P(t)$ as the fraction of the cases where the elementary event t occurs. The equation $\sum_{t \in U} P(t) = 1$ follows from the fact that exactly one elementary event occurs each time we do our experiment. Think of $P(E)$ as the fraction of time an elementary event in E occurs.

Theorem 9 (Disjoint events) *Suppose that (U, P) is a probability space and that X and Y are disjoint subsets of U; that is, $X \cap Y = \emptyset$. Then $P(X \cup Y) = P(X) + P(Y)$.*

Proof: By definition, $P(X \cup Y)$ is the sum of $P(t)$ over all $t \in X \cup Y$. If $t \in X \cup Y$, then either $t \in X$ or $t \in Y$, but not both because $X \cap Y = \emptyset$. Thus we can split the sum into two sums, one over $t \in X$ and the other over $t \in Y$. These two sums are $P(X)$ and $P(Y)$, respectively. Thus $P(X \cup Y) = P(X) + P(Y)$.

We could rephrase this using summation notation:

$$P(X \cup Y) = \sum_{t \in X \cup Y} P(t) = \sum_{t \in X} P(t) + \sum_{t \in Y} P(t) = P(X) + P(Y),$$

where we could split the sum into two sums because $t \in X \cup Y$ means that either $t \in X$ or $t \in Y$, but not both because $X \cap Y = \emptyset$. □

Example 25 (Dealing a full house) What is the probability of being dealt a full house? There are $\binom{52}{5}$ distinct hands of cards so we could simply divide the answer 3,744 from Example 15 by this number. That gives the correct answer, but there is another way to think about the problem.

When a hand of cards is dealt, the *order* in which you receive the cards matters: Thus receiving 3♠ 6◇ 2♡ in that order is a different dealing of the cards than receiving 2♡ 3♠ 6◇ in that order. Thus, we regard each of the $52 \times 51 \times 50 \times 49 \times 48$ ways of dealing five cards from 52 as equally likely. Thus each hand has probability $1/52 \times 51 \times 50 \times 49 \times 48$. Since all the cards in a hand of five cards are different, they can be ordered in 5! ways. Hence the probability of being dealt a full house is $\frac{3{,}774 \times 5!}{52 \times 51 \times 50 \times 49 \times 48}$, which does indeed equal 3,744 divided by $\binom{52}{5}$.

If cards are not all distinct and if we are not careful, the two approaches give different answers. The first approach gives the wrong answer. We now explain why. Be prepared to think carefully, because this is a difficult concept for beginning students.

To illustrate consider a deck of 4 cards that contains two aces of spades and two jacks of diamonds. There are 3 possible two card hands: 2 aces, 1 ace and 1 jack, or 2 jacks, but the probability of getting two aces is only 1/6. Can you see how to calculate that correctly?

[29]

Basic Counting and Listing

We can look at this in at least two ways. Suppose we are being dealt the top two cards. The probability of getting two aces equals the fraction of ways to assign positions to cards so that the top two are given to aces. There are $\binom{4}{2}$ ways to assign positions to aces and only one of those results in the aces being in the top two positions.

Here's the other way to look at it: Mark the cards so that the aces can be told apart, and the jacks can be told apart, say A_1, A_2, J_1, and J_2. Since the cards are distinct each hand can be ordered in the same number of ways, namely 2!, and so we can count ordered or unordered hands. There are now $\binom{4}{2}$ unordered hands (or 4×3 ordered ones) and only one of these (or 2×1 ordered ones) contain A_1 and A_2. $\square$

Example 26 (Venn diagrams and probability) A "Venn diagram" shows the relationship between elements of sets. The interior of the rectangle in the following figure represents the sample space U. The interior of each of the circular regions represents the events A and B.

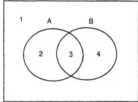

Let's list what each of the regions in the figure are:

$$1 \text{ is } (A \cup B)^c \quad 2 \text{ is } A - B \quad 3 \text{ is } A \cap B \quad 4 \text{ is } B - A.$$

We can compute either set cardinalities or probabilities. For example, $U \setminus A$ is all of U except what is in the region labeled A. Thus $|U \setminus A| = |U| - |A|$. On the other hand, A and A^c partition the sample space and so $P(A) + P(A^c) = 1$. Rewriting this as

$$P(A^c) = 1 - P(A)$$

puts it in the same form as $|U \setminus A| = |U| - |A|$ since $U \setminus A = A^c$. Notice that the only difference between the set and probability equations is the presence of the function P and the fact that $P(U) = 1$. Also notice that the probability form did *not* assume that the probability was uniformly at random.

What about $A \cup B$? It corresponds to the union of the disjoint regions labeled 2, 3 and 4 in the Venn diagram. Thus

$$P(A \cup B) = P(A - B) + P(A \cap B) + P(B - A)$$

by Theorem 9. We can express $P(A - B)$ in terms of $P(A)$ and $P(A \cap B)$ because A is the disjoint union of $A - B$ and $A \cap B$: $P(A) = P(A - B) + P(A \cap B)$. Solving for $P(A - B)$ and writing a similar expression for $P(B - A)$:

$$P(A - B) = P(A) - P(A \cap B) \qquad P(B - A) = P(B) - P(A \cap B).$$

Section 4: Probability and Basic Counting

Combining our previous results

$$P(A \cup B) = P(A - B) + P(A \cap B) + P(B - A) = P(A) + P(B) - P(A \cap B).$$

There is a less formal way of saying this. If we take A and B we get the region labeled 3 twice — once in A and once in B. The region labeled 3 corresponds to $A \cap B$ since it is the region that belongs to both A and B. Thus $|A| + |B|$ gives us regions 2, 3 and 4 (which is $|A \cup B|$) and a second copy of 3, (which is $|A \cap B|$). We have shown that

$$|A| + |B| = |A \cup B| + |A \cap B|.$$

The probability form is $P(A) + P(B) = P(A \cup B) + P(A \cap B)$. We can rewrite this as

$$P(A \cup B) = P(A) + P(B) - P(A \cap B).$$

(This is the two set case of the *Principle of Inclusion and Exclusion*.)

One more example: Using DeMorgan's Rule from Theorem 6, $(A \cup B)^c = A^c \cap B^c$. (Check this out in the Venn diagram.) Combining the results of the two previous paragraphs,

$$P(A^c \cap B^c) = 1 - P(A \cup B) = 1 - \Big(P(A) + P(B) - P(A \cap B)\Big)$$
$$= 1 - P(A) - P(B) + P(A \cap B).$$

This is another version of the Principle of Inclusion and Exclusion. ☐

Example 27 (Combining events) Let U be a sample space with probability function P. Let A and B be events. Suppose we know that

- A occurs with probability 7/15,
- B occurs with probability 6/15, and
- the probability that neither of the events occurs is 3/15.

What is the probability that both of the events occur?

Let's translate the given information into mathematical notation. The first two data are easy: $P(A) = 7/15$ and $P(B) = 6/15$. What about the last? What is the event corresponding to neither of A and B occurring? One person might say $A^c \cap B^c$; another might say $(A \cup B)^c$. Both are correct by DeMorgan's Rule. Thus the third datum can be written $P((A \cup B)^c) = P(A^c \cap B^c) = 3/15$. We are asked to find $P(A \cap B)$.

What do we do now? A Venn diagram can help. The situation is shown in the following Venn diagram for A and B. The rectangle stands for U, the whole sample space. (We've put in some numbers that we haven't computed yet, so you should ignore them.)

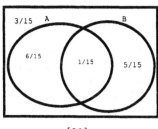

[31]

CL-31

Basic Counting and Listing

We have been given just partial information, namely $P(A) = 7/15$, $P(B) = 6/15$, and $P((A \cup B)^c) = 3/15$. The best way to work such problems is to use the information given to, if possible, find the probabilities of the four fundamental regions associated with A and B, namely the regions

$$(A \cup B)^c \qquad A - B \qquad B - A \qquad A \cap B.$$

(You should identify and label the regions in the figure.) Recall that $P(E^c) = 1 - P(E)$ for any event E. Thus

$$P(A \cup B) = 1 - P((A \cup B)^c) = 1 - 3/15 = 12/15.$$

From this we get (check the Venn diagram)

$$P(A - B) = P(A \cup B) - P(B) = 12/15 - 6/15 = 6/15.$$

Similarly, $P(B - A) = 12/15 - 7/15 = 5/15$. Finally,

$$P(A \cap B) = P(A) - P(A - B) = 7/15 - 6/15 = 1/15.$$

The answer to the question we were asked at the beginning is that $P(A \cap B) = 1/15$. ☐

Example 28 (Odds and combining events) Let U be a sample space and let A and B be events where the odds of A occurring are 1:2, the odds of B occurring are 5:4 and the odds of both A and B occurring are 1:8. Find the odds of neither A nor B occurring. In popular culture, probabilities are often expressed as *odds*. If an event E occurs with odds $a : b$, then it occurs with $P(E) = a/(a+b)$. Thus, $P(A) = 1/3$, $P(B) = 5/9$, and $P(A \cap B) = 1/9$. From the equation $P(A \cup B) = P(A) + P(B) - P(A \cap B)$ in Example 26, $P(A \cup B) = 7/9$. From the equation $P(A^c) = 1 - P(A)$ in that example, with $A \cup B$ replacing A, we have $P((A \cup B)^c) = 2/9$. The odds of neither A nor B occurring are 2:7.

Caution: It is not always clear what odds mean. If someone says that the odds on Beatlebomb in a horse race are 100:1, this means that the probability is $100/(100+1)$ that Beatlebomb will *lose*. The probability that he will win is $1/(100+1)$. ☐

Example 29 (Hypergeometric probabilities) Six light bulbs are chosen at random from 18 light bulbs of which 8 are defective. What is the probability that exactly two of the chosen bulbs are defective? We'll do the general situation. Let

- B denote the total number of bulbs,
- D the total number of defective bulbs, and
- b the number of bulbs chosen.

Let the probability space be the set of all $\binom{B}{b}$ ways to choose b bulbs from B and let the probability be uniform. Let $E(B, D, b, d)$ be the event consisting of all selections of b from B when a total of D bulbs are defective and d of the selected bulbs are defective. We want $P(E(B, D, b, d))$. The total number of ways to choose b, of which exactly d are defective,

Section 4: Probability and Basic Counting

is $\binom{D}{d}\binom{B-D}{b-d}$. To see this, first choose d bulbs from the D defective bulbs and then choose $b-d$ bulbs from the $B-D$ good bulbs. Thus,

$$P(E(B,D,b,d)) = \frac{\binom{D}{d}\binom{B-D}{b-d}}{\binom{B}{b}}.$$

Substituting $B = 18$, $b = 6$, $D = 8$, and $d = 2$ gives $P(E(18,8,6,2)) = 0.32$, the answer to our original question.

The function $P(E(B,D,b,d))$ occurs frequently. It is called the *hypergeometric probability distribution*. ☐

Example 30 (Sampling with replacement from six cards) First one card and then a second card are selected at random, with replacement, from 6 cards numbered 1 to 6. What is the probability that the sum of the values on the cards equals 7? That the sum of the values of the cards is divisible by 5? Since both cards are selected from the same set of cards numbered one to six, this process is called "sampling with replacement." The idea is that one can choose the first card, write down its number, *replace* it and repeat the process a second (or more) times. The basic sample space is $S \times S = \{(i,j) \mid 1 \leq i \leq 6,\ 1 \leq j \leq 6\}$. Every point in this sample space is viewed as equally likely. Call the two events of interest E_7 (sum equals 7) and D_5 (sum divisible by 5). It is helpful to have a way of visualizing $S \times S$. This can be done as follows:

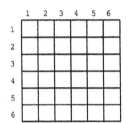

The 6×6 rectangular array has 36 squares. The square with row label i and column label j corresponds to $(i,j) \in S \times S$. The rectangular array on the right has the sum $i+j$ in square (i,j). Thus

$$E_7 = \{(i,j) : 1 \leq i \leq 6,\ 1 \leq j \leq 6,\ i+j = 7\}$$

corresponds to six points in the sample space and so $P(E_7) = |E_7|/36 = 6/36$.

A number k is divisible by 5 if $k = 5j$ for some integer j. In that case, we write $5|k$. Thus

$$D_5 = \{(i,j) : 1 \leq i \leq 6,\ 1 \leq j \leq 6,\ 5|(i+j)\}$$

and so $D_5 = E_5 \cup E_{10}$. Finally, $|D_5| = 4 + 3 = 7$ and $P(D_5) = 7/36$. ☐

Basic Counting and Listing

Example 31 (Girls and boys sit in a row) Four girls and two boys sit in a row. Find the probability that each boy has a girl to his left and to his right. Suppose that the girls are g_1, g_2, g_3, g_4 and the boys are b_1, b_2. There are $6! = 720$ ways of putting these six people in a row. This set of 720 such permutations is the sample space S, and we assume each permutation is equally likely. Let S_g denote the set of such permutations where each boy has a girl on his left and one on his right. There are three patterns where each boy has a girl on both his left and his right: *gbgbgg*, *gbggbg*, and *ggbgbg*. For each pattern, there are $2!\,4! = 48$ ways of placing the girls and boys into that pattern. Thus, $(3 \times 48)/6! = 144/720 = 1/5$ is the required probability. Note that we could have also taken the sample space to be the set of $\binom{6}{2}$ patterns. Each pattern is equally likely since each arises from the same number of arrangements of the 6 children. The probability would then be computed as $3/\binom{6}{2} = 3/15 = 1/5$. ☐

Example 32 (Dealing cards from a standard deck of 52 cards) A man is dealt 4 spade cards from an ordinary deck of 52 cards. If he is given five more cards, what is the probability that three of them are spades? This is another example of the hypergeometric probability distribution. There are $B = 48$ cards remaining, $D = 9$ of them spades. We ask for the probability that, from $b = 5$ cards selected, $d = 3$ are spades.

$$P(E(B,D,b,d)) = \frac{\binom{D}{d}\binom{B-D}{b-d}}{\binom{B}{b}} = \frac{\binom{9}{3}\binom{39}{2}}{\binom{48}{5}} = 0.036\,. \qquad \square$$

Example 33 (Selecting points at random from a square) Suppose we have a square with side s and inside it is a circle of diameter $d \leq 1$. A point is selected uniformly at random from the square. What is the probability that the point selected lies inside the circle?

We haven't defined probability for infinite sample spaces. The intuition is that probability is proportional to area — a "geometric probability" problem. Thus we have

$$P(E) = \frac{\text{area}(E)}{\text{area}(U)},$$

where U is the sample space, which is the set of points in the square. Computing areas, we obtain $P = \pi d^2/(4s^2)$. This is the correct answer. Clearly, this answer doesn't depend on the figure being a circle. It could be any figure of area $\pi d^2/4$ that fits inside the square. ☐

The next example deals with the following question: If k items are randomly put one at a time into n boxes what are the chances that no box contains more than one item? A related problem that can be dealt with in a similar manner is the following: If I choose k_1 items at random from n items and you choose k_2 items from the same n, what are the chances that our choices contain no items in common? These problems arise in the analysis of some algorithms.

Section 4: Probability and Basic Counting

*__Example 34 (The birthday problem)__ Assume that all days of the year are equally likely to be birthdays and ignore leap years. If k people are chosen at random, what is the probability that they all have different birthdays? While we're at it, let's replace the number of days in a year with n.

Here's one way we can think about this. Arrange the people in a line. Their birthdays, listed in the same order as the people, are $b_1, b_2, \ldots, b_k$. The probability space is $\underline{n} \times \underline{n} \times \cdots \times \underline{n}$, where there are k copies of $\underline{n}$. Each of the n^k possible k-long lists are equally likely. We are interested in $P(A)$, where A consists of those lists without repeats. Thus $|A| = n(n-1)\cdots(n-(k-1))$ and so

$$P(A) = \frac{n(n-1)\cdots(n-(k-1))}{n^k} = \prod_{i=1}^{k-1} \frac{n-i}{n} = \prod_{i=1}^{k-1}\left(1 - \frac{i}{n}\right).$$

While this answer is perfectly correct, it does not give us any idea how large $P(A)$ is. Of course, if k is very small, $P(A)$ will be nearly 1, and, if k is very large, $P(A)$ will be nearly 0. (In fact $P(A) = 0$ if $k > n$. This can be proved by using our formula. You should do it.) Where does this transition from near 1 to near 0 occur and how does $P(A)$ behave during the transition? Our goal is to answer this question.

We now suppose that $k \leq n^{3/5}$. Why 3/5? Just accept that this is a good choice. We need the following fact which will be proved at the end of the example:

If $0 \leq x \leq 1/2$, then $e^{-x-x^2} \leq 1 - x \leq e^{-x}$.

First, we get an upper bound for $P(A)$. Using $1 - x \leq e^{-x}$ with $x = i/n$,

$$P(A) = \prod_{i=1}^{k-1}\left(1 - \frac{i}{n}\right) \leq \prod_{i=1}^{k-1} e^{-i/n} = \exp\left(-\sum_{i=1}^{k-1} i/n\right).$$

Using the formula[7] $1 + 2 + \cdots + N = N(N+1)/2$ with $N = k - 1$, we have

$$\sum_{i=1}^{k-1} \frac{i}{n} = \frac{(k-1)k}{2n} = \frac{k^2}{2n} - \frac{k}{2n}.$$

Thus $P(A) \leq e^{-k^2/2n} e^{k/2n}$. Since $0 \leq k/n \leq n^{3/5}/n = n^{-2/5}$, which is small when n is large, $e^{k/2n}$ is close to 1. Thus, we have an upper bound for $P(A)$ that is close to $e^{-k^2/2n}$.

Next, we get our lower bound for $P(A)$ From the other inequality in our fact, namely $1 - x \geq e^{-x-x^2}$, we have

$$P(A) = \prod_{i=1}^{k-1}\left(1 - \frac{i}{n}\right) \geq \prod_{i=1}^{k-1} e^{-i/n - (i/n)^2} = \left(\prod_{i=1}^{k-1} e^{-i/n}\right)\left(\prod_{i=1}^{k-1} e^{-(i/n)^2}\right).$$

Let's look at the last product. It is less than 1. Since $i < k$, all of the factors in the product are greater than $e^{-(k/n)^2}$. Since there are less than k factors, the product is greater than

---
[7] This is a formula you should have learned in a previous class.

Basic Counting and Listing

e^{-k^3/n^2}. Since $k \leq n^{3/5}$, $k^3/n^2 \leq n^{9/5}/n^2 = n^{-1/5}$, which is small when n is large. Thus the last product is close to 1 when n is large. This shows that $P(A)$ has a lower bound which is close to $\prod_{i=1}^{k-1} e^{-i/n}$, which is the upper bound estimate we got in the previous paragraph. Since our upper and lower bounds are close together, they are both close to $P(A)$. In the previous paragraph, we showed that the upper bound is close to $e^{-k^2/2n}$.

To summarize, we have shown that

If n is large and $k \leq n^{3/5}$, then $P(A)$ is close to $e^{-k^2/2n}$.

What happens when $k > n^{3/5}$?

- First note that $P(A)$ decreases as k increases. You can see this by thinking about the original problem. You can also see it by looking at the product we obtained for $P(A)$, noting that each factor is less than 1 and noting that we get more factors as k increases.

- Second note that, when k is near $n^{3/5}$ but smaller than $n^{3/5}$, then $k^2/2n$ is large and so $P(A)$ is near 0 since e to a large negative power is near 0.

Putting these together, we see that $P(A)$ must be near 0 when $k \geq n^{3/5}$.

How does $e^{-k^2/2n}$ behave? When k is much smaller than $\sqrt{n}$, $k^2/2n$ is close to 0 and so $e^{-k^2/2n}$ is close to 1. When k is much larger than $\sqrt{n}$, $k^2/2n$ is large and so $e^{-k^2/2n}$ is close to 0. Put in terms of birthdays, for which $n = 365$ and $\sqrt{365} \approx 19$:

- When k is much smaller than 19, the probability of distinct birthdays is nearly 1.

- When k is much larger than 19, the probability of distinct birthdays is nearly 0.

- In between, the probability of distinct birthdays is close to $e^{-k^2/(2 \times 365)}$.

Here's a graph of $P(A)$ ("staircase" curve) and the approximation function $e^{-k^2/2n}$ (smooth curve) for various values of k when $n = 365$.[8] As you can see, the approximation is quite accurate.

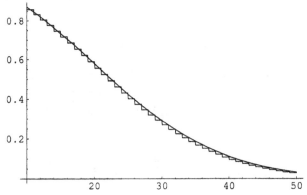

---

[8] Since $P(A)$ is defined only for k an integer, it should be a series of dots. To make it more visible, we've plotted it as a step function (staircase). The approximation is given by the function $e^{-x^2/2n}$, which is a smooth curve.

Section 4: Probability and Basic Counting

We now prove our fact. It requires calculus. By Taylor's theorem, the series for the natural logarithm is

$$\ln(1-x) = -\sum_{i=1}^{\infty} \frac{x^k}{k}.$$

Since $x > 0$ all the terms in the sum are negative. Throwing away all but the first term, $\ln(1-x) < -x$. Exponentiating, we have $1 - x < e^{-x}$, which is half of our fact.

Note that

$$\sum_{k=1}^{\infty} x^k/k = x + x^2 \sum_{k=2}^{\infty} x^{k-2}/k.$$

Since $0 \leq x \leq 1/2$ and $k \geq 2$ in the second sum, we have $x^{k-2}/k \leq (1/2)^{k-2}/2$. By the formula for the sum of a geometric series,[9]

$$\sum_{k=2}^{\infty} (1/2)^{k-2}/2 = 1/2 + (1/2)^2 + (1/2)^3 + \cdots = \frac{1/2}{1 - 1/2} = 1.$$

Thus

$$\sum_{k=1}^{\infty} x^k/k = x + x^2 \sum_{k=2}^{\infty} x^{k-2}/k \leq x + x^2,$$

and so $\ln(1-x) \geq -x - x^2$, which gives us the other half of our fact. □

Exercises for Section 4

4.1. Six horses are in a race. You pick two of them at random and bet on them both. Find the probability that you picked the winner. State clearly what your probability space is.

4.2. A roulette wheel consists of 38 containers numbered 0 to 36 and 00. In a fair wheel the ball is equally likely to fall into each container. A special wheel is designed in which all containers are the same size except that 00 is 5% larger than any of the others so that 00 has a 5% greater chance of occurring than any of the other values. What is the probability that 00 will occur on a spin of the wheel?

4.3. Alice and Bob have lost a key at the beach. They each get out their metal detectors and hunt until the key is found. If Alice can search 20% faster than Bob, what are the odds that she finds the key? What is the probability that Alice finds the key?

---

[9] Recall that the sum of the geometric series $a + ar + ar^2 + \cdots$ is $a/(1-r)$. You should be able to see that here $a = 1/2$ and $r = 1/2$.

Basic Counting and Listing

4.4. Six horses are in a race. You pick two of them at random and bet on them both. Find the probability that you picked a horse that won or placed (came in second). This should include the possibility that one of your picks won and the other placed.

4.5. Suppose 4 different balls are placed into 4 labeled boxes at random. (This can be done in 4^4 ways.)

  (a) What is the probability that no box is empty?

  (b) What is the probability that exactly one box is empty?

  (c) What is the probability that at least one box is empty?

  (d) Repeat (a)–(c) if there are 5 balls and 4 boxes.

4.6. For each event E determine $P(E)$.

  (a) Suppose a fair die is thrown k times and the values shown are recorded. What is the sample space? What is the probability of the event E that the sum of the values is even?

  (b) A card is drawn uniformly at random from a regular deck of cards. This process is repeated n times, with replacement. What is the sample space? What is the probability that a king, K, doesn't appear on any of the draws? What is the probability that at least one K appears in n draws?

  (c) An urn contains 3 white, 4 red, and 5 blue marbles. Two marbles are drawn without replacement. What is the sample space? What is the probability that both marbles are red?

4.7. Six light bulbs are chosen at random from 15 bulbs of which 5 are defective. What is the probability that exactly 3 are defective?

4.8. An urn contains ten labeled balls, labels $1, 2, \ldots, 10$.

  (a) Two balls are drawn together. What is the sample space? What is the probability that the sum of the labels on the balls is odd?

  (b) Two balls are drawn one after the other without replacement. What is the sample space? What is the probability that the sum is odd?

  (c) Two balls are drawn one after the other with replacement. What is the sample space? What is the probability that the sum is odd?

4.9. Let A and B be events with $P(A) = 3/8$, $P(B) = 1/2$, and $P((A \cup B)^c) = 3/8$. What is $P(A \cap B)$?

4.10. Of the students at a college, 20% are computer science majors and 58% of the entire student body are women. 430 of the 5,000 students at the college are women majoring in computer science.

  (a) How many women are not computer science majors?

Section 4: Probability and Basic Counting

(b) How many men are not computer science majors?

(c) What is the probability that a student selected at random is a woman computer science major?

(d) What is the probability that a female student selected at random is a computer science major?

4.11. The odds on the horse Beatlebomb in the Kentucky Derby are 100 to 1. A man at the races tells his wife that he is going to flip a coin. If it comes up heads he will bet on Beatlebomb, otherwise he will skip this race and not bet. What is the probability that he bets on Beatlebomb and wins?

4.12. Four persons, called North, South, East, and West, are each dealt 13 cards from an ordinary deck of 52 cards. If South has exactly two aces, what is the probability that North has the other two aces?

4.13. You have been dealt 4 cards and discover that you have 3 of a kind; that is, 3 cards have the same face value and the fourth is different. For example, you may have been dealt 4♠ 4♡ 10♠ 4♣. The other three players each receive four cards, but you do not know what they have been dealt. What is the probability that the fifth card will improve your hand by making it 4 of a kind or a full house (3 of a kind and a pair)?

4.14. Three boys and three girls are lined up in a row.

(a) What is the probability of all three girls being together?

(b) Suppose they are then seated around a circular table with six seats in the same order they were lined up. What is the probability that all three girls sit together?

4.15. Prove the *principle of inclusion exclusion*, for three sets namely that

$$P(A^c \cap B^c \cap C^c) = 1 - P(A) - P(B) - P(C) + P(A \cap B) + P(A \cap C) + P(B \cap C) - P(A \cap B \cap C).$$

(The formula extends in a fairly obvious way to any number of sets.)
Hint: Recall that, that for two sets, $P(A^c \cap B^c) = 1 - P(A) - P(B) + P(A \cap B)$.

4.16. A point is selected uniformly at random on a stick. This stick is broken at this point. What is the probability that the longer piece is at least twice the length of the shorter piece?

4.17. Two points are selected uniformly at random on a stick of unit length. The stick is broken at these two points. What is the probability that the three pieces form a triangle?

*4.18. What is the probability that a coin of diameter $d \leq 1$ when tossed onto the Euclidean plane (i.e., $\mathbb{R} \times \mathbb{R}$, $\mathbb{R}$ the real numbers) covers a lattice point of the plane

Basic Counting and Listing

(i.e., a point (p, q), where p and q are integers)?
Hint: Compare this problem with Example 33.

*4.19. Three points are selected at random on a circle C. What is the probability that all three points lie on a common semicircle of C? What if 3 is replaced by k?

Review Questions

Multiple Choice Questions for Review

1. Suppose there are 12 students, among whom are three students, M, B, C (a Math Major, a Biology Major, a Computer Science Major). We want to send a delegation of four students (chosen from the 12 students) to a convention. How many ways can this be done so that the delegation includes *exactly* two (not more, not less) students from $\{M, B, C\}$?

   (a) 32   (b) 64   (c) 88   (d) 108   (e) 144

2. The permutations of $\{a, b, c, d, e, f, g\}$ are listed in lex order. What permutations are just before and just after $bacdefg$?

   (a) Before: $agfedbc$, After: $bacdfge$

   (b) Before: $agfedcb$, After: $badcefg$

   (c) Before: $agfebcd$, After: $bacedgf$

   (d) Before: $agfedcb$, After: $bacdfge$

   (e) Before: $agfedcb$, After: $bacdegf$

3. Teams A and B play in a basketball tournament. The first team to win two games in a row or a total of three games wins the tournament. What is the number of ways the tournament can occur?

   (a) 8   (b) 9   (c) 10   (d) 11   (e) 12

4. The number of four letter words that can be formed from the letters in BUBBLE (each letter occurring at most as many times as it occurs in BUBBLE) is

   (a) 72   (b) 74   (c) 76   (d) 78   (e) 80

5. The number of ways to seat 3 boys and 2 girls in a row if each boy must sit next to at least one girl is

   (a) 36   (b) 48   (c) 148   (d) 184   (e) 248

6. Suppose there are ten balls in an urn, four blue, four red, and two green. The balls are also numbered 1 to 10. How many ways are there to select an ordered sample of four balls without replacement such that there are two blue balls and two red balls in the sample?

   (a) 144   (b) 256   (c) 446   (d) 664   (e) 864

7. How many different rearrangements are there of the letters in the word BUBBLE?

   (a) 40   (b) 50   (c) 70   (d) 80   (e) 120

8. The English alphabet has 26 letters of which 5 are vowels (A,E,I,O,U). How many seven letter words, with all letters distinct, can be formed that start with B, end with the letters ES, and have exactly three vowels? The "words" for this problem are just strings of letters and need not have linguistic meaning.

   (a) $2^3 \times 3^4 \times 17$

   (b) $2^3 \times 3^4 \times 19$

[41]

Basic Counting and Listing

(c) $2^4 \times 3^4 \times 19$

(d) $2^4 \times 3^3 \times 19$

(e) $2^4 \times 3^3 \times 17$

9. The permutations on $\{a,b,c,d,e,f,g\}$ are listed in lex order. All permutations $x_1 x_2 x_3 x_4 x_5 x_6 x_7$ with $x_4 = a$ or $x_4 = c$ are kept. All others are discarded. In this reduced list what permutation is just after $dagcfeb$?

(a) $dbacefg$

(b) $dbcaefg$

(c) $dbacgfe$

(d) $dagcfbe$

(e) $dcbaefg$

10. The number of four letter words that can be formed from the letters in SASSABY (each letter occurring at most as many times as it occurs in SASSABY) is

(a) 78   (b) 90   (c) 108   (d) 114   (e) 120

11. How many different rearrangements are there of the letters in the word TATARS if the two A's are never adjacent?

(a) 24   (b) 120   (c) 144   (d) 180   (e) 220

12. Suppose there are ten balls in an urn, four blue, four red, and two green. The balls are also numbered 1 to 10. How many ways are there to select an *ordered* sample of four balls without replacement such that the number $B \geq 0$ of blue balls, the number $R \geq 0$ of red balls, and the number $G \geq 0$ of green balls are all different?

(a) 256   (b) 864   (c) 1152   (d) 1446   (e) 2144

13. Suppose there are ten balls in an urn, four blue, four red, and two green. The balls are also numbered 1 to 10. You are asked to select an *ordered* sample of four balls without replacement. Let $B \geq 0$ be the number of blue balls, $R \geq 0$ be the number of red balls, and $G \geq 0$ be the number of green balls in your sample. How many ways are there to select such a sample if *exactly* one of B, R, or G must be zero?

(a) 256   (b) 1152   (c) 1446   (d) 2144   (e) 2304

14. The number of partitions of $X = \{a,b,c,d\}$ with a and b in the same block is

(a) 4   (b) 5   (c) 6   (d) 7   (e) 8

15. Let W_{ab} and W_{ac} denote the set of partitions of $X = \{a,b,c,d,e\}$ with a and b belonging to the same block and with a and c belonging to the same block, respectively. Similarly, let W_{abc} denote the set of partitions of $X = \{a,b,c,d,e\}$ with a, b, and c belonging to the same block. What is $|W_{ab} \cup W_{ac}|$? (Note: $B(3) = 5$, $B(4) = 15$, $B(5) = 52$, where $B(n)$ is the number of partitions of an n-element set).

(a) 25   (b) 30   (c) 35   (d) 40   (e) 45

16. The number of partitions of $X = \{a,b,c,d,e,f,g\}$ with a, b, and c in the same block and c, d, and e in the same block is

(a) 2   (b) 5   (c) 10   (d) 15   (e) 52

17. Three boys and four girls sit in a row with all arrangements equally likely. Let x be the probability that no two boys sit next to each other. What is x?

    (a) 1/7   (b) 2/7   (c) 3/7   (d) 4/7   (e) 5/7

18. A man is dealt 4 spade cards from an ordinary deck of 52 cards. He is given 2 more cards. Let x be the probability that they both are the same suit. Which is true?

    (a) $.2 < x \le .3$

    (b) $0 < x \le .1$

    (c) $.1 < x \le .2$

    (d) $.3 < x \le .4$

    (e) $.4 < x \le .5$

19. Six light bulbs are chosen at random from 15 bulbs of which 5 are defective. What is the probability that exactly one is defective?

    (a) $C(5,1)C(10,6)/C(15,6)$

    (b) $C(5,1)C(10,5)/C(15,6)$

    (c) $C(5,1)C(10,1)/C(15,6)$

    (d) $C(5,0)C(10,6)/C(15,6)$

    (e) $C(5,0)C(10,5)/C(15,6)$

20. A small deck of five cards are numbered 1 to 5. First one card and then a second card are selected at random, with replacement. What is the probability that the sum of the values on the cards is a prime number?

    (a) 10/25   (b) 11/25   (c) 12/25   (d) 13/25   (e) 14/25

21. Let A and B be events with $P(A) = 6/15$, $P(B) = 8/15$, and $P((A \cup B)^c) = 3/15$. What is $P(A \cap B)$?

    (a) 1/15   (b) 2/15   (c) 3/15   (d) 4/15   (e) 5/15

22. Suppose the odds of A occurring are 1:2, the odds of B occurring are 5:4, and the odds of both A and B occurring are 1:8. The odds of $(A \cap B^c) \cup (B \cap A^c)$ occurring are

    (a) 2:3   (b) 4:3   (c) 5:3   (d) 6:3   (e) 7:3

23. A pair of fair dice is tossed. Find the probability that the greatest common divisor of the two numbers is one.

    (a) 12/36   (b) 15/36   (c) 17/36   (d) 19/36   (e) 23/36

24. Three boys and three girls sit in a row. Find the probability that exactly two of the girls are sitting next to each other (the remaining girl separated from them by at least one boy).

    (a) 4/20   (b) 6/20   (c) 10/20   (d) 12/20   (e) 13/20

25. A man is dealt 4 spade cards from an ordinary deck of 52 cards. If he is given five more, what is the probability that none of them are spades?

Basic Counting and Listing

(a) $\binom{39}{1}/\binom{48}{5}$ (b) $\binom{39}{2}/\binom{48}{5}$ (c) $\binom{39}{3}/\binom{48}{5}$ (d) $\binom{39}{5}/\binom{48}{5}$ (e) $\binom{39}{6}/\binom{48}{5}$

Answers: **1** (d), **2** (e), **3** (c), **4** (a), **5** (a), **6** (e), **7** (e), **8** (c), **9** (a), **10** (d), **11** (b), **12** (c), **13** (e), **14** (b), **15** (a), **16** (b), **17** (b), **18** (a), **19** (b), **20** (b), **21** (b), **22** (d), **23** (e), **24** (d), **25** (d).

Unit Fn

Functions

Section 1: Some Basic Terminology

Functions play a fundamental role in nearly all of mathematics. Combinatorics is no exception. In this section we review the basic terminology and notation for functions. Permutations are special functions that arise in a variety of ways in combinatorics. Besides studying them for their own interest, we'll see them as a central tool in other topic areas.

Except for the real numbers $\mathbb{R}$, rational numbers $\mathbb{Q}$ and integers $\mathbb{Z}$, our sets are normally finite. The set of the first n positive integers, $\{1, 2, \ldots, n\}$ will be denoted by $\underline{n}$.

Recall that $|A|$ is the number of elements in the set A. When it is convenient to do so, we linearly order the elements of a set A. In that case we denote the ordering by $a_1, a_2, \ldots, a_{|A|}$ or by $(a_1, a_2, \ldots, a_{|A|})$. Unless clearly stated otherwise, the ordering on a set of numbers is the numerical ordering. For example, the ordering on $\underline{n}$ is $1, 2, 3, \ldots, n$.

A review of the terminology concerning sets will be helpful. When we speak about sets, we usually have a "universal set" U in mind, to which the various sets of our discourse belong. Let U be a set and let A and B be subsets of U.

- The sets $A \cap B$ and $A \cup B$ are the *intersection* and *union* of A and B.
- The set $A \setminus B$ or $A - B$ is the *set difference* of A and B; that is, the set $\{x : x \in A, x \notin B\}$.
- The set $U \setminus A$ or A^c is the *complement* of A (relative to U). The complement of A is also written A' and $\sim A$.
- The set $A \oplus B = (A \setminus B) \cup (B \setminus A)$ is *symmetric difference* of A and B; that is, those x that are in *exactly one* of A and B. We have $A \oplus B = (A \cup B) \setminus (A \cap B)$.
- $\mathcal{P}(A)$ is the set of all subsets of A. (The notation for $\mathcal{P}(A)$ varies from author to author.)
- $\mathcal{P}_k(A)$ the set of all subsets of A of size (or cardinality) k. (The notation for $\mathcal{P}_k(A)$ varies from author to author.)
- The Cartesian product $A \times B$ is the set of all ordered pairs built from A and B:

$$A \times B = \{\, (a,b) \mid a \in A \text{ and } b \in B \,\}.$$

We also call $A \times B$ the *direct product* of A and B.

If $A = B = \mathbb{R}$, the real numbers, then $\mathbb{R} \times \mathbb{R}$, written $\mathbb{R}^2$, is frequently interpreted as coordinates of points in the plane. Two points are the same if and only if they have the same coordinates, which says the same thing as our definition, $(a, b) = (a', b')$ if $a = a'$ and $b = b'$. Recall that the direct product can be extended to any number of sets. How can $\mathbb{R} \times \mathbb{R} \times \mathbb{R} = \mathbb{R}^3$ be interpreted?

Definition 1 (Function) *If A and B are sets, a function from A to B is a rule that tells us how to find a **unique** $b \in B$ for each $a \in A$. We write $f: A \to B$ to indicate that f is a function from A to B.*

[45]

Functions

We call the set A the *domain of f* and the set B the *range*[1] or, equivalently, *codomain of f*. To specify a function completely you must give its domain, range and rule.

The set of all functions from A to B is written B^A, for a reason we will soon explain. Thus $f: A \to B$ and $f \in B^A$ say the same thing.

In calculus you dealt with functions whose ranges were $\mathbb{R}$ and whose domains were contained in $\mathbb{R}$; for example, $f(x) = 1/(x^2 - 1)$ is a function from $\mathbb{R} - \{-1, 1\}$ to $\mathbb{R}$. You also studied functions of functions! The derivative is a function whose domain is all differentiable functions and whose range is all functions. If we wanted to use functional notation we could write $D(f)$ to indicate the function that the derivative associates with f.

Definition 2 (One-line notation) *When A is ordered, a function can be written in one-line notation as $(f(a_1), f(a_2), \ldots, f(a_{|A|}))$. Thus we can think of a function as an element of $B \times B \times \ldots \times B$, where there are $|A|$ copies of B. Instead of writing $B^{|A|}$ to indicate the set of all functions, we write B^A. Writing $B^{|A|}$ is incomplete because the domain A is not specified. Instead, only its size $|A|$ is given.*

Example 1 (Using the notation) To get a feeling for the notation used to specify a function, it may be helpful to imagine that you have an envelope or box that contains a function. In other words, this envelope contains all the information needed to completely describe the function. Think about what you're going to see when you open the envelope.

You might see

$$P = \{a, b, c\}, \qquad g: P \to \underline{4}, \qquad g(a) = 3, \quad g(b) = 1 \quad \text{and} \quad g(c) = 4.$$

This tells you that name of the function is g, the domain of g is P, which is $\{a, b, c\}$, and the range of g is $\underline{4} = \{1, 2, 3, 4\}$. It also tells you the values in $\underline{4}$ that g assigns to each of the values in its domain. Someone else may have put

$$g: \underline{4}^{\{a,b,c\}}, \qquad \text{ordering: } a, b, c, \qquad g = (3, 1, 4).$$

in the envelope instead. This describes the same function. It doesn't give a name for the domain, but we don't need a name like P for the set $\{a, b, c\}$ — we only need to know what is in the set. On the other hand, it gives an order on the domain so that the function can be given in one-line form. Can you describe other possible envelopes for the same function?

What if the envelope contained only $g = (3, 1, 4)$? You've been cheated! You *must* know the domain of g in order to known what g is. What if the envelope contained

$$\text{the domain of } g \text{ is } \{a, b, c\}, \qquad \text{ordering: } a, b, c, \qquad g = (3, 1, 4)?$$

We haven't specified the range of g, but is it necessary since we know the values of the function? Our definition included the requirement that the range be specified, so this is not a complete definition. On the other hand, in some discussions the range may not be important; for example, if $g = (3, 1, 4)$ all that may matter is that the range is large enough to contain 1, 3 and 4. In such cases, we'll be sloppy and accept this as if it were a complete specification. ☐

---

[1] Some people define "range" to be the values that the function actually takes on. Most people call that the *image*, a concept we will discuss a bit later.

Section 1: Some Basic Terminology

Example 2 (Counting functions) By the Rule of Product, $|B^A| = |B|^{|A|}$. We can represent a subset S of A by a unique function $f: A \to \{0,1\}$ where $f(x) = 0$ if $a \notin S$ and $f(x) = 1$ if $x \in S$. This proves that there are $2^{|A|}$ such subsets. For example, if $A = \{a, b, d\}$, then the number of subsets of A is $2^{|\{a,b,d\}|} = 2^3 = 8$.

We can represent a multiset S formed from A by a unique function $f: A \to \mathbb{N} = \{0, 1, 2, \ldots\}$ where $f(x)$ is the number of times x appears in S. If no element is allowed to appear more than k times, then we can restrict the codomain of f to be $\{0, 1, \ldots, k\}$ and so there are $(k+1)^{|A|}$ such multisets. For example, the number of multisets of $A = \{a, b, d\}$ where each element can appear at most 4 times is $(4 + 1)^{|A|} = 5^3 = 125$. The particular multiset $\{a, a, a, d, d\}$ is represented by the function $f(a) = 3$, $f(b) = 0$ and $f(d) = 2$.

We can represent a k-list of elements drawn from a set B, with repetition allowed, by a unique function $f: \underline{k} \to B$. In this representation, the list corresponds to the function written in one-line notation. (Recall that the ordering on $\underline{k}$ is the numerical ordering.) This proves that there are exactly $|B|^k$ such lists. For example, the number of 4-lists that can be formed from $B = \{a, b, d\}$ is $|B|^4 = 3^4 = 81$. The 4-list (b, d, d, a) corresponds to the function $f = (b, d, d, a)$ in 1-line notation, where the domain is $\underline{4}$. ☐

Definition 3 (Types of functions) Let $f: A \to B$ be a function. (Specific examples of these concepts are given after the definition.)

- If for every $b \in B$ there is an $a \in A$ such that $f(a) = b$, then f is called a *surjection* (or an *onto function*). Another way to describe a surjection is to say that it takes on each value in its range at least once.

- If $f(x) = f(y)$ implies $x = y$, then f is called an *injection* or a *one-to-one function*). Another way to describe an injection is to say that it takes on each value in its range at most once. The injections in $S^{\underline{k}}$ correspond to k-lists without repetitions.

- If f is both an injection and a surjection, it is a called a *bijection*.

- The bijections of A^A are called the *permutations* of A.

- If $f: A \to B$ is a bijection, we may talk about the *inverse* of f, written f^{-1}, which reverses what f does. Thus $f^{-1}: B \to A$ and $f^{-1}(b)$ is that unique $a \in A$ such that $f(a) = b$. Note that $f(f^{-1}(b)) = b$ and $f^{-1}(f(a)) = a$.[2]

Example 3 (Types of functions) Let $A = \{1, 2, 3\}$ and $B = \{a, b\}$ be the domain and range of the function $f = (a, b, a)$. The function is a surjection because every element of the range is "hit" by the function. It is not an injection because a is hit twice.

Now consider the function g with domain B and range A given by $g(a) = 3$ and $g(b) = 1$. It is not a surjection because it misses 2; however, it is an injection because each element of A is hit at most once.

Neither f nor g is a bijection because some element of the range is either hit more than once or is missed. The function h with domain B and range $C = \{1, 3\}$ given by $h(a) = 3$ and $h(b) = 1$ is a bijection. At first, it may look like g and h are the same function. They

---

[2] Do not confuse f^{-1} with $1/f$. For example, if $f: \mathbb{R} \to \mathbb{R}$ is given by $f(x) = x^3 + 1$, then $1/f(x) = 1/(x^3 + 1)$ and $f^{-1}(y) = (y - 1)^{1/3}$.

Functions

are not because they have different ranges. You can tell if a function is an injection without knowing its range, but you *must* know its range to decide if it is a surjection.

The inverse of the bijection h has domain C and range B it is given by $h^{-1}(1) = b$ and $h^{-1}(a) = 3$.

The function f with domain and range $\{a, b, c, d\}$ given in 2-line form by

$$f = \begin{pmatrix} a & b & c & d \\ b & c & a & d \end{pmatrix}$$

is a permutation. You can see this immediately because the domain equals the range and the bottom line of the 2-line form is a rearrangement of the top line. The 2-line form is convenient for writing the inverse—just switch the top and bottom lines. In this example,

$$f^{-1} = \begin{pmatrix} b & c & a & d \\ a & b & c & d \end{pmatrix}. \quad \square$$

Example 4 (Functions as relations) There is another important set-theoretic way of defining functions. Let A and B be sets. A *relation from A to B* is a subset of $A \times B$. For example:

If $A = \underline{3}$ and $B = \underline{4}$, then $R = \{(1,4), (1,2), (3,3), (2,3)\}$ is a relation from A to B.

If the relation R satisfies the condition that, for all $x \in A$ there is a *unique* $y \in B$ such that $(x, y) \in R$, then the relation R is called a *functional relation*. In the notation from logic, this can be written

$$\forall x \in A \; \exists! y \in B \ni (x, y) \in R.$$

This mathematical shorthand is well worth knowing:

- "$\forall$" means "for all",
- "$\exists$" means "there exists",
- "$\exists!$" means "there exists a unique", and
- "$\ni$" means "such that."

In algebra or calculus, when you draw a graph of a real-valued function $f : \mathbb{R} \to \mathbb{R}$ (such as $f(x) = x^3$), you are attempting a pictorial representation of the set $\{(x, f(x)) : x \in \mathbb{R}\}$, which is the subset of $\mathbb{R} \times \mathbb{R}$ that is the "functional relation from $\mathbb{R}$ to $\mathbb{R}$." In general, if $R \subset A \times B$ is a functional relation, then the function f corresponding to R has domain A and codomain B and is given by the ordered pairs $\{(x, f(x)) \mid x \in A\} = R$.

If you think of the "envelope game," Example 1, you will realize that a functional relation is yet another thing you might find in the envelope that describes a function. When a subset is defined it is formally required in mathematics that the "universal set" from which it has been extracted to form a subset also be described. Thus, in the envelope, in addition to R, you must also find enough information to describe completely $A \times B$. As you can see, a function can be described by a variety of different "data structures."

Given any relation $R \subseteq A \times B$, the inverse relation R^{-1} from B to A is defined to be $\{(y, x) : (x, y) \in R\}$. Recall the example in the previous paragraph where $A = \underline{3}$, $B = \underline{4}$, and

Section 1: Some Basic Terminology

$R = \{(1,4), (1,2), (3,3), (2,3)\}$, The inverse relation is $R^{-1} = \{(4,1), (2,1), (3,3), (3,2)\}$. Notice that all we've had to do is reverse the order of the elements in the ordered pairs $(1,4), \ldots, (2,3)$ of R to obtain the ordered pairs $(4,1), \ldots, (3,2)$ of R^{-1}.

Note that neither R nor R^{-1} is a functional relation in the example in the previous paragraph. You should make sure you understand why this statement is true (Hint: R fails the "∃!" test and R^{-1} fails the "∀" part of the definition of a functional relation). Note also that if both R and R^{-1} are functional relations then $|A| = |B|$. In this case, R (and R^{-1}) are bijections in the sense of Definition 3. □

Example 5 (Two-line notation) Since one-line notation is a simple, brief way to specify functions, we'll use it frequently. If the domain is not a set of numbers, the notation is poor because we must first pause and order the domain. There are other ways to write functions which overcome this problem. For example, we could write $f(a) = 4$, $f(b) = 3$, $f(c) = 4$ and $f(d) = 1$. This could be shortened up somewhat to

$$a \to 4,\ b \to 3,\ c \to 4 \text{ and } d \to 1.$$

By turning each of these sideways, we can shorten it even more: $\begin{pmatrix} a & b & c & d \\ 4 & 3 & 4 & 1 \end{pmatrix}$. For obvious reasons, this is called *two-line notation*. Since x always appears directly over $f(x)$, there is no need to order the domain; in fact, we need not even specify the domain separately since it is given by the top line. If the function is a bijection, its inverse function is obtained by interchanging the top and bottom lines.

The arrows we introduced in the last paragraph can be used to help visualize different properties of functions. Imagine that you've listed the elements of the domain A in one column and the elements of the range B in another column to the right of the domain. Draw an arrow from a to b if $f(a) = b$. Thus the heads of arrows are labeled with elements of B and the tails with elements of A. Here are some arrow diagrams.

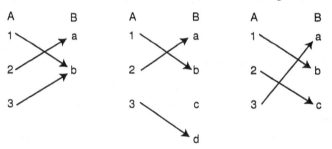

In all three functions, the domain $A = \{1, 2, 3\}$; however, the range B is different for each function. Since each diagram represents a function f, no two arrows have the same tail. If f is an injection, no two arrows have the same head. Thus the second and third diagrams are injections, but the first is not. If f is a surjection, every element of B is on the head of some arrow. Thus the first and third diagrams are surjections, but the second is not. Since the third diagram is both an injection and a surjection, it is a bijection. You should be able to describe the situation with the arrowheads when f is a bijection. How can you tell if a diagram represents a permutation? □

[49]

Functions

Exercises for Section 1

1.1. This exercise lets you check your understanding of the definitions. In each case below, some information about a function is given to you. Answer the following questions and give reasons for your answers:

- Have you been given enough information to specify the function?
- Can you tell whether or not the function is an injection? a surjection? a bijection?
- If possible, give the function in two-line form.

(a) $f \in \underline{3}^{\{>,<,+,?\}}$, $f = (3, 1, 2, 3)$.

(b) $f \in \{>,<,+,?\}^{\underline{3}}$, $f = (?, <, +)$.

(c) $f \in \underline{4}^{\underline{3}}$, $2 \to 3$, $1 \to 4$, $3 \to 2$.

1.2. Let A and B be finite sets and $f: A \to B$. Prove the following claims. Some are practically restatements of the definitions, some require a few steps.

(a) If f is an injection, then $|A| \leq |B|$.

(b) If f is a surjection, then $|A| \geq |B|$.

(c) If f is a bijection, then $|A| = |B|$.

(d) If $|A| = |B|$, then f is an injection if and only if it is a surjection.

(e) If $|A| = |B|$, then f is a bijection if and only if it is an injection or it is a surjection.

1.3. Let S be the set of students attending a large university, let I be the set of student ID numbers for those students, let D be the set of dates for the past 100 years (month/day/year), let G be the set of 16 possible grade point averages between 2.0 and 3.5, rounded to the nearest tenth. For each of the following, decide whether or not it is a function. If it is, decide whether it is an injection, bijection or surjection. Give reasons for your answers.

(a) The domain is S, the codomain is I and the function maps each student to his or her ID number.

(b) The domain is S, the codomain is D and the function maps each student to his or her birthday.

(c) The domain is D, the codomain is I and the function maps each date to the ID number of a student born on that date. If there is more than one such student, the lexicographically least ID number is chosen.

(d) The domain is S, the codomain is G and the function maps each student to his or her GPA rounded to the nearest tenth.

(e) The domain is G, the codomains is I and the function maps each GPA to the ID number of a student with that GPA. If there is more than one such student, the lexicographically least ID number is chosen.

1.4. Let $A = \{1, 2, 3\}$ and $B = \{a, b, d\}$. Consider the following subsets of sets.

$$\{(3,a), (2,b), (1,a)\}, \quad \{(1,a), (2,b), (3,c)\},$$
$$\{(1,a), (2,b), (1,d)\}, \quad \{(1,a), (2,b) (3,d), (1,b)\}.$$

Which of them are relations on $A \times B$? Which of the are functional relations? Which of their inverses are functional relations?

Section 2: Permutations

Before beginning our discussion, we need the notion of composition of functions. Suppose that f and g are two functions such that the values f takes on are contained in the domain of g. We can write this as $f: A \to B$ and $g: C \to D$ where $f(a) \in C$ for all $a \in A$. We define the *composition* of g and f, written $gf: A \to D$ by $(gf)(x) = g(f(x))$ for all $x \in A$. The notation $g \circ f$ is also used to denote composition. Suppose that f and g are given in two-line notation by

$$f = \begin{pmatrix} p & q & r & s \\ P & R & T & U \end{pmatrix} \qquad g = \begin{pmatrix} P & Q & R & S & T & U & V \\ 1 & 3 & 5 & 2 & 4 & 6 & 7 \end{pmatrix}.$$

Then $gf = \begin{pmatrix} p & q & r & s \\ 1 & 5 & 4 & 6 \end{pmatrix}$. To derive $(gf)(p)$, we noted that $f(p) = P$ and $g(P) = 1$. The other values of gf were derived similarly.

The set of permutations on a set A is denoted in various ways in the literature. Two notations are $\text{PER}(A)$ and $\mathcal{S}(A)$. Suppose that f and g are permutations of a set A. Recall that a permutation is a bijection from a set to itself and so it makes sense to talk about f^{-1} and fg. We claim that fg and f^{-1} are also permutations of A. This is easy to see if you write the permutations in two-line form and note that the second line is a rearrangement of the first if and only if the function is a permutation. You may want to look ahead at the next example which illustrates these ideas.

The permutation f given by $f(a) = a$ for all $a \in A$ is called the *identity permutation*. Notice that $f \circ f^{-1}$ and $f^{-1} \circ f$ both equal the identity permutation. You should be able to show that, if f is any permutation of A and e is the identity permutation of A, then $f \circ e = e \circ f = f$.

Again suppose that f is a permutation. Instead of $f \circ f$ or ff we write f^2. Note that $f^2(x)$ is not $(f(x))^2$. (In fact, if multiplication is not defined in A, $(f(x))^2$ has no meaning.) We could compose three copies of f. The result is written f^3. In general, we can compose k copies of f to obtain f^k. A cautious reader may be concerned that $f \circ (f \circ f)$ may not be the same as $(f \circ f) \circ f$. They are equal. In fact, $f^{k+m} = f^k \circ f^m$ for all nonnegative integers k and m, where f^0 is defined by $f^0(x) = x$ for all x in the domain. This is based

Functions

on the "associative law" which states that $f \circ (g \circ h) = (f \circ g) \circ h$ whenever the compositions make sense. We'll prove these results.

To prove that the two functions are equal, it suffices to prove that they take on the same values for all x in the domain. Let's use this idea for $f \circ (g \circ h)$ and $(f \circ g) \circ h$. We have

$$(f \circ (g \circ h))(x) = f((g \circ h)(x)) \qquad \text{by the definition of } \circ,$$
$$= f(g(h(x))) \qquad \text{by the definition of } \circ.$$

Similarly

$$((f \circ g) \circ h)(x) = (f \circ g)(h(x)) \qquad \text{by the definition of } \circ,$$
$$= f(g(h(x))) \qquad \text{by the definition of } \circ.$$

More generally, one can use this approach to prove by induction that $f_1 \circ f_2 \circ \cdots \circ f_n$ is well defined. This result then implies that $f^{k+m} = f^k \circ f^m$. Note that we have proved that the associative law for any three functions f, g and h for which the domain of f contains the values taken on by g and the domain of g contains the values taken on by h.

Example 6 (Composing permutations) We'll use the notation. Let f and g be the permutations

$$f = \begin{pmatrix} 1 & 2 & 3 & 4 & 5 \\ 2 & 1 & 4 & 5 & 3 \end{pmatrix} \qquad g = \begin{pmatrix} 1 & 2 & 3 & 4 & 5 \\ 2 & 3 & 4 & 5 & 1 \end{pmatrix}.$$

We can compute fg by calculating all the values. This can be done fairly easily from the two-line form: For example, $(fg)(1)$ can be found by noting that the image of 1 under g is 2 and the image of 2 under f is 1. Thus $(fg)(1) = 1$. You should be able to verify that

$$fg = \begin{pmatrix} 1 & 2 & 3 & 4 & 5 \\ 1 & 4 & 5 & 3 & 2 \end{pmatrix} \qquad gf = \begin{pmatrix} 1 & 2 & 3 & 4 & 5 \\ 3 & 2 & 5 & 1 & 4 \end{pmatrix} \neq fg$$

and that

$$f^2 = \begin{pmatrix} 1 & 2 & 3 & 4 & 5 \\ 1 & 2 & 5 & 3 & 4 \end{pmatrix} \quad f^3 = \begin{pmatrix} 1 & 2 & 3 & 4 & 5 \\ 2 & 1 & 3 & 4 & 5 \end{pmatrix} \quad g^5 = f^6 = \begin{pmatrix} 1 & 2 & 3 & 4 & 5 \\ 1 & 2 & 3 & 4 & 5 \end{pmatrix}.$$

Note that it is easy to get the inverse, simply interchange the two lines. Thus

$$f^{-1} = \begin{pmatrix} 2 & 1 & 4 & 5 & 3 \\ 1 & 2 & 3 & 4 & 5 \end{pmatrix} \quad \text{which is the same as } f^{-1} = \begin{pmatrix} 1 & 2 & 3 & 4 & 5 \\ 2 & 1 & 5 & 3 & 4 \end{pmatrix},$$

since the order of the columns in two-line form does not matter. □

Let f be a permutation of the set A and let $n = |A|$. If $x \in A$, we can look at the sequence

$$x, f(x), f(f(x)), \ldots, f^k(x), \ldots,$$

which is often written as

$$x \to f(x) \to f(f(x)) \to \cdots \to f^k(x) \to \cdots.$$

Section 2: Permutations

Since the range of f has n elements, this sequence will contain a repeated element in the first $n + 1$ entries. Suppose that $f^s(x)$ is the first sequence entry that is ever repeated and that $f^{s+p}(x)$ is the first time that it is repeated. Thus $f^s(x) = f^{s+p}(x)$. Apply $(f^{-1})^s$ to both sides of this equality to obtain $x = f^p(x)$ and so, in fact, $s = 0$. It follows that the sequence cycles through a pattern of length p forever since

$$f^{p+1}(x) = f(f^p(x)) = f(x), \quad f^{p+2}(x) = f^2(f^p(x)) = f^2(x), \quad \text{and so on.}$$

We call $(x, f(x), \ldots, f^{p-1}(x))$ the *cycle* containing x and call p the *length of the cycle*. If a cycle has length p, we call it a p-cycle.[3] Cyclic shifts of a cycle are considered the same; for example, if (1,2,6,3) is the cycle containing 1 (as well as 2, 3 and 6), then (2,6,3,1), (6,3,1,2) and (3,1,2,6) are other ways of writing the cycle (1,2,6,3). A cycle looks like a function in one-line notation. How can we tell them apart? Either we will be told or it will be clear from the context.

Example 7 (Using cycle notation) Consider the permutation

$$f = \begin{pmatrix} 1 & 2 & 3 & 4 & 5 & 6 & 7 & 8 & 9 \\ 2 & 4 & 8 & 1 & 5 & 9 & 3 & 7 & 6 \end{pmatrix}.$$

Since $1 \to 2 \to 4 \to 1$, the cycle containing 1 is (1,2,4). We could equally well write it (2,4,1) or (4,1,2); however, (1,4,2) is a different cycle since it corresponds to $1 \to 4 \to 2 \to 1$. The usual convention is to list the cycle starting with its smallest element. The cycles of f are (1,2,4), (3,8,7), (5) and (6,9). We write f in *cycle form* as

$$f = (1, 2, 4)\ (3, 8, 7)\ (5)\ (6, 9).$$

It is common practice to omit the cycles of length one and write $f = (1, 2, 4)(3, 8, 7)(6, 9)$. The inverse of f is obtained by reading the cycles backwards because $f^{-1}(x)$ is the lefthand neighbor of x in a cycle. Thus

$$f^{-1} = (4, 2, 1)(7, 8, 3)(9, 6) = (1, 4, 2)(3, 7, 8)(6, 9). \qquad \square$$

Cycle form is useful in certain aspects of the branch of mathematics called "finite group theory." Here's an application.

---

[3] If $(a_1, a_2, \ldots, a_p)$ is a cycle of f, then

$$f(a_1) = a_2, \quad f(a_2) = a_3, \quad \ldots, \quad f(a_{p-1}) = a_p, \quad f(a_p) = a_1.$$

Functions

Example 8 (Powers of permutations) With a permutation in cycle form, it's very easy to calculate a power of the permutation. For example, suppose we want the tenth power of the permutation whose cycle form (including cycles of length 1) is $(1,5,3)(7)(2,6)$. To find the image of 1, we take ten steps: $1 \to 5 \to 3 \to 1 \cdots$. Where does it stop after ten steps? Since three steps bring us back to where we started (because 1 is in a cycle of length three), nine steps take us around the cycle three times and the tenth takes us to 5. Thus $1 \to 5$ in the tenth power. Similarly, $5 \to 3$ and $3 \to 1$. Clearly $7 \to 7$ regardless of the power. Ten steps take us around the cycle $(2,6)$ exactly five times, so $2 \to 2$ and $6 \to 6$. Thus the tenth power is $(1,5,3)(7)(2)(6)$. □

Suppose we have a permutation in cycle form whose cycle lengths all divide k. The reasoning in the previous example shows that the kth power of that permutation will be the identity permutation; that is, all the cycles will be 1-long and so every element is mapped to itself (i.e., $f(x) = x$ for all x). In particular, if we are considering permutations of an n-set, every cycle has length at most n and so we can take $k = n!$, regardless of the permutation. We have shown

Theorem 1 (A fixed power of n-permutations is the identity) *Given a set S, there are $k > 0$ depending on $|S|$ such that f^k is the identity permutation for every permutation f of S. Furthermore, $k = |S|!$ is one such k.*

*Example 9 (Involutions)** An *involution* is a permutation which is equal to its inverse. Since $f(x) = f^{-1}(x)$, we have $f^2(x) = f(f^{-1}(x)) = x$. Thus involutions are those permutations which have all their cycles of lengths one and two. How many involutions are there on $\underline{n}$? Let's count the involutions with exactly k 2-cycles and use the Rule of Sum to add up the results. We can build such an involution as follows:

- Select $2k$ elements for the 2-cycles AND
- partition these $2k$ elements into k blocks that are all of size 2 AND
- put the remaining $n - 2k$ elements into 1-cycles.

Since there is just one 2-cycle on two given elements, we can interpret each block as 2-cycle. This specifies f. The number of ways to carry out the first step is $\binom{n}{2k}$. For the second step, we might try the multinomial coefficient $\binom{2k}{2,\ldots,2} = (2k)!/2^k$. This is almost right! In using the multinomial coefficient, we're assuming an ordering on the pairs even though they don't have one. For example, with $k = 3$ and the set $\underline{6}$, there are just 15 possible partitions as follows.

$$\{\{1,2\},\{3,4\},\{5,6\}\} \quad \{\{1,2\},\{3,5\},\{4,6\}\} \quad \{\{1,2\},\{3,6\},\{4,5\}\}$$
$$\{\{1,3\},\{2,4\},\{5,6\}\} \quad \{\{1,3\},\{2,5\},\{4,6\}\} \quad \{\{1,3\},\{2,6\},\{4,5\}\}$$
$$\{\{1,4\},\{2,3\},\{5,6\}\} \quad \{\{1,4\},\{2,5\},\{3,6\}\} \quad \{\{1,4\},\{2,6\},\{3,5\}\}$$
$$\{\{1,5\},\{2,3\},\{4,6\}\} \quad \{\{1,5\},\{2,4\},\{3,6\}\} \quad \{\{1,5\},\{2,6\},\{3,4\}\}$$
$$\{\{1,6\},\{2,3\},\{4,5\}\} \quad \{\{1,6\},\{2,4\},\{3,5\}\} \quad \{\{1,6\},\{2,5\},\{3,4\}\}$$

This is smaller than $\binom{6}{2,2,2} = 6!/2!2!2! = 90$ because all 3! ways to order the three blocks in each partition are counted differently to obtain the number 90. This is because we've chosen a first, second and third block instead of simply dividing $\underline{6}$ into three blocks of size two.

How can we solve the dilemma? Actually, the discussion of what went wrong contains the key to the solution: The multinomial coefficient counts ordered collections of k blocks and we want unordered collections. Since the blocks in a partition are all distinct, there are $k!$ ways to order the blocks and so the multinomial coefficient counts each unordered collection $k!$ times. Thus we must simply divide the multinomial coefficient by $k!$. If this dividing by $k!$ bothers you, try looking at it this way. Let $f(k)$ be the number of ways to carry out the second step, partition the $2k$ elements into k blocks that are all of size 2. Since the k blocks can be permuted in $k!$ ways, the Rule of Product tells us that there are $f(k)\,k!$ ways to select k ordered blocks of 2 elements each. Thus $f(k)\,k! = \binom{2k}{2,\ldots,2}$.

Since there is just one way to carry out Step 3, the Rule of Product tells us that the number of involutions with exactly k 2-cycles is

$$\binom{n}{2k} \frac{1}{k!} \binom{2k}{2,\ldots,2} = \frac{n!}{(2k)!(n-2k)!} \frac{1}{k!} \frac{(2k)!}{(2!)^k}.$$

Simplifying and using the Rule of Sum to combine the various possible values of k, we obtain a formula for involutions.

We have just proved: The number of involutions of $\underline{n}$ is

$$\sum_{k=0}^{\lfloor n/2 \rfloor} \frac{n!}{(n-2k)!\,2^k k!}$$

where $\lfloor n/2 \rfloor$ denotes the largest integer less than or equal to $n/2$. Let's use this to compute the number of involutions when $n=6$. Since $\lfloor 6/2 \rfloor = 3$, the sum has four terms:

$$\frac{6!}{(6-0)!\,2^0 0!} + \frac{6!}{(6-2)!\,2^1 1!} + \frac{6!}{(6-4)!\,2^2 2!} + \frac{6!}{(6-6)!\,2^3 3!}$$
$$= 1 + \frac{6 \times 5}{2} + \frac{6 \times 5 \times 4 \times 3}{4 \times 2!} + \frac{6 \times 5 \times 4}{8}$$
$$= 1 + 15 + 45 + 15 = 76.$$

The last term in the sum, namely $k=3$ corresponds to those involutions with three 2-cycles (and hence no 1-cycles). Thus it counts the 15 partitions listed earlier in this example. $\square$

If you're familiar with the basic operations associated with matrices, the following example gives a correspondence between matrix multiplication and composition of permutations.

*Example 10 (Permutation matrices) Suppose f and g are permutations of $\underline{n}$. We can define an $n \times n$ matrix F to consist of zeroes except that the $(i,j)^{\text{th}}$ entry, $F_{i,j}$, equals one whenever $f(j) = i$. Define G similarly. Then

$$(FG)_{i,j} = \sum_{k=1}^{n} F_{i,k} G_{k,j} = F_{i,g(j)},$$

Functions

since $G_{k,j} = 0$ except when $g(j) = k$. By the definition of F, this entry of F is zero unless $f(g(j)) = i$. Thus $(FG)_{i,j}$ is zero unless $(fg)(j) = i$, in which case it is one. We've proven that FG corresponds to fg. In other words:

Composition of permutations corresponds to multiplication of matrices.

It is also easy to prove that f^{-1} corresponds to F^{-1}. Using this correspondence, we can prove things such as $(fg)^{-1} = g^{-1}f^{-1}$ and $(f^k)^{-1} = (f^{-1})^k$ by noting that they are true for matrices F and G.

As an example, let f and g be the permutations

$$f = \begin{pmatrix} 1 & 2 & 3 & 4 & 5 \\ 2 & 1 & 4 & 5 & 3 \end{pmatrix} \qquad g = \begin{pmatrix} 1 & 2 & 3 & 4 & 5 \\ 2 & 3 & 4 & 5 & 1 \end{pmatrix}.$$

We computed fg in Example 6. We obtained

$$fg = \begin{pmatrix} 1 & 2 & 3 & 4 & 5 \\ 1 & 4 & 5 & 3 & 2 \end{pmatrix}.$$

Using our correspondence, we obtain

$$F = \begin{pmatrix} 0 & 1 & 0 & 0 & 0 \\ 1 & 0 & 0 & 0 & 0 \\ 0 & 0 & 0 & 0 & 1 \\ 0 & 0 & 1 & 0 & 0 \\ 0 & 0 & 0 & 1 & 0 \end{pmatrix} \qquad G = \begin{pmatrix} 0 & 0 & 0 & 0 & 1 \\ 1 & 0 & 0 & 0 & 0 \\ 0 & 1 & 0 & 0 & 0 \\ 0 & 0 & 1 & 0 & 0 \\ 0 & 0 & 0 & 1 & 0 \end{pmatrix}$$

You should multiply these two matrices together and verify that you get the matrix FG corresponding to

$$fg = \begin{pmatrix} 1 & 2 & 3 & 4 & 5 \\ 1 & 4 & 5 & 3 & 2 \end{pmatrix}. \qquad \square$$

---

*Example 11 (Derangements) A *derangement* is a permutation f with no fixed points; i.e., $f(x) \neq x$ for all x. We first show that the probability is $1/n$ that a permutation f, selected uniformly at random from all permutations of $\underline{n}$, has $f(k) = k$. If $f(k) = k$, then the elements of $\underline{n} - \{k\}$ can be permuted in any fashion. This can be done in $(n-1)!$ ways. Thus, $(n-1)!$ is the cardinality of the set of all permutations of $\underline{n}$ that satisfy $f(k) = k$. Since there are $n!$ permutations, the probability that $f(k) = k$ is $(n-1)!/n! = 1/n$. Hence the probability that $f(k) \neq k$ is $1 - 1/n$. If we toss a coin with probability p of heads for n tosses, the probability that no heads occurs in n tosses is $(1-p)^n$. This is because each toss is "independent" of the prior tosses. If we, incorrectly, treat the n events $f(1) \neq 1, \ldots, f(n) \neq n$ as independent in this same sense, the probability that $f(k) \neq k$ for $k = 1, \ldots, n$, would be $(1-1/n)^n$. One of the standard results in calculus is that $(1-1/n)^n$ approaches $1/e$ as $n \to \infty$. (You can prove it by writing $(1-1/n)^n = \exp(\ln(1-1/n)/(1/n))$, setting $1/n = x$ and using l'Hôpital's Rule.) Thus, we might expect approximately $n!/e$ derangements of $\underline{n}$ for large n. Although our argument is wrong, the result is right! We get partial credit for this example. $\square$

Section 2: Permutations

Exercises for Section 2

2.1. This exercise lets you check your understanding of cycle form. A permutation is given in one-line, two-line or cycle form. Convert it to the other two forms. Give its inverse in all three forms.

(a) (1,5,7,8) (2,3) (4) (6).

(b) $\begin{pmatrix} 1 & 2 & 3 & 4 & 5 & 6 & 7 & 8 \\ 8 & 3 & 7 & 2 & 6 & 4 & 5 & 1 \end{pmatrix}.$

(c) (5,4,3,2,1), which is in one-line form.

(d) (5,4,3,2,1), which is in cycle form.

2.2. A carnival barker has four cups upside down in a row in front of him. He places a pea under the cup in the first position. He quickly interchanges the cups in the first and third positions, then the cups in the first and fourth positions and then the cups in the second and third positions. This entire set of interchanges is done a total of five times. Where is the pea?
Hint: Write one entire set of interchanges as a permutation in cycle form.

2.3. Let f be a permutation of $\underline{n}$. The cycle of f that contains 1 is call the *cycle generated by* 1.

(a) Prove that the number of permutations in which the cycle generated by 1 has length n is $(n-1)!$.

(b) For how many permutations does the cycle generated by 1 have length k? (Remember that a permutation must be defined on *all* elements of its domain $\underline{n}$.)

(c) If your answer to (b) is correct, when it is summed over $1 \leq k \leq n$ it should equal $n!$, the total number of permutations of $\underline{n}$. Why? Show that your answer to (b) has the correct sum.

2.4. This exercise deals with powers of permutations. All our permutations will be written in cycle form.

(a) Compute $(1,2,3)^{300}$.

(b) Compute $\big((1,3)(2,5,4)\big)^{300}$.

(c) Show that for every permutation f of $\underline{5}$, we have f^{60} is the identity permutation. What is f^{61}?

Functions

Section 3: Other Combinatorial Aspects of Functions

This section contains two independent parts. The first deals with the concept of the inverse of a general function. The second deals with functions related to computer storage of unordered lists.

The Inverse of an Arbitrary Function

Again, let $f: A \to B$ be a function. The *image* of f is the set of values f actually takes on: Image$(f) = \{ f(a) \mid a \in A \}$. The definition of a surjection can be rewritten Image$(f) = B$ because a surjection was defined to be a function $f : A \to B$ such that, for every $b \in B$ there is an $a \in A$ with $f(a) = b$.

For each $b \in B$, the *inverse image* of b, written $f^{-1}(b)$ is the set of those elements in A whose image is b; i.e.,
$$f^{-1}(b) = \{ a \mid a \in A \text{ and } f(a) = b \}.$$
This extends our earlier definition of f^{-1} from bijections to all functions; however, such an f^{-1} can't be thought of as a function from B to A unless f is a bijection because it will not give a unique $a \in A$ for each $b \in B$.[4]

Suppose f is given by the functional relation $R \subset A \times B$. Then $f^{-1}(b)$ is all those a such that $(a, b) \in R$. Equivalently, $f^{-1}(b)$ is all those a such that $(b, a) \in R^{-1}$.

Definition 4 (Coimage) *Let $f: A \to B$ be a function. The collection of nonempty inverse images of elements of B is called the coimage of f. In mathematical notation*
$$\text{Coimage}(f) = \{ f^{-1}(b) \mid b \in B, f^{-1}(b) \neq \emptyset \} = \{ f^{-1}(b) \mid b \in \text{Image}(f) \}.$$

We claim that the coimage of f is the partition of A whose blocks[5] are the maximal subsets of A on which f is constant. For example, if $f \in \{a, b, c\}^{\underline{5}}$ is given in one line form as (a, c, a, a, c), then
$$\text{Coimage}(f) = \{f^{-1}(a), f^{-1}(c)\} = \{\{1,3,4\}, \{2,5\}\},$$
f is a on $\{1,3,4\}$ and is c on $\{2,5\}$.

We now prove the claim. If $x \in A$, let $y = f(x)$. Then $x \in f^{-1}(y)$ and the set $f^{-1}(y)$ is an element of Coimage(f). Hence the union of the nonempty inverse images contains A. Clearly it does not contain anything which is not in A. If $y_1 \neq y_2$, then we cannot have $x \in f^{-1}(y_1)$ and $x \in f^{-1}(y_2)$ because this would imply $f(x) = y_1$ and $f(x) = y_2$, a contradiction of the definition of a function. Thus Coimage(f) is a partition of A. Since the value of $f(x)$ determines the block to which x belongs, x_1 and x_2 belong to the same block if and only if $f(x_1) = f(x_2)$. Hence a block is a maximal set on which f is constant.

---

[4] There is a slight abuse of notation here: If $f: A \to B$ is a bijection, our new notation is $f^{-1}(b) = \{a\}$ and our old notation is $f^{-1}(b) = a$.
[5] Recall that a partition of a set S is an unordered collection of disjoint nonempty subsets of S whose union is S. These subsets are called the blocks of the partition.

Section 3: Other Combinatorial Aspects of Functions

Example 12 (f^{-1} **as a function**) Let $f: A \to B$ be a function. For each $b \in B$,

$$f^{-1}(b) = \{\, a \mid a \in A \text{ and } f(a) = b \,\}.$$

Thus, for each $b \in B$, $f^{-1}(b) \in \mathcal{P}(A)$. Hence f^{-1} is a function with domain B and range (codomain) $\mathcal{P}(A)$, the set of all subsets of A. This is true for any function f and does not require f to be bijection. For example, if $f \in \{a,b,c\}^{\underline{5}}$ is given in one-line form as (a,c,a,a,c), then, f^{-1}, in two-line notation is

$$\begin{pmatrix} a & b & c \\ \{1,3,4\} & \emptyset & \{2,5\} \end{pmatrix}$$

If, we take the domain of f^{-1} to be Image(f), instead of all of B, then f^{-1} is a bijection from Image(f) to Coimage(f). In the case of our example (a,c,a,a,c), we get, in two-line notation

$$\begin{pmatrix} a & c \\ \{1,3,4\} & \{2,5\} \end{pmatrix}$$

for the image–coimage bijection associated with f^{-1}. If we are only given the coimage of a function then we don't have enough information to specify the function. For example, suppose we are given only that $\{\{1,3,4\},\{2,5\}\}$ is the coimage of some function g with codomain $\{a,b,c\}$. We can see immediately that the domain of g is $\underline{5}$. But what is g? To specify g we need to know the elements x and y in $\{a,b,c\}$ that make

$$\begin{pmatrix} x & y \\ \{1,3,4\} & \{2,5\} \end{pmatrix}$$

the correct two-line description of g^{-1} (restricted to its image). There are $(3)_2 = 6$ choices[6] for xy, namely, ab, ac, bc, ba, ca, and cb. In general, suppose $f: A \to B$ and we are given that a particular partition of A with k blocks is the coimage of f. Then, by comparison with our example ($A = \underline{5}$, $B = \{a,b,c\}$), it is easy to see that there are exactly $(|B|)_k$ choices for the function f. □

We can describe the image and coimage of a function by the arrow pictures introduced in Example 5. Image(f) is the set of those $b \in B$ which appear as labels of arrowheads. A block in Coimage(f) is the set of labels on the tails of those arrows that all have their heads pointing to the same value; for example, the block of Coimage(f) arising from $b \in$ Image(f) is the set of labels on the tails of those arrows pointing to b.

Example 13 (Counting functions with specified image size) How many functions in B^A have an image with exactly k elements? You will need to recall that the symbol $S(n,k)$, stands for the number of partitions of set of size n into k blocks. (The $S(n,k)$ are called the Stirling numbers of the second kind and are discussed in Unit CL. See the index for page numbers.) If $f \in B^A$ has k elements in its image, then this means that the coimage of f is a partition of A having exactly k blocks. Suppose that $|A| = a$ and $|B| = b$. There

---

[6] Recall that $(n)_k = n(n-1)\cdots(n-k+1) = n!/(n-k)!$ is the number of k-lists without repeats that can be made from an n-set.

Functions

are $S(a,k)$ ways to choose the blocks of the coimage. The partition of A does not fully specify a function $f \in B^A$. To complete the specification, we must specify the image of the elements in each block (Example 12). In other words, an injection from the set of k blocks to B must be specified. This is an ordered selection of size k without replacement from B. There are $(b)_k = b!/(b-k)!$ such injections, independent of which k block partition of A we are considering. By the Rule of Product, there are $S(a,k)(b)_k$ functions $f \in B^A$ with $|\text{Image}(f)| = k$. For example, when the domain is $\underline{5}$ and the range is $\{a,b,c\}$, the number of functions with $|\text{Image}(f)| = 2$ is $S(5,2)(3)_2 = 15 \times 6 = 90$, where the value of $S(5,2)$ was obtained from the table in the discussion of Stirling numbers in Unit CL. Example 12 gave one of these 90 possibilities.

We consider some special cases.

- Suppose $k = a$.
    - If $b < a$, there are no functions f with $|\text{Image}(f)| = a$ because the size a of the image is at most the size b of the codomain.
    - If $b \geq a$ there are $(b)_a$ functions with $|\text{Image}(f)| = a$.
    - If $b = a$, the previous formula, $(b)_a$, reduces to $a!$ and the functions are injections from A to B.
- Suppose $k = b$.
    - If $b > a$ there are no functions f with $|\text{Image}(f)| = b$ because the size of the image is at most the size of the domain.
    - If $b \leq a$ then there are $S(a,b)(b)_b = S(a,b)\,b!$ functions $f \in B^A$ with $|\text{Image}(f)| = b$. These functions are exactly the surjections. $\square$

Monotonic Lists and Unordered Lists

In computers, all work with data structures requires that the parts of the data structure be ordered. The most common orders are arrays and linked lists.

Sometimes the order relates directly to an order associated with the corresponding mathematical objects. For example, the one-line notation for a function is simply an ordered list, which is an array. Thus there is a simple correspondence (i.e., bijection) between lists and functions: A k-list from S is a function $f : \underline{k} \to S$. Thus functions (mathematical objects) are easily stored as ordered lists (computer objects).

Sometimes the order is just an artifact of the algorithm using the structures. In other words, the order is imposed by the designer of the algorithm. Finding such a "canonical" ordering[7] is essential if one wants to work with unordered objects efficiently in a computer.

---

[7] In mathematics, people refer to a unique thing (or process or whatever) that has been selected as *canonical*.

Section 3: Other Combinatorial Aspects of Functions

Since sets and multisets[8] are basic unordered mathematical objects, it is important to have ways of representing them in a computer. We'll discuss a canonical ordering for k-sets and k-multisets whose elements lie in an n-set.

We need to think of a unique way to order the set or multiset, say $s_1, s_2, \ldots, s_k$ so that we have an ordered list. (A mathematician would probably speak of a canonical ordering of the multiset rather than a unique ordering; however, both terms are correct.)

Let's look at a small example, the 3-element multisets whose elements are chosen from $\underline{5}$. Here are the $\binom{5+3-1}{3} = 35$ such multisets.[9] An entry like 2,5,5 stands for the multiset containing one 2 and two 5's.

1,1,1	1,1,2	1,1,3	1,1,4	1,1,5	1,2,2	1,2,3	1,2,4	1,2,5	1,3,3
1,3,4	1,3,5	1,4,4	1,4,5	1,5,5	2,2,2	2,2,3	2,2,4	2,2,5	2,3,3
2,3,4	2,3,5	2,4,4	2,4,5	2,5,5	3,3,3	3,3,4	3,3,5	3,4,4	3,4,5
3,5,5	4,4,4	4,4,5	4,5,5	5,5,5					

We've simply arranged the elements in each 3-multiset to be in "weakly increasing order." Let $(b_1, b_2, \ldots, b_k)$ be an ordered list. We say the list is in *weakly increasing order* if the values are not decreasing as we move from one element to the next; that is, if $b_1 \leq b_2 \leq \cdots \leq b_k$. The list of lists we've created can be thought of as a bijection from

(i) the 3-multisets whose elements lie in $\underline{5}$ to

(ii) the weakly increasing functions in $\underline{5}^{\underline{3}}$ written in one-line notation.

Thus, 3-multisets with elements in $\underline{5}$ correspond to weakly increasing functions in $\underline{5}^{\underline{3}}$. For example the multiset $\{2, 5, 5\}$ corresponds to the weakly increasing function $f = (2, 5, 5)$ in 1-line form.

Since we have seen that functions with domain $\underline{k}$ can be viewed as k-lists, we say that $f \in \underline{n}^{\underline{k}}$ is a *weakly increasing function* if its one-line form is weakly increasing; that is, $f(1) \leq f(2) \leq \cdots \leq f(k)$. In a similar fashion we say that the list $b_1, b_2, \ldots, b_k$ is in

$$\left. \begin{array}{l} \text{weakly decreasing} \\ \text{strictly decreasing} \\ \text{strictly increasing} \end{array} \right\} \text{order if} \left\{ \begin{array}{l} b_1 \geq b_2 \geq \cdots \geq b_k; \\ b_1 > b_2 > \cdots > b_k; \\ b_1 < b_2 < \cdots < b_k. \end{array} \right.$$

Again, this leads to similar terminology for functions. All such functions are also called *monotone functions*.

In the bijection we gave for our 35 lists, the lists without repetition correspond to the strictly increasing functions. Thus 3-subsets of $\underline{5}$ correspond to strictly increasing functions. From our previous list of multisets, we can read off these functions in 1-line form:

$$(1,2,3) \quad (1,2,4) \quad (1,2,5) \quad (1,3,4) \quad (1,3,5)$$
$$(1,4,5) \quad (2,3,4) \quad (2,3,5) \quad (2,4,5) \quad (3,4,5)$$

We can interchange the strings "decreas" and "increas" in the previous paragraphs and read the functions in the list backwards. For example, the bijection between 3-subsets of $\underline{5}$ and strictly decreasing functions is given by

$$3,2,1 \quad 4,2,1 \quad 4,3,1 \quad 4,3,2 \quad 5,2,1 \quad 5,3,1 \quad 5,3,2 \quad 5,4,1 \quad 5,4,2 \quad 5,4,3.$$

---

[8] Recall that a multiset is like a set except that repeated elements are allowed.

[9] A later example explains how we got this number.

Functions

The function $(4,3,1)$ corresponds to the 3-subset $\{4,3,1\} = \{1,3,4\}$ of $\underline{5}$.

All these things are special cases of the following 2-in-1 theorem.

Theorem 2 (Sets, unordered lists and monotone functions) *Read either the top lines in all the braces or the bottom lines in all the braces. There are bijections between each of the following:*

- $\left\{ \begin{matrix} k\text{-multisets} \\ k\text{-sets} \end{matrix} \right\}$ whose elements lie in $\underline{n}$,

- the $\left\{ \begin{matrix} \text{weakly} \\ \text{strictly} \end{matrix} \right\}$ increasing ordered k-lists made from $\underline{n}$,

- the $\left\{ \begin{matrix} \text{weakly} \\ \text{strictly} \end{matrix} \right\}$ increasing functions in $\underline{n}^{\underline{k}}$.

In these correspondences, the items in the list are the elements of the (multi)-set and are the values of the function.
In the correspondences, "increasing" can be replaced by "decreasing."

For example, reading the top lines in the braces with $k=3$ and $n=5$ and, in the last one, replacing "increasing" with "decreasing," we have: There are bijections between

(a) 3-multisets whose elements lie in $\underline{5}$,

(b) the weakly increasing ordered 3-lists in $\underline{5}$ and

(c) the weakly decreasing functions in $\underline{5}^{\underline{3}}$.

In these bijections, $\{2,5,5\}$ corresponds to list $(2,5,5)$ and the function $f=(5,5,2)$.

Example 14 (Counting multisets) Earlier we said there were $\binom{5+3-1}{3} = 35$ different 3-element multisets whose elements come from $\underline{5}$ and gave the list

```
1,1,1   1,1,2   1,1,3   1,1,4   1,1,5   1,2,2   1,2,3   1,2,4   1,2,5   1,3,3
1,3,4   1,3,5   1,4,4   1,4,5   1,5,5   2,2,2   2,2,3   2,2,4   2,2,5   2,3,3
2,3,4   2,3,5   2,4,4   2,4,5   2,5,5   3,3,3   3,3,4   3,3,5   3,4,4   3,4,5
3,5,5   4,4,4   4,4,5   4,5,5   5,5,5
```

How did we know this? To see the trick, do the following to each 3-list:

- add 0 to the first item,
- add 1 to the second item, and
- add 2 to the third item.

Thus the first ten become

$$1,2,3 \quad 1,2,4 \quad 1,2,5 \quad 1,2,6 \quad 1,2,7 \quad 1,3,4 \quad 1,3,5 \quad 1,3,6 \quad 1,3,7 \quad 1,4,5$$

You should be able to see that you've created strictly increasing 3-lists from $\underline{7}$. In other words, you have listed all subsets of $\underline{7}$ of size 3. We know there are $\binom{7}{3} = 35$ such subsets and hence there were 35 multisets in the original list. In general, suppose we have listed

Section 3: Other Combinatorial Aspects of Functions

all weakly increasing k-lists from $\underline{n}$. Suppose each such k-list is in weakly increasing order. If, as in the above example, we add the list $0, 1, \ldots, k-1$ to each such k-list element by element, we get the strictly increasing k-lists from $\underline{n+k-1}$. By Theorem 2, this is a list of all k-subsets of $\underline{n+k-1}$. Thus, the number of weakly increasing k-lists from $\underline{n}$ is $\binom{n+k-1}{k}$. By Theorem 2, this is also the number of k-multisets from $\underline{n}$. ☐

Exercises for Section 3

3.1. This exercise lets you check your understanding of the definitions. In each case below, some information about a function is given to you. Answer the following questions and give reasons for your answers:

- Have you been given enough information to specify the function; i.e., would this be enough data for a function envelope?
- Can you tell whether or not the function is an injection? a surjection? a bijection? If so, what is it?

(a) $f \in \underline{4}^{\underline{5}}$, Coimage$(f) = \{\{1, 3, 5\}, \{2, 4\}\}$.

(b) $f \in \underline{5}^{\underline{5}}$, Coimage$(f) = \{\{1\}, \{2\}, \{3\}, \{4\}, \{5\}\}$.

(c) $f \in \underline{4}^{\underline{5}}$, $f^{-1}(2) = \{1, 3, 5\}$, $f^{-1}(4) = \{2, 4\}\}$.

(d) $f \in \underline{4}^{\underline{5}}$, $|\text{Image}(f)| = 4$.

(e) $f \in \underline{4}^{\underline{5}}$, $|\text{Image}(f)| = 5$.

(f) $f \in \underline{4}^{\underline{5}}$, $|\text{Coimage}(f)| = 5$.

3.2. Let A and B be finite sets and let $f: A \to B$ be a function. Prove the following claims.

(a) $|\text{Image}| = |\text{Coimage}(f)|$.

(b) f is an injection if and only if $|\text{Image}| = |A|$.

(c) f is a surjection if and only if $|\text{Coimage}(f)| = |B|$.

3.3. In each case we regard the specified functions in one-line form. (As strings of integers, such functions are ordered lexicographically.)

(a) List all strictly decreasing functions in $\underline{5}^{\underline{3}}$ in lexicographic order. Note that this lists all subsets of size 3 from $\underline{5}$ in lexicographic order.

(b) In the list for part (a), for each string $x_1 x_2 x_3$ compute $\binom{x_1-1}{3} + \binom{x_2-1}{2} + \binom{x_3-1}{1} + 1$. What do these integers represent in terms of the list of part (a)?

(c) List all strictly increasing functions in $\underline{5}^{\underline{3}}$ in lexicographic order. Note that this also lists all subsets of size 3 from $\underline{5}$ in lexicographic order.

Functions

(d) What is the analog, for part (c) of the formula of part (b)?
Hint: For each $x_1\, x_2\, x_3$ in the list of part (c), form the list $(6-x_1)(6-x_2)(6-x_3)$.

(e) In the lexicographic list of all strictly decreasing functions in $\underline{9}^{\underline{5}}$, find the successor and predecessor of 98321.

(f) How many elements are there before 98321 in this list?

3.4. We study the problem of listing all ways to put 5 balls (unlabeled) into 4 boxes (labeled 1 to 4). Consider ten consecutive points, labeled $0\cdots 9$. Some points are to be converted to box boundaries, some to balls. Points 0 and 9 are always box boundaries (call these *exterior box boundaries*). From the remaining points labeled $1\cdots 8$, we can arbitrarily pick three points to convert to *interior* box boundaries, four points to convert to balls. Here are two examples:

```
0 1 2 3 4 5 6 7 8 9        points      0 1 2 3 4 5 6 7 8 9
|• •|• •|•  |  |           conversion  |• •|  •|  |• • •|
  1    2  3  4             box no.       1    2  3    4
```

In this way, the placements of 5 balls into 4 boxes are made to correspond to subsets of size 3 (the box boundaries we can select) from $\underline{8}$. Lexicographic order on subsets of size 3 from $\underline{8}$, where the subsets are listed as strictly decreasing strings of length 3 from $\underline{8}$, correspondingly lex orders all placements of 5 balls into three boxes.

(a) Find the successor and predecessor of each of the above placements of balls into boxes.

(b) In the lex order of 5 balls into 4 boxes, which placement of balls into boxes is the last one in the first half of the list? The first one in the second half of the list?
Hint: The formula $p(x_1, x_2, x_3) = \binom{x_1-1}{3} + \binom{x_2-1}{2} + \binom{x_3-1}{1} + 1$ gives the position of the string $x_1x_2x_3$ in the list of decreasing strings of length three from $\underline{8}$. Try to solve the equation $p(x_1,x_2,x_3) = \binom{8}{3}/2 = 28$ for the variables x_1, x_2, x_3.

3.5. Listing all of the partitions of an n set can be tricky and, of course, time consuming as there are lots of them. This exercise shows you a useful trick for small n. We define a class of functions called *restricted growth* functions that have the property that their collection of coimages is exactly the set of all partitions of $\underline{n}$.

(a) Call a function $f \in \underline{n}^{\underline{n}}$ a restricted growth function if $f(1) = 1$ and $f(i) - 1$ is at most the maximum of $f(k)$ over all $k < i$. Which of the following functions in one-line form are restricted growth functions? Give reasons for your answers.

$$2,2,3,3 \quad 1,2,3,3,2,1 \quad 1,1,1,3,3 \quad 1,2,3,1.$$

(b) List, in lexicographic order, all restricted growth functions for $n = 4$. Use one-line form and, for each one, list its coimage partition.

Section 4: Functions and Probability

(c) For $n = 5$, list in lexicographic order the first fifteen restricted growth functions. Use one-line form. For the functions in positions 5, 10, and 15, list their coimage partitions.

3.6. How many functions f are there from $\underline{6}$ to $\underline{5}$ with $|\text{Image}(f)| = 3$?

3.7. How many ways can 6 different balls be placed into 5 labeled cartons in such a way that exactly 2 of the cartons contain no balls?

3.8. Count each of the following

(a) the number of multisets of size 6 whose elements lie in $\{a, b, c, d\}$,

(b) the number of weakly increasing functions from $\underline{6}$ to $\underline{4}$,

(c) the number of weakly decreasing ordered 6-lists made from $\underline{4}$,

(d) the number of strictly increasing functions from $\underline{6}$ to $\underline{9}$.

Section 4: Functions and Probability

In this section we look at various types of functions that occur in elementary probability theory. For the most part, we deal with finite sets. The functions we shall define are not difficult to understand, but they do have special names and terminology common to the subject of probability theory. We describe these various functions with a series of examples.

Probability functions have already been encountered in our studies. To review this idea, let U be a finite sample space (that is, a finite set) and let P be a function from U to $\mathbb{R}$ such that $P(t) \geq 0$ for all $t \in U$ and $\sum_{t \in U} P(t) = 1$. Then P is called a *probability function* on U. For any event $E \subseteq U$, define $P(E) = \sum_{t \in E} P(t)$. $P(E)$ is called the *probability of the event* E. The pair (U, P) is called a *probability space*.

Functions

Random Variables

Consider tossing a coin. The result is either heads or tails, which we denote by H and T. Thus the sample space is $\{H,T\}$. Sometimes we want to associate numerical values with elements of the sample space. Such functions are called "random variables." The function $X : \{H,T\} \to \mathbb{R}$, defined by $X(H) = 1$, $X(T) = 0$, is a random variable. Likewise the function $Y(H) = 1$, $Y(T) = -1$, same domain and range, is a random variable.

As another example, consider the sample space $U = \times^4\{H,T\}$, which contains the possible results of four coin tosses. The random variable $X(t_1, t_2, t_3, t_4) = |\{i \mid t_i = H\}|$ counts the number of times H appears in the sequence t_1, t_2, t_3, t_4. The function X has Image$(X) = \{0, 1, 2, 3, 4\}$, which is a subset of the set of real numbers.

Definition 5 (Random variable) Let (U, P) be a probability space, and let $g : U \to \mathbb{R}$ be a function with domain U and range (codomain) $\mathbb{R}$, the real numbers. Such a function g is called a *random variable* on (U, P). The term "random variable" informs us that the range is the set of real numbers and that, in addition to the domain U, we also have a probability function P.

Random variables are usually denoted by capital letters near the end of the alphabet. Thus, instead of g in the definition of random variable, most texts would use X.

By combining the two concepts, random variable and probability function, we obtain one of the most important definitions in elementary probability theory, that of a distribution function.

Definition 6 (Distribution function of a random variable) Let $X : U \to \mathbb{R}$ be a random variable on a sample space U with probability function P. For each real number $t \in \text{Image}(X)$, let $X^{-1}(t)$ be the inverse image of t. Define a function $f_X : \text{Image}(X) \to \mathbb{R}$ by $f_X(t) = P(X^{-1}(t))$. The function f_X is called the *probability distribution function* of the random variable X. The distribution function is also called the *density function*.

Since $P(X^{-1}(t))$ is the probability of the set of events $E = \{e \mid X(e) = t\}$, one often writes $P(X = t)$ instead of $P(X^{-1}(t))$.

Example 15 (Some distribution functions) Suppose we roll a fair die. Then $U = \{1, 2, 3, 4, 5, 6\}$ and $P(i) = 1/6$ for all i.

- If $X(t) = t$, then $f_X(t) = P(X^{-1}(t)) = P(t) = 1/6$ for $t = 1, 2, 3, 4, 5, 6$.
- If $X(t) = 0$ when t is even and $X(t) = 1$ when t is odd, then $f_X(0) = P(X^{-1}(0)) = P(\{2, 4, 6\}) = 1/2$ and $f_X(1) = 1/2$.
- If $X(t) = -1$ when $t \leq 2$ and $X(t) = 1$ when $t > 2$, then $f_X(-1) = P(\{1, 2\}) = 1/3$ and $f_X(1) = 2/3$. ∎

The function f_X, in spite of its fancy sounding name, is nothing but a probability function on the set Image(X). Why is that?

Section 4: Functions and Probability

- Since P is a nonnegative function with range $\mathbb{R}$, so is f_X.
- Since $\text{Coimage}(X) = \{X^{-1}(t) \mid t \in \text{Image}(X)\}$ is a partition of the set U,

$$1 = P(U) = \sum_{t \in \text{Image}(X)} P(X^{-1}(t)) = \sum_{t \in \text{Image}(X)} f_X(t).$$

Thus, f_X is a nonnegative function from $\text{Image}(X)$ to $\mathbb{R}$ such that

$$\sum_{t \in \text{Image}(X)} f_X(t) = 1.$$

This is exactly the definition of a probability function on the set $\text{Image}(X)$.

Example 16 (Distribution of six cards with replacement) First one card and then a second card are selected at random, with replacement, from 6 cards numbered 1 to 6. The basic sample space is $S \times S = \{(i,j) : 1 \leq i \leq 6, 1 \leq j \leq 6\}$. Every point (i,j) in this sample space is viewed as equally likely: $P(i,j) = 1/36$. Define a random variable X on $S \times S$ by $X(i,j) = i+j$. $S \times S$ can be visualized as a 6×6 rectangular array and X can be represented by inserting $X(i,j)$ in the position represented by row i and column j. This can be done as follows:

	1	2	3	4	5	6
1	2	3	4	5	6	7
2	3	4	5	6	7	8
3	4	5	6	7	8	9
4	5	6	7	8	9	10
5	6	7	8	9	10	11
6	7	8	9	10	11	12

It is evident that $\text{Image}(X) = \{2, 3, 4, 5, 6, 7, 8, 9, 10, 11, 12\}$. The blocks of the $\text{Coimage}(X)$ are the sets of pairs (i,j) for which $i+j$ is constant. Thus,

$$X^{-1}(5) = \{(1,4), (2,3), (3,2), (4,1)\}$$

and

$$X^{-1}(8) = \{(2,6), (3,5), (4,4), (5,3), (6,2)\}.$$

Since every point in $S \times S$ has probability $1/36$, $P(X^{-1}(i)) = |X^{-1}(i)|/36$. Thus, with a little counting in the previous figure, the distribution function of X, $f_X = |X^{-1}(i)|/36$, in two-line form is

$$f_X = \begin{pmatrix} 2 & 3 & 4 & 5 & 6 & 7 & 8 & 9 & 10 & 11 & 12 \\ 1/36 & 2/36 & 3/36 & 4/36 & 5/36 & 6/36 & 5/36 & 4/36 & 3/36 & 2/36 & 1/36 \end{pmatrix}. \quad \square$$

Suppose we have two random variables X and Y defined on the same sample space U. Since the range of both X and Y is the real numbers, it makes sense to add the two random variables to get a new random variable: $Z = X + Y$; that is, $Z(t) = X(t) + Y(t)$ for all

[67]

Functions

$t \in U$. Likewise, if r is a real number, then $W = rX$ is a random variable, $W(t) = rX(t)$ for all $t \in U$. Thus we can do basic arithmetic with random variables.

For any random variable, X on U, we have the following important definition:

Definition 7 (Expectation of a random variable) Let X be a random variable on a sample space U with probability function P. The expectation $E(X)$, or expected value of X, is defined by
$$E(X) = \sum_{t \in U} X(t) P(t).$$
The $E(X)$ is often denoted by μ_X and referred to as the mean of X.

If we collect terms in preceding sum according to the value of $X(t)$, we obtain another formula for the expectation:
$$E(X) = \sum_{t \in U} X(t) P(t) = \sum_{r} r \left(\sum_{\substack{t \in U \\ X(t) = r}} P(t) \right) = \sum_{r} r f_X(r).$$

The expectation E is a function whose arguments are functions. Another example of such a function is differentiation in calculus. Such functions, those whose arguments are themselves functions, are sometimes called "operators." Sometimes you will see a statement such as $E(2) = 2$. Since the arguments of the expectation operator E are functions, the "2" inside the parentheses is interpreted as the constant function whose value is 2 on all of U. The second 2 in $E(2) = 2$ is the number 2.

If X and Y are two random variables on a sample space U with probability function P, and $Z = X + Y$, then
$$E(Z) = \sum_{t \in U} Z(t) P(t)$$
$$= \sum_{t \in U} (X(t) + Y(t)) P(t)$$
$$= \sum_{t \in U} X(t) P(t) + \sum_{t \in U} Y(t) P(t) = E(X) + E(Y).$$

Similarly, for any real number r, $E(rX) = rE(X)$. Putting these observations together, we have proved the following theorem:

Theorem 3 (Linearity of expectation) If X and Y are two random variables on a sample space U with probability function P and if a and b are real numbers, then $E(aX + bY) = aE(X) + bE(Y)$.

We now introduce some additional functions defined on random variables, the covariance, variance, standard deviation, and correlation.

Definition 8 (Covariance, variance, standard deviation, correlation) Let U be a sample space with probability function P, and let X and Y be random variables on U.

Section 4: Functions and Probability

- Then the covariance of X and Y, $\text{Cov}(X,Y)$, is defined by
  $\text{Cov}(X,Y) = E(XY) - E(X)E(Y)$.

- The variance of X, $\text{Var}(X)$ (also denoted by σ_X^2), is defined by
  $\text{Var}(X) = \text{Cov}(X,X) = E(X^2) - (E(X))^2$.

- The standard deviation of X is $\sigma_X = (\text{Var}(X))^{1/2}$.

- Finally, the correlation, $\rho(X,Y)$ of X and Y is $\rho(X,Y) = \text{Cov}(X,Y)/\sigma_X\sigma_Y$, provided $\sigma_X \neq 0$ and $\sigma_Y \neq 0$.

Example 17 (Sampling from a probability space) Consider the probability space (U,P) consisting of all possible outcomes from tossing a fair coin three times. Let the random variable X be the number of heads. Suppose we now actually toss a fair coin three times and record the result. This is called "sampling from the distribution." Given our sample $e \in U$, we can compute $X(e)$. Suppose we sample many times and average the values of $X(e)$ that we compute. As we increase the number of samples, our average moves around; e.g., if our values of $X(e)$ are $1, 2, 0, 1, 2, 3, \ldots$ our averages of the first one, first two, first three, etc., are $1, 3/2, 3/3, 4/4, 6/5, 9/6, \ldots$, which reduce to $1, 1.5, 1, 1, 1.2, 1.5, \ldots$. What can we say about the average? When we take a large number of samples the average will tend to be close to μ_X, which is 1.5 in this case. Thus without knowing P we can estimate μ_x by sampling many times computing X and averaging the results.

The same idea applies to other functions of random variables: Sample many times, compute the function of the random variable(s) for each sample and average the results. In this way, we can estimate $E(X^2)$. In the previous paragraph, we estimated $E(X)$. Combining these estimates, we obtain an estimate for $\text{Var}(X) = E(X^2) - (E(X))^2$. Refinements of this procedure are discussed in statistics classes. ∎

Theorem 4 (General properties of covariance and variance) Let X, Y and Z be random variables on a probability space (U, P) and let a, b and c be real numbers. The covariance and variance satisfy the following properties:

(1) *(symmetry)* $\text{Cov}(X,Y) = \text{Cov}(Y,X)$

(2) *(bilinearity)* $\text{Cov}(aX + bY, Z) = a\text{Cov}(X,Z) + b\text{Cov}(Y,Z)$ and
   $\text{Cov}(X, aY + bZ) = a\text{Cov}(X,Y) + b\text{Cov}(X,Z)$.

Thinking of a as the constant function equal to a and likewise for b, we have

(3) $\text{Cov}(a, X) = 0$ and $\text{Cov}(X + a, Y + b) = \text{Cov}(X,Y)$.

In particular, $\text{Cov}(X,Y) = E((X - \mu_X)(Y - \mu_Y))$ and $\text{Var}(X) = E((X - \mu_X)^2)$.

(4) $\text{Var}(aX + bY + c) = \text{Var}(aX + bY)$
   $= a^2\text{Var}(X) + 2ab\,\text{Cov}(X,Y) + b^2\text{Var}(Y)$.

The last two formulas in (3) are sometimes taken as the definition of the covariance and variance. In that case, the formulas for them in Definition 8 would be proved as a theorem. Note that the formula $\text{Var}(X) = E((X - \mu_X)^2)$ says that $\text{Var}(X)$ tells us something about how far X is likely to be from its mean.

Functions

Proof: Property (1) follows immediately from the fact that $XY = YX$ and $E(X)E(Y) = E(Y)E(X)$. The following calculations prove (2).

$$\begin{aligned}
\text{Cov}(aX + bY, Z) &= E((aX + bY)Z) - E(aX + bY)E(Z) & \text{by definition} \\
&= aE(XZ) + bE(YZ) - (aE(X)E(Z) + bE(Y)E(Z)) & \text{by Theorem 3} \\
&= a(E(XZ) - E(X)E(Z)) + b(E(YZ) - E(Y)E(Z)) \\
&= a\text{Cov}(X, Z) + b\text{Cov}(Y, Z) & \text{by definition.}
\end{aligned}$$

We now turn to (3). By definition, $\text{Cov}(a, X) = E(aX) - E(a)E(X)$. Since $E(aX) = aE(X)$ and $E(a) = a$, $\text{Cov}(a, X) = 0$. Using the various parts of the theorem,

$$\begin{aligned}
\text{Cov}(X + a, Y + b) &= \text{Cov}(X, Y + b) + \text{Cov}(a, Y + b) = \text{Cov}(X, Y + b) \\
&= \text{Cov}(X, Y) + \text{Cov}(X, b) = \text{Cov}(X, Y).
\end{aligned}$$

The particular results follow when we set $a = -\mu_X$ and $b = -\mu_Y$.

You should be able to prove (4) by using (1), (2), (3) and the definition $\text{Var}(Z) = \text{Cov}(Z, Z)$. □

Example 18 (General properties of correlation) Recall that the correlation of X and Y is $\rho(X, Y) = \text{Cov}(X, Y)/\sigma_X \sigma_Y$, provided $\sigma_x \neq 0$ and $\sigma_Y \neq 0$. Since

$$\text{Cov}(X + c, Y + d) = \text{Cov}(X, Y) \quad \text{and} \quad \text{Var}(X + c) = \text{Var}(X)$$

for any constant functions c and d, we have $\rho(X + c, Y + d) = \rho(X, Y)$ for any constant functions c and d. Suppose that X and Y both have mean zero. Note that

$$0 \leq E((X + tY)^2) = E(X^2) + 2E(XY)t + E(Y^2)t^2 = f(t)$$

defines a nonnegative polynomial of degree 2 in t. From high school math or from calculus, the minimum value of a polynomial of the form $A + 2Bt + Ct^2$ $(C > 0)$ is

$$A - B^2/C = E(X^2) - (E(XY))^2/E(Y^2).$$

Since for all t, $f(t) \geq 0$, we have $E(X^2) - (E(XY))^2/E(Y^2) \geq 0$. Since X and Y have mean zero,

$$\text{Cov}(X, Y) = E(XY), \quad \text{Var}(X) = E(X^2), \quad \text{and} \quad \text{Var}(Y) = E(Y^2).$$

Thus,

$$(E(XY))^2/E(X^2)E(Y^2) = (\rho(X, Y))^2 \leq 1$$

or, equivalently, $-1 \leq \rho(X, Y) \leq +1$. If the means of X and Y are not zero then replace X and Y by $X - \mu_X$ and $Y - \mu_Y$ respectively, to obtain $-1 \leq \rho(X, Y) \leq +1$. □

We have just proved the following theorem about the correlation of random variables.

Theorem 5 (Bounds on the correlation of two random variables) Let X and Y be random variables on a sample space U and let $\rho(X, Y)$ be their correlation. Then $-1 \leq \rho(X, Y) \leq +1$.

Section 4: Functions and Probability

The intuitive interpretation of the correlation $\rho(X,Y)$ is that values close to 1 mean points in U where X is large also tend to have Y large. Values close to -1 mean the opposite. As extreme examples, take Y=X. Then $\rho(X,Y) = 1$. If we take $Y = -X$ then $\rho(X,Y) = -1$.

Tchebycheff's inequality[10] is another easily proved inequality for random variables. It relates the tendency of a random variable to be far from its mean to the size of its variance. Such results are said to be "measures of central tendency."

Theorem 6 (Tchebycheff's inequality) Let X be a random variable on a probability space (U, P). Suppose that $E(X) = \mu$ and $\text{Var}(X) = \sigma^2$. Let $\epsilon > 0$ be a real number. Then

$$P\Big(\{u \mid |X(u) - \mu| \geq \epsilon\}\Big) \leq \frac{\sigma^2}{\epsilon^2}.$$

The left side of the inequality contains the set of all u for which $|X(u) - \mu| \geq \epsilon$. Thus it can be thought of as the probability that the random variable X satisfies $|X - \mu| \geq \epsilon$.

The most important aspect of Tchebycheff's inequality is the universality of its applicability: the random variable X is arbitrary.

Proof: Let's look carefully at the computation of the variance:

$$\text{Var}(X) = E[(X - \mu)^2] = \sum_{\{u \mid |X-\mu|\geq\epsilon\}} (X(u) - \mu)^2 P(u) + \sum_{\{u \mid |X-\mu|<\epsilon\}} (X(u) - \mu)^2 P(u).$$

In breaking down the variance into these two sums, we have partitioned U into two disjoint sets $\{u \mid |X - \mu| \geq \epsilon\}$ and $\{u \mid |X - \mu| < \epsilon\}$. Since all terms are positive, $\text{Var}(X)$ is greater than or equal to either one of the above sums. In particular,

$$\text{Var}(X) = E[(X - \mu)^2] \geq \sum_{\{u \mid |X-\mu|\geq\epsilon\}} (X(u) - \mu)^2 P(u).$$

Note that

$$\sum_{\{u \mid |X-\mu|\geq\epsilon\}} (X(u) - \mu)^2 P(u) \geq \epsilon^2 \left(\sum_{\{u \mid |X-\mu|\geq\epsilon\}} P(u)\right) = \epsilon^2 P\Big(\{u \mid |X - \mu| \geq \epsilon\}\Big).$$

Putting all of this together proves the theorem. ∎

---

[10] Tchebycheff is also spelled Chebyshev, depending on the system used for transliterating Russian.

Joint Distributions

A useful concept for working with pairs of random variables is the *joint distribution function*:

Definition 9 (Joint distribution function) Let X and Y be random variables on a sample space U with probability function P. For each $(i,j) \in \text{Image}(X) \times \text{Image}(Y)$, define $h_{X,Y}(i,j) = P(X^{-1}(i) \cap Y^{-1}(j))$. The function $h_{X,Y}$ is called the *joint distribution function* of X and Y.

Recalling the meaning of the distribution functions f_X and f_Y (Definition 6) you should be able to see that

$$f_X(i) = \sum_{j \in \text{Image}(Y)} h_{X,Y}(i,j) \quad \text{and} \quad f_Y(j) = \sum_{i \in \text{Image}(X)} h_{X,Y}(i,j).$$

Example 19 (A joint distribution for coin tosses) A fair coin is tossed three times, recording H if heads, T if tails. The sample space is

$$U = \{HHH, HHT, HTH, HTT, THH, THT, TTH, TTT\}.$$

Let X be the random variable defined by $X(t_1 t_2 t_3) = 0$ if $t_1 = T$, $X(t_1 t_2 t_3) = 1$ if $t_1 = H$. Let Y be the random variable that counts the number of times T occurs in the three tosses. $\text{Image}(X) \times \text{Image}(Y) = \{0,1\} \times \{0,1,2,3\}$. We compute $h_{X,Y}(0,2)$. $X^{-1}(0) = \{THH, THT, TTH, TTT\}$. $Y^{-1}(2) = \{HTT, THT, TTH\}$. $X^{-1}(0) \cap Y^{-1}(2) = \{THT, TTH\}$. Thus, $h_{X,Y}(0,2) = 2/8$. Computing the rest of the values of $h_{X,Y}$, we can represent the results in the following table:

$h_{X,Y}$	Y=0	Y=1	Y=2	Y=3	f_X
X=0	0	1/8	2/8	1/8	1/2
X=1	1/8	2/8	1/8	0	1/2
f_Y	1/8	3/8	3/8	1/8	

In this table, the values of $h_{X,Y}(i,j)$ are contained in the submatrix

$$\begin{matrix} 0 & 1/8 & 2/8 & 1/8 \\ 1/8 & 2/8 & 1/8 & 0 \end{matrix}$$

The last column gives the distribution function f_X and the last row gives the distribution function f_Y. These distributions are called the *marginal distributions* of the joint distribution $h_{X,Y}$. You should check that $E(X) = 1/2$, $E(Y) = 3/2$, $\text{Var}(X) = 1/4$, $\text{Var}(Y) = 3/4$. Compute also that $E(XY) = \sum ij h_{X,Y}(i,j) = 1/2$, where the sum is over $(i,j) \in \{0,1\} \times \{0,1,2,3\}$. Thus, $\text{Cov}(X,Y) = -1/4$. Putting this all together gives $\rho(X,Y) = -3^{-1/2}$. It should be no surprise that the correlation is negative. If X is "large" (i.e., 1) then Y should be "small," since the first toss came up H, making the total number of T's at most 2. □

Section 4: Functions and Probability

Example 20 (Another joint distribution for coin tosses) As before, a fair coin is tossed three times, recording H if heads, T if tails. The sample space is

$$U = \{HHH, HHT, HTH, HTT, THH, THT, TTH, TTT\}.$$

Again, let X be the random variable defined by $X(t_1t_2t_3) = 0$ if $t_1 = T$, $X(t_1t_2t_3) = 1$ if $t_1 = H$. However, now let Y be the random variable that counts the number of times T occurs in the last two tosses. Image$(X) \times$ Image$(Y) = \{0,1\} \times \{0,1,2\}$. We compute $h_{X,Y}(0,2)$. $X^{-1}(0) = \{THH, THT, TTH, TTT\}$. $Y^{-1}(2) = \{HTT, TTT\}$. $X^{-1}(0) \cap Y^{-1}(2) = \{TTT\}$. Thus, $h_{X,Y}(0,2) = 1/8$. Computing the rest of the values of $h_{X,Y}$, we can represent the results in the following table:

$h_{X,Y}$	Y=0	Y=1	Y=2	f_X
X=0	1/8	2/8	1/8	1/2
X=1	1/8	2/8	1/8	1/2
f_Y	1/4	1/2	1/4	

You should compute that $E(X) = 1/2$, $E(Y) = 1$, $\text{Var}(X) = 1/4$, $\text{Var}(Y) = 1/2$ and $\text{Cov}(X,Y) = 0$. Since all the previous numbers were nonzero, it is rather surprising that the covariance is zero. This is a consequence of "independence," which we will study next. ∎

Independence

If the expectation of the sum of random variables is the sum of the expectations, then maybe the expectation of the product is the product of the expectations? Not so. We'll look a simple example with $Y(T) = X(T)$. Let $U = \{H, T\}$ and $P(H) = P(T) = 1/2$ and let $X(T) = 0$ and $X(H) = 1$ be a random variable on (U, P). Then $E(X) = X(T)(1/2) + X(H)(1/2) = 1/2$. Since $X^2 = X$, we have $E(XX) = E(X^2) = E(X) = 1/2$. This does not equal $E(X)^2$. In order to give some general sufficient conditions for when $E(XY) = E(X)E(Y)$ we need the following definition.

Definition 10 (Independence) Let (U, P) be a probability space.

- If $A \subseteq U$ and $B \subseteq U$ are events (subsets) of U such that $P(A \cap B) = P(A)P(B)$. Then A and B are said to be *a pair of independent events*.

- If X and Y are random variables on U such that

$$\text{for \underline{all} } s \in \text{Image}(X) \text{ and \underline{all} } t \in \text{Image}(Y),$$
$$X^{-1}(s) \text{ and } Y^{-1}(t) \text{ are a pair of independent events},$$

then X and Y are said to be *a pair of independent random variables*.

[73]

Functions

The definition of independence for events and random variables sounds a bit technical. In practice we will use independence in a very intuitive way.

Recall the definitions of the marginal and joint distributions f_X, f_Y and $h_{X,Y}$ in Definition 9. Using that notation, the definition of independence can be rephrased as

X and Y are independent random variables
if and only if
$h_{X,Y}(i,j) = f_X(i) f_Y(j)$ for all $(i,j) \in \text{Image}(X) \times \text{Image}(Y)$.

You should verify that X and Y are independent in Example 20. Intuitively, knowing what happens on the first toss gives us no information about the second and third tosses. We explore this a bit in the next example.

Example 21 (Independent coin tosses) Suppose a fair coin is tossed twice, one after the other. In everyday language, we think of the tosses as being independent. We'll see that this agrees with our mathematical definition of independence.

The sample space is $U = \{(H,H), (H,T), (T,H), (T,T)\}$. Note that $U = \{H,T\} \times \{H,T\}$ and p is the uniform probability function on U; i.e., $P(e) = 1/4$ for each $e \in U$.

Let A be the event that the first toss is H and let B be the event that the second is H. Thus $A = \{HH, HT\}$ and $B = \{HH, TH\}$. You should be able to see that $P(A) = 1/2$, $P(B) = 1/2$ and

$$P(A \cap B) = P(HH) = 1/4 = (1/2)(1/2)$$

and so A and B are independent.

What about independent random variables? Let

$$X(t_1, t_2) = \begin{cases} 0, & \text{if } t_1 = T, \\ 1, & \text{if } t_1 = H. \end{cases}$$

and

$$Y(t_1, t_2) = \begin{cases} 0, & \text{if } t_2 = T, \\ 1, & \text{if } t_2 = H. \end{cases}$$

Thus X "looks at" just the first toss and Y "looks at" just the second. You should be able to verify that $X^{-1}(1) = A$, $X^{-1}(0) = A^c$, $Y^{-1}(1) = B$ and $Y^{-1}(0) = B^c$. To see that X and Y are independent, we must verify that each of the following 4 pairs of events is independent

A and B A and B^c A^c and B A^c and B^c.

In the previous paragraph, we saw that A and B are independent. You should be able to do the other 3.

This seems like a lot of work to verify the mathematical notion of independence, compared with the obvious intuitive notion. Why bother? There are two reasons. First, we want to see that the two notions of independence are the same. Second, we can't do any calculations with an intuitive notion, but the mathematical definition will allow us to obtain useful results. ∎

The preceding example can be generalized considerably. The result is an important method for building up new probability spaces from old ones.

Section 4: Functions and Probability

*Example 22 (Product spaces and independence)   Let (U_1, P_1) and (U_2, P_2) be probability spaces. Define the *product space* (U, P) by

$$U = U_1 \times U_2 \qquad \text{(Cartesian product)}$$
$$P(e_1, e_2) = P_1(e_1) \times P_2(e_2) \qquad \text{(multiplication of numbers)}$$

Suppose $A = A_1 \times U_2$ and $B = U_1 \times B_2$. We claim that A and B are independent events. Before proving this, let's see how it relates to the previous example.

Suppose $U_1 = U_2 = \{H, T\}$ and that P_1 and P_2 are the uniform probability functions. Then (U_1, P_1) describes the first toss and (U_2, P_2) describes the second. Also, you should check that (U, P) is the same probability space as in the previous example. Check that, if $A_1 = \{H\}$ and $B_2 = \{H\}$, then A and B are the same as in the previous example.

We now prove that A and B are independent in our general setting. We have

$$\begin{aligned}
P(A) &= \sum_{e \in A} P(e) & &\text{definition of } P(A) \\
&= \sum_{e \in A_1 \times U_2} P(e) & &\text{definition of } A \\
&= \sum_{\substack{e_1 \in A_1 \\ e_2 \in U_2}} P(e_1, e_2) & &\text{definition of Cartesian product} \\
&= \sum_{\substack{e_1 \in A_1 \\ e_2 \in U_2}} P_1(e_1) P_2(e_2) & &\text{definition of } P \\
&= \left(\sum_{e_1 \in A_1} P_1(e_1) \right) \times \left(\sum_{e_2 \in U_2} P_2(e_2) \right) & &\text{algebra} \\
&= \sum_{e_1 \in A_1} P_1(e_1) \times 1 & &\text{definition of probability} \\
&= P_1(A_1) & &\text{definition of } P_1(A_1).
\end{aligned}$$

Similarly, $P(B) = P_2(B_2)$. You should verify that $A \cap B = A_1 \times B_2$. By doing calculations similar to what we did for $P(A)$, you should show that $P(A_1 \times B_2) = P_1(A_1) P_2(B_2)$. This proves independence.

Suppose X is a random variable such that $X(e_1, e_2)$ depends only on e_1. What does $X^{-1}(r)$ look like? Suppose $X(e_1, e_2) = r$ so that $(e_1, e_2) \in X^{-1}(r)$. Since X does not depend on e_2, $X(e_1, u_2) = r$ for all $u_2 \in U_2$. Thus $\{e_1\} \times U_2 \subseteq X^{-1}(r)$. Proceeding in this way, one can show that $X^{-1}(r) = A_1 \times U_2$ for some $A_1 \subseteq U_1$.

Suppose Y is a random variable such that $Y(e_1, e_2)$ depends only on e_2. We then have $Y^{-1} = U_1 \times B_2$ for some $B_2 \subseteq U_2$. By our earlier work in this example, it follows that $X^{-1}(r)$ and $Y^{-1}(s)$ are independent events and so X and Y are independent.

What made this work? X "looked at" just the first component and Y "looked at" just the second.

This can be generalized to the product of any number of probability spaces. Random variables X and Y will be independent if the components that X "looks at" are disjoint

Functions

from the set of components that Y "looks at." For example, suppose a coin is tossed "independently" 20 times. Let X count the number of heads in the first 10 tosses and let Y count the number of tails in the last 5 tosses. ☐

We now return to $E(XY)$, with the assumption that X and Y are independent random variables on a sample space U with probability function P. We also look at variance, covariance, and correlation.

Theorem 7 (Properties of independent random variables) Suppose that U is a sample space with probability function P and that X and Y are independent random variables on U. Then the following are true

- $E(XY) = E(X)E(Y)$,
- $\text{Cov}(X,Y) = 0$ and $\rho(X,Y) = 0$,
- $\text{Var}(X+Y) = \text{Var}(X) + \text{Var}(Y)$,
- if $f, g : \mathbb{R} \to \mathbb{R}$, then $f(X)$ and $g(Y)$ are independent random variables.

Proof: First note that there are two ways to compute $E(Z)$ for a random variable Z:

$$E(Z) = \sum_{u \in U} Z(u) P(u) \qquad \text{(the first way)}$$

$$= \sum_{k \in \text{Image}(Z)} k P(Z^{-1}(k)) = \sum_{k \in \text{Image}(Z)} k f_Z(k) \qquad \text{(the second way)}.$$

For $Z = XY$, we use the second way. If $Z = k$, then $X = i$ and $Y = j$ for some i and j with $ij = k$. Thus

$$E(Z) = \sum_{k \in \text{Image}(Z)} k P(Z^{-1}(k))$$

$$= \sum_{\substack{i \in \text{Image}(X) \\ j \in \text{Image}(Y)}} ij P(X^{-1}(i) \cap Y^{-1}(j)).$$

From the definition of independence, $P(X^{-1}(i) \cap Y^{-1}(j)) = P(X^{-1}(i)) P(Y^{-1}(j))$, and hence

$$\sum_{\substack{i \in \text{Image}(X) \\ j \in \text{Image}(Y)}} ij P(X^{-1}(i) \cap Y^{-1}(j)) = \sum_{\substack{i \in \text{Image}(X) \\ j \in \text{Image}(Y)}} ij P(X^{-1}(i)) P(Y^{-1}(j))$$

$$= \left(\sum_{i \in \text{Image}(X)} i P(X^{-1}(i)) \right) \times \left(\sum_{j \in \text{Image}(Y)} j P(Y^{-1}(j)) \right).$$

The right hand side of the above equation is just $E(X)E(Y)$. This proves the first part of the theorem.

By Definition 8 and the fact that we have just proved $E(XY) = E(X)E(Y)$, it follows that $\text{Cov}(X,Y) = 0$. It then follows from Definition 8 that $\rho(X,Y) = 0$. You should use some of the results in Theorem 4 to show that $\text{Var}(X+Y) = \text{Var}(X) + \text{Var}(Y)$.

We omit the proof of the last part of the theorem; however, you should note that $f(X)$ is a *function* on U because, for $u \in U$, $f(X)(u) = f(X(u))$. ☐

Section 4: Functions and Probability

*Example 23 (Generating random permutations) Suppose we want to generate permutations of $\underline{n}$ randomly so that each permutation is equally likely to occur. How can we do so? We now look at a simple, efficient method for doing this. The procedure is based on a bijection between two sets:

- Seq_n, the set of all sequences $a_2, a_3, \ldots, a_n$ of integers with $1 \le a_k \le k$ for all k and
- Perm_n, the set of all permutations of $\underline{n} = \{1, 2, \ldots, n\}$.

We can tell there must be such a bijection even without giving one! Why is this? Since there are k choices for a_k, the number of sequences $a_2, a_3, \ldots, a_n$ is $2 \times 3 \times \cdots \times n = n!$. We know that there are $n!$ permutations. Thus $|\text{Seq}_n| = |\text{Perm}_n|$. Because Seq_n and Perm_n have the same size, there must be a bijection between them. Since there's a bijection, what's the problem? The problem is to find one that is easy to use.

Before providing the bijection, let's look at how to use it to generate permutations uniformly at random. It is easy to generate the sequences uniformly at random: Choose a_k uniformly at random from $\underline{k}$ and choose each of the a_k independently. This makes Seq_n into a probability space with the uniform probability distribution. Once we have a sequence, we use the bijection to construct the permutation that corresponds to the sequence. This makes Perm_n into a probability space with the uniform probability distribution.

Now we want to specify a bijection. There are lots of choices. (You should be able to show that there are $(n!)!$ bijections.) Here's one that is easy to use:

Step 1. Write out the sequence $1, 2, \ldots, n$ and set $k = 2$

Step 2. If $a_k \ne k$, swap the elements in positions k and a_k in the sequence. If $a_k = k$, do nothing.

Step 3. If $k < n$, increase k by 1 and go to Step 2. If $k = n$, stop.

The result is a permutation in one-line form.

For example, suppose $n = 5$ and the sequence of a_k's is 2,1,3,3. Here's what happens, where "next step" tells which step to use to produce the permutation on the next line:

action	permutation	information	next step
the start (Step 1)	1,2,3,4,5	$k = 2$, $a_k = 2$	Step 3
do nothing	1,2,3,4,5	$k = 3$, $a_k = 1$	Step 2
swap at 3 and 1	3,2,1,4,5	$k = 4$, $a_k = 3$	Step 2
swap at 4 and 3	3,2,4,1,5	$k = 5$, $a_k = 3$	Step 2
swap at 5 and 3	3,2,5,1,4		all done

Thus, the sequence $2, 1, 3, 3$ corresponds to the permutation $3, 2, 5, 1, 4$.

How can we prove that this is a bijection? We've described a function F from Seq_n to Perm_n. Since $|\text{Seq}_n| = |\text{Perm}_n|$, we can prove that F is a bijection if we can show that it is a surjection. (This is just Exercise 1.2(c).) In other words, we want to show that, for every permutation p in Perm_n, there is a sequence $\mathbf{a}$ in Seq_n such that $F(\mathbf{a}) = p$.

Let's try an example. Suppose we have the permutation $4, 1, 3, 2, 6, 5$. What sequence does it come from? This is a permutation of $\underline{6}$. The only way 6 could move from the last place is because $a_6 \ne 6$. In fact, since 6 is in the fifth place, we must have had $a_6 = 5$.

Functions

Which caused us to swap the fifth and sixth positions. So, just before we used $a_6 = 5$, we had the permutation $4, 1, 3, 2, 5, 6$. None of $a_2, \ldots, a_5$ can move what's in position 6 since they are all less than 6, so $a_2, \ldots, a_5$ must have rearranged $1, 2, 3, 4, 5, (6)$ to give $4, 1, 3, 2, 5, (6)$. (We've put 6 in parentheses to remember that it's there and that none of $a_2, \ldots, a_5$ can move it.) How did this happen? Well, only a_5 could have affected position 5. Since 5 is there, it didn't move and so $a_5 = 5$. Now we're back to $4, 1, 3, 2, (5, 6)$ and trying to find a_4. Since 4 is in the first position, $a_4 = 1$. So, just before using, a_4 we had $2, 1, 3, (4, 5, 6)$. Thus $a_3 = 3$ and we're back to $2, 1, (3, 4, 5, 6)$. Finally $a_2 = 1$. We've found that the sequence $1, 3, 1, 5, 5$ gives the permutation $4, 1, 3, 4, 6, 5$. You should apply the algorithm described earlier in Steps 1, 2, and 3 and so see for yourself that the sequence gives the permutation.

The idea in the previous paragraph can be used to give a proof by induction on n. For those of you who would like to see it, here's the proof. The fact that F is a surjection is easily checked for $n = 2$: There are two sequences, namely 1 and 2. These correspond to the two permutations $2, 1$ and $1, 2$, respectively. Suppose $n > 2$ and $p_1, \ldots, p_n$ is a permutation of $\underline{n}$. We need to find the position of n in the permutation. The position is that k for which $p_k = n$. So we set $a_n = k$ and define a new permutation $q_1, \ldots, q_{n-1}$ of $\{1, 2, \ldots, n-1\}$ to correspond to the situation just before using $a_n = k$:

- If $k = n$, then $q_i = p_i$ for $1 \leq i \leq n - 1$.

- If $k \neq n$, the $q_k = p_n$ and $q_i = p_i$ for $1 \leq i < k$ and for $k < i \leq n - 1$.

You should be able to see that $q_1, \ldots, q_{n-1}$ is a permutation of $\underline{n-1}$. By induction, there is a sequence $a_2, \ldots, a_{n-1}$ that gives $q_1, \ldots, q_{n-1}$ when we apply our 3-step procedure to $1, 2, 3, \ldots, (n-1)$. After that, we must apply $a_n = k$ to $q_1, \ldots, q_{n-1}, n$. What happens? You should be able to see that it gives us $p_1, \ldots, p_n$. This completes the proof. □

Some Standard Distributions

We now take a look at some examples of random variables and their distributions that occur often in applications. The first such distribution is the *binomial distribution*.

Example 24 (Binomial distribution) Suppose we toss a coin, sequentially and independently, n times, recording H for heads and T for tails. Suppose the probability of H in a single toss of the coin is p. Define

$$P^*(t) = \begin{cases} p, & \text{if } t = H, \\ q = 1 - p, & \text{if } t = T. \end{cases}$$

Our sample space is $U = \times^n \{H, T\}$ and the probability function P is given by $P(t_1, \ldots, t_n) = P^*(t_1) \cdots P^*(t_n)$ because of independence. This is an example of a product space. We discussed product spaces in Example 22.

Define the random variable $X(t_1, \ldots, t_n)$ to be the number of H's in the sequence $(t_1, \ldots, t_n)$. This is a standard example of a *binomial random variable*.

Section 4: Functions and Probability

We want to compute $P(X = k)$ for $k \in \mathbb{R}$. Note that $\text{Image}(X) = \{0, \ldots, n\}$. Hence $P(x = k) = 0$ if k is not in $\{0, \ldots, n\}$. Note that $(t_1, \ldots, t_n) \in X^{-1}(k)$ if and only if $(t_1, \ldots, t_n)$ contains exactly k heads (H's). In this case, $P(t_1, \ldots, t_n) = p^k q^{n-k}$. Since all elements of $X^{-1}(k)$ have the same probability $p^k q^{n-k}$, it follows that $f_X(k) = |X^{-1}(k)| p^k q^{n-k}$. What is the value of $|X^{-1}(k)|$. It is the number of sequences with exactly k heads. Since the positions for k heads must be chosen from among the n tosses, $|X^{-1}(k)| = \binom{n}{k}$. Thus $f_X(k) = \binom{n}{k} p^k q^{n-k}$. This is the *binomial distribution* function. A common alternative notation for this distribution function is $b(k; n, p)$. This notation has the advantage of explicitly referencing the parameters, n and p.

An alternative way of thinking about the random variable X is to write it as a sum, $X = X_1 + \cdots + X_n$, of n independent random variables. The random variable X_i is defined on the sample space $U = \times^n \{H, T\}$ by the rule

$$X_i(t_1, \ldots, t_n) = \begin{cases} 1, & \text{if } t_i = H, \\ 0, & \text{if } t_i = T. \end{cases}$$

Using this representation of X, we can compute $E(X) = E(X_1) + \cdots + E(X_n)$, and $\text{Var}(X) = \text{Var}(X_1) + \cdots + \text{Var}(X_n)$. Computation gives

$$E(X_i) = 1 \times P(X_i = 1) + 0 \times P(X_i = 0) = p$$

and

$$\text{Var}(X_i) = E(X_i^2) - E(X_i)^2 = p - p^2 = p(1-p),$$

where we have used $X_i^2 = X_i$ because X_i must be 0 or 1. Thus, we obtain $E(X) = np$ and $\text{Var}(X) = np(1-p) = npq$. □

Of course, the binomial distribution is not restricted to coin tosses, but is defined for any series of outcomes that

- are restricted to two possibilities,
- are independent, and
- have a fixed probability p of one outcome, $1 - p$ of the other outcome.

Our next example is a random variable X that is defined on a countably infinite sample space U. This distribution, the Poisson, is associated with random distributions of objects.

Example 25 (Poisson distribution and its properties) Suppose a 500 page book has 2,000 misprints. If the misprints are distributed randomly, what is the probability of exactly k misprints appearing on page 95? (We want the answers for $k = 0, 1, 2, \ldots$.)

Imagine that the misprints are all in a bag. When we take out a misprint, it appears on page 95 with probability $1/500$. Call the case in which a misprint appears on page 95 a "success" and the case when it does not a "failure." We have just seen that, for a randomly selected misprint, the probability of success is $p = 1/500$. Since we have assumed the misprints are independent, we can use the binomial distribution. Our answer is therefore that the probability of exactly k misprints on page 95 is $b(k; 2000, 1/500)$.

Functions

Thus we have our answer: $b(k; 2000, 1/500) = \binom{2000}{k}(1/500)^k(1-1/500)^{2000-k}$. Unfortunately, its hard to use: for large numbers the binomial distribution is awkward to work with because there is a lot of calculation involved and numbers can be very large or very small. Can we get a more convenient answer? Yes. There is a nice approximation which we will now discuss.

The function $f_X(k) = e^{-\lambda}\frac{\lambda^k}{k!}$ is also denoted by $p(k; \lambda)$ and is called the *Poisson distribution*. Clearly the $p(k; \lambda)$ are positive. Also, they sum to one:

$$\sum_{k=0}^{\infty} e^{-\lambda}\frac{\lambda^k}{k!} = e^{-\lambda}\sum_{k=0}^{\infty}\frac{\lambda^k}{k!} = e^{-\lambda}e^{\lambda} = 1.$$

We have used the Taylor Series expansion, obtained in calculus courses, $\sum_{k=0}^{\infty}\frac{\lambda^k}{k!} = e^{\lambda}$. In a similar manner, it can be shown that

$$E(X) = \lambda \quad \text{and} \quad \text{Var}(X) = \lambda.$$

Thus, a Poisson distributed random variable X has the remarkable property that $E(X) = \lambda$ and $\text{Var}(X) = \lambda$ where $\lambda > 0$ is the parameter in the distribution function $P(X = k) = p(k; \lambda) = e^{-\lambda}\lambda^k/k!$.

We now return to our binomial distribution $b(k; 2000, 1/500)$. The Poisson is a good approximation to $b(k; n, p)$ when n is large and np is not large. In this case, take $\lambda = np$, the mean of the binomial distribution. For our problem, $\lambda = 2000(1/500) = 4$, which is not large when compared to the other numbers in the problem, namely 2,000 and 500. Let's compute some estimates for P_k, the probability of exactly k errors on page 95.

$$P_0 = e^{-4} = 0.0183, \quad P_1 = 4e^{-4} = 0.0733, \quad P_3 = 4^3e^{-4}/3! = 0.1954,$$

and so on. ☐

Our final example of a random variable X has its underlying sample space $U = \mathbb{R}$, the real numbers. Rather than starting with a description of X itself, we start with the distribution function $f_X(x) = \phi_{\mu,\sigma}(x)$, called the *normal distribution function with mean μ and standard deviation σ*.

$$\phi_{\mu,\sigma}(x) = \frac{1}{\sigma\sqrt{2\pi}}e^{-\frac{1}{2}(\frac{x-\mu}{\sigma})^2}.$$

For computations concerning the normal distribution, it suffices in most problems, to work with the special case when $\mu = 0$ and $\sigma = 1$. In this case, we use the notation

$$\phi(x) = \frac{1}{\sqrt{2\pi}}e^{-\frac{1}{2}x^2}$$

where $\phi(x) = \phi_{0,1}(x)$ is called the *standard normal distribution*.

The function $\phi(x)$ is defined for $-\infty < x < \infty$ and is symmetric about $x = 0$. The maximum of $\phi(x)$ occurs at $x = 0$ and is about 0.4. Here is a graph of $\phi(x)$ for $-2 \leq x \leq t$:

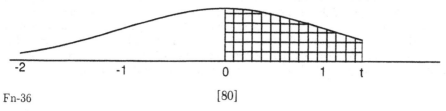

Section 4: Functions and Probability

In this graph of $\phi(x)$ shown above, the area between the curve and the interval from 0 to t on the x-axis is shaded. This area, as we shall discuss below, represents the probability that a random variable with distribution function ϕ lies between 0 and t. For $t = 1$ the probability is about 0.34, for $t = 1.5$ the probability is about 0.43, and for $t = 2$, the probability is about 0.48. Since this is a probability distribution, the area under the whole curve is 1. Also, since the curve is symmetric, the area for $x < 0$ is $1/2$. We'll use these values in the examples and problems, so you will want to refer back to them.

Example 26 (The normal distribution and probabilities) The way the normal curve relates to probability is more subtle than in the finite or discrete case. If a random variable X has $\phi_{\mu,\sigma}(x) = \frac{1}{\sqrt{2\pi}} e^{-\frac{1}{2}(\frac{x-\mu}{\sigma})^2}$ as its distribution function then we compute the probability of any event of the form $[a, b] = \{x \mid a \leq x \leq b\}$ by computing the area under the curve $\phi_{\mu,\sigma}(x)$ and above the interval $[a, b]$.

How can we compute this area? Tables and computer programs for areas below $y = \phi(x)$ are available. Unfortunately $\phi_{\mu, sigma}$ and ϕ are different functions unless $\mu = 0$ and $\sigma = 1$. Fortunately, there is a simple recipe for converting one to the other. Let $h(t) = (t - \mu)/\sigma$. The area below $\phi_{\mu,\sigma}(x)$ above the interval $[a, b]$ equals the area below ϕ above the interval $[h(a), h(b)]$.

A farmer weighs some oranges from his crop and comes to you for help. From his data you notice that the mean weight is 8 ounces and the standard deviation is 0.67 ounces. You've read somewhere (Was it here?) that for such things a normal distribution is a good approximation to the weight. The farmer can sell oranges that weigh at least 9 ounces at a higher price per ounce, so he wants to estimate what fraction of his crop weighs at least 9 ounces. Using our recipe, $h(9) = (9 - 8)/0.67 = 1.5$. We know that the area under $\phi(x)$ for the interval $[0, 1.5]$ is 0.43. Since the area under $\phi(x)$ for $x < 0$ is $1/2$, the area for $x \leq 1.5$ is $0.43 + 0.5 = 0.93$. Since these are the "underweight" oranges, the farmer can expect about 7% of his crop to be at least 9 ounces. $\square$

Example 27 (Approximating the binomial distribution) Recall the binomial distribution from Example 24: $b(k; n, p)$ is the probability of exactly k heads in n tosses and p is the probability of a head on one toss. We derived the formula $b(k; n, p) = \binom{n}{k} p^k q^{n-k}$, where $q = 1 - p$. We also found, that for a binomial random variable X, $E(X) = np$ and $\text{Var}(X) = npq$. How does the random variable behave when n is large? We already saw in Example 25 how to use the Poisson approximation when $E(X)$ is not large. When $E(X)$ and $\text{Var}(X)$ are large, a better approximation is given by the normal distribution $\phi_{\mu,\sigma}$ with $\mu = np$ and $\sigma = \sqrt{npq}$.

Suppose that our book in Example 25 is a lot worse: About one word in ten is wrong. How can we estimate the probability of at most 30 errors on page 95? If the errors are independent, the distribution is a binomial with $p = 0.1$ and n equal to the number of words on page 95. We estimate that n is about 400. Thus we are dealing with $b(k; 400, 0.1)$. We have
$$\mu = 400 \times 0.1 = 40 \quad \text{and} \quad \sigma = \sqrt{400 \times 0.1 \times 0.9} = \sqrt{36} = 6.$$
Thus we want the area under $\phi(x)$ for $x < h(30) = (30 - 40)/6 \approx -1.5$. By the symmetry of ϕ, this is the area under $\phi(x)$ for $x > 1.5$, which is $0.5 - 0.43 = 7\%$.

Functions

We've done some rounding off here, which is okay since our estimates are rather crude. There are ways to improve the estimates, but we will not discuss them. ☐

Approximations like those in the preceding example are referred to as "limit theorems" in probability theory. The next example discusses the use of an important limit theorem, the Central Limit Theorem, for estimating how close an average of measurements is to the true value of a number. This is often used in experimental science when estimating a physical constant.

*Example 28 (The Central Limit Theorem and the normal distribution) Suppose a student must estimate a quantity, say the distance between two buildings on campus. The student makes a number n of measurements. Each measurement can be thought of as a sample of a random variable. Call the random variable for measurement i X_i. If the student is not influenced by the previous measurements, we can think of the random variables as being independent and identically distributed. The obvious thing to do is average these measurements. How accurate is the result?

Let's phrase this in probabilistic terms. We have a new random variable given by $X = (X_1 + \cdots + X_n)/n$ and our average is a sample of the value of the random variable X. What can we say about X?

We can approximate X with a normal distribution. This approximation is a consequence of the *Central Limit Theorem*. Let A_1 be the average of the n measurements and let A_2 the average of the squares of the n measurements. Then we estimate μ and σ by A_1 and $\sqrt{(A_2 - (A_1)^2)/(n-1)}$, respectively.[11] We could now use $\phi_{\mu,\sigma}$ to estimate the distribution of the random variable X.

This can be turned around, $\phi_{\mu,\sigma}$ can also be used to estimate the true mean of the random variable X. You might have thought that A_1 was the mean. No. It is just the average of some observed values. Thus, the probability that the mean of X lies in $[\mu - \sigma, \mu + \sigma]$ equals $0.34 + 0.34 = 0.68$. ☐

We've looked at several different distributions: binomial, normal, Poisson and marginal. What do we use when? How are they related?

The binomial distribution occurs when you have a sequence of repeated independent events and want to know how many times a certain event occurred. For example, the probability of k heads in n tosses of a coin. The coin tosses are the repeated independent events and the heads are the events we are interested in.

The normal distribution is usually an approximation for estimating a number whose value is the sum of a lot of (nearly) independent random variables. For example, let X_i be 1 or 0 according as the i-th coin toss is a head or tail. We want to know the probability that $X_1 + X_2 + \ldots + X_n$ equals k. The exact answer is the binomial distribution. The normal distribution gives an approximation.

The Poisson distribution is associated with rare events. For example, if light bulbs fail at random (we're not being precise here) and have an average lifetime L, then the number

---

[11] The estimate for σ is a result from statistics. We cannot derive it here.

Section 4: Functions and Probability

of failures in a time interval T is roughly Poisson if $\lambda = T/L$ is not too big or too small. Another example is errors in a text, which are rare and have a distribution associated with them that is like the binomial.

Unlike the previous three distributions, which exist by themselves, a marginal distribution is always derived from some given distribution. In our coin toss experiment, let X be the number of heads and let Y be the number of times two or more tails occur together. We could ask for the distribution given by $P(X = k$ and $Y = j)$. This is called a "joint distribution" for the random variables X and Y. *Given* the joint distribution, we could ask for the distribution of just one of the random variables. These are "marginal distributions" associated with the joint distribution. In this example, $P(X = k)$ and $P(Y = j)$ are marginal distributions. The first one (the probability of k heads) is the sum of $P(X = k$ and $Y = j)$ over all j and the second (the probability of two or more tails together happening j times) is the sum of $P(X = k$ and $Y = j)$ over all k.

Exercises for Section 4

4.1. A fair coin is tossed four times, recording H if heads, T if tails. Let X be the random variable defined by $X(t_1 t_2 t_3 t_4) = |\{i \mid t_i = H\}|$. Let Y be the random variable defined by

$$Y(t_1 t_2 t_3 t_4) = \begin{cases} 0, & \text{if } t_i = T \text{ for all } i = 1, 2, 3, 4; \\ \max\{k \mid H = t_i = t_{i+1} = \cdots = t_{i+k-1}, i = 1,2,3,4\}, & \text{otherwise.} \end{cases}$$

The random variable X equals the number of H's. The random variable Y equals the length of the longest consecutive string of H's. Compute

(a) the joint distribution function $h_{X,Y}$,

(b) the marginal distributions f_X and f_Y,

(c) the covariance $\text{Cov}(X,Y)$, and

(d) the correlation $\rho(X,Y)$.

Give an intuitive explanation of the value of $\rho(X,Y)$.

4.2. Let X and Y be random variables on a sample space U and let a and b be real numbers.

(a) Show that $\text{Cov}(aX + bY, aX - bY)$ is $a^2 \text{Var}(X) - b^2 \text{Var}(Y)$.

(b) What is $\text{Var}((aX - bY)(aX + bY))$?

4.3. Let X be random variable on a sample space U and let a and b be real numbers. What is $E((aX + b)^2)$ if

(a) X has the binomial distribution $b(k; n, p)$?

Functions

(b) X has the Poisson distribution $e^{-\lambda}\lambda^k/k!$?

4.4. An 100 page book has 200 misprints. If the misprints are distributed uniformly throughout the book, show how to use the Poisson approximation to the binomial distribution to calculate the probability of there being less than 4 misprints on page 8.

4.5. Let X and Y be independent random variables and let a and b be real numbers. Let $Z = aX + bY$. Then, for all $\epsilon > 0$, Tchebycheff's inequality gives an upper bound for $P(|Z - E(Z)| \geq \epsilon)$. Give this upper bound for the cases where

(a) X and Y have Poisson distribution $p(k;\gamma)$ and $p(k;\delta)$ respectively.

(b) X and Y have binomial distribution $p(k;n,r)$ and $p(k;n,s)$ respectively.

4.6. Each time a customer checks out at Super Save Groceries, a wheel with nine white and one black dot, symmetrically placed around the wheel, is spun. If the black dot is uppermost, the customer gets the least expensive item in their grocery cart for free. Assuming the probability of any dot being uppermost is 1/10, what is the probability that out of the first 1000 customers, between 85 and 115 customers get a free item? Write the formula for the exact solution and show how the normal distribution can be used to approximate this solution. You need not compute the values of the normal distribution.

4.7. Let $X_1, \ldots, X_n$ be independent random variables each having mean μ and variance σ^2. (These could arise by having one person repeat n times an experiment that produces an estimate of a number whose value is μ. See Example 28.) Let $X = (X_1 + \cdots + X_n)/n$.

(a) Compute the mean and variance of X.

(b) Explain why an observed value of X could be used as an estimate of μ.

(c) It turns out that the error we can expect in approximating μ with X is proportional to the value of σ_X. Suppose we want to reduce this expected error by a factor of 10. How much would we have to increase n. (In other words, how many more measurements would be needed.)

Review Questions

Multiple Choice Questions for Review

1. In each case some information is given about a function. In which case is the information *not* sufficient to define a function?

   (a) $f \in \underline{4}^{\underline{3}}$, $2 \to 3$, $1 \to 4$, $3 \to 2$.

   (b) $f \in \{>, <, +, ?\}^{\underline{3}}$, $f = (?, <, +)$.

   (c) $f \in \underline{3}^{\{>,<,+,?\}}$, $f = (3, 1, 2, 3)$.

   (d) $f \in \underline{3}^{\{>,<,+,?\}}$, $f = (3, 1, 2, 3)$. Domain ordered as follows: $>, <, +, ?$.

   (e) $f \in \{>, <, +, ?\}^{\underline{3}}$, $f = (?, <, +)$. Domain ordered as follows: $3, 2, 1$.

2. The following function is in two line form: $\begin{pmatrix} 1 & 2 & 3 & 4 & 5 & 6 & 7 & 8 & 9 \\ 9 & 3 & 7 & 2 & 6 & 4 & 5 & 1 & 8 \end{pmatrix}$. Which of the following is a correct cycle form for this function?

   (a) $(1, 8, 9)(2, 3, 7, 5, 6, 4)$

   (b) $(1, 9, 8)(2, 3, 5, 7, 6, 4)$

   (c) $(1, 9, 8)(2, 3, 7, 5, 4, 6)$

   (d) $(1, 9, 8)(2, 3, 7, 5, 6, 4)$.

   (e) $(1, 9, 8)(3, 2, 7, 5, 6, 4)$

3. In each case some information about a function is given to you. Based on this information, which function is an injection?

   (a) $f \in \underline{6}^{\underline{5}}$, $\text{Coimage}(f) = \{\{1\}, \{2\}, \{3\}, \{4\}, \{5\}\}$

   (b) $f \in \underline{6}^{\underline{6}}$, $\text{Coimage}(f) = \{\{1\}, \{2\}, \{3\}, \{4\}, \{5, 6\}\}$

   (c) $f \in \underline{5}^{\underline{5}}$, $f^{-1}(2) = \{1, 3, 5\}$, $f^{-1}(4) = \{2, 4\}$

   (d) $f \in \underline{4}^{\underline{5}}$, $|\text{Image}(f)| = 4$

   (e) $f \in \underline{5}^{\underline{5}}$, $\text{Coimage}(f) = \{\{1, 3, 5\}, \{2, 4\}\}$

4. The following function is in two line form: $f = \begin{pmatrix} 1 & 2 & 3 & 4 & 5 & 6 & 7 & 8 & 9 \\ 8 & 5 & 9 & 2 & 4 & 1 & 3 & 6 & 7 \end{pmatrix}$. Which of the following is a correct cycle form for $h = f^3 \circ f^{-1}$?

   (a) $(1, 6, 8)(2, 3, 7)(5, 6, 4)$

   (b) $(1, 6, 8)(2, 4, 5)(3, 7, 9)$

   (c) $(1, 8, 6)(2, 3, 7)(5, 9, 4)$

   (d) $(1, 9, 8)(2, 3, 5)(7, 6, 4)$

   (e) $(8, 5, 9, 2, 4, 1, 3, 6, 7)$

5. The following permutation is in two line form: $f = \begin{pmatrix} 1 & 2 & 3 & 4 & 5 & 6 & 7 & 8 & 9 \\ 8 & 6 & 4 & 7 & 2 & 9 & 1 & 3 & 5 \end{pmatrix}$. The permutation $g = (1, 2, 3)$ is in cycle form. Let $h = f \circ g$ be the composition of f and g. Which of the following is a correct cycle form for h?

Functions

   (a) (1, 6, 9, 5, 2, 4, 7)(3, 8)

   (b) (1, 8, 3, 4, 7, 2, 6)(5, 9)

   (c) (1, 8, 3, 7, 4, 2, 6)(9, 5)

   (d) (1, 8, 4, 3, 7, 2, 6)(9, 5)

   (e) (8, 6, 4, 7, 9, 1, 2)(3, 5)

6. We want to find the smallest integer $n > 0$ such that, for every permutation f on $\underline{4}$, the function f^n is the identity function on $\underline{4}$. What is the value of n?

   (a) 4   (b) 6   (c) 12   (d) 24   (e) It is impossible.

7. In the lexicographic list of all strictly decreasing functions in $\underline{9}^{\underline{5}}$, find the successor of 98432.

   (a) 98431   (b) 98435   (c) 98521   (d) 98532   (e) 98543

8. The 16 consecutive points $0, 1, \ldots, 14, 15$ have 0 and 15 converted to exterior box boundaries. The interior box boundaries correspond to points $1, 5, 7, 9$. This configuration corresponds to

   (a) 9 balls into 5 boxes

   (b) 9 balls into 6 boxes

   (c) 10 balls into 5 boxes

   (d) 10 balls into 6 boxes

   (e) 11 balls into 4 boxes

9. The 16 consecutive points $0, 1, \ldots, 14, 15$ have 0 and 15 converted to exterior box boundaries. The interior box boundaries correspond to the strictly increasing functions $1 \leq x_1 < x_2 < x_3 < x_4 \leq 14$ in lex order. How many configurations of balls into boxes come before the configuration •||||•••••••••? (Exterior box boundaries are not shown.)

   (a) $\binom{13}{3}$ (b) $\binom{13}{4}$ (c) $\binom{14}{3}$ (d) $\binom{14}{4}$ (e) $\binom{15}{3}$

10. Suppose $f \in \underline{7}^{\underline{6}}$. How many such functions have $|\text{Image}(f)| = 4$?

    (a) $S(7, 4)$ (b) $S(7, 4)(6)_4$ (c) $S(6, 4)(7)_4$ (d) $S(4, 7)(6)_4$ (e) $S(7, 4)\, 6!$

11. Let X be a random variable with distribution $b(k; n, p)$, $q = 1 - p$. Let $Y = (X + 1)^2$. Then $E(Y) = ?$

    (a) $npq + (np + 1)^2$

    (b) $2npq + (np + 1)^2$

    (c) $npq + 2(np + 1)^2$

    (d) $(npq)^2 + (np + 1)^2$

    (e) $2npq(np + 1)^2$

12. Let X and Y be independent random variables with distribution $b(k; n, a)$ and $b(k; n, b)$ respectively. Let $Z = X + 2Y$. Then, for all $\epsilon > 0$, Tchebycheff's inequality guarantees that $P(|Z - na - 2nb| \geq \epsilon)$ is always less than or equal to what?

(a) $(na(1-a) + nb(1-b))/\epsilon^2$

(b) $(na(1-a) + 2nb(1-b))/\epsilon^2$

(c) $(na(1-a) + 4nb(1-b))/\epsilon^2$

(d) $(na(1-a) + 2nb(1-b))/\epsilon^3$

(e) $(na(1-a) + 4nb(1-b))/\epsilon^3$

13. An 800 page book has 400 misprints. If the misprints are distributed uniformly throughout the book, and the Poisson approximation to the binomial distribution is used to calculate the probability of exactly 2 misprints on page 16, which of the following represents the correct use of the Poisson approximation?

    (a) $e^{0.5}/8$ (b) $e^{-0.5}/8$ (c) $e^{0.5}/16$ (d) $e^{-0.5}/16$ (e) $e^{-0.5}/32$

14. For 40 weeks, once per hour during the 40 hour work week, an employee of Best Cars draws a ball from an urn that contains 1 black and 9 white balls. If black is drawn, a $10 bill is tacked to a bulletin board. At the end of the 40 weeks, the money is given to charity. What is the expected amount of money given?

    (a) 1000   (b) 1200   (c) 1400   (d) 1600   (e) 1800

15. For 40 weeks, once per hour during the 40 hour work week, an employee of Best Cars draws a ball from an urn that contains 1 black and 9 white balls. If black is drawn, $10 is tacked to a bulletin board. At the end of the 40 weeks, the money is given to charity. Using the normal approximation, what interval under the standard normal curve should be used to get the area which equals the probability that $1800 or more is given?

    (a) from 1.67 to ∞

    (b) from 0 to 1.67

    (c) from 0.6 to ∞

    (d) from 0 to 0.6

    (e) from 0.6 to 1.67

16. A fair coin is tossed three times. Let X be the random variable which is one if the first throw is T (for tails) and the third throw is H (for heads), zero otherwise. Let Y denote the random variable that is one if the second and third throws are both H, zero otherwise. The covariance, $\text{Cov}(X, Y)$ is

    (a) $1/8$ (b) $-1/8$ (c) $1/16$ (d) $-1/16$ (e) $1/32$

17. A fair coin is tossed three times. Let X be the random variable which is one if the first throw is T (for tails) and the third throw is H (for heads), zero otherwise. Let Y denote the random variable that is one if the second and third throws are both H, zero otherwise. The correlation, $\rho(X, Y)$ is

    (a) 0   (b) $1/3$ (c) $-1/3$ (d) $1/8$ (e) $-1/8$

18. A fair coin is tossed three times and a T (for tails) or H (for heads) is recorded, giving us a 3-long list. Let X be the random variable which is zero if no T has another T adjacent to it, and is one otherwise. Let Y denote the random variable that counts

Functions

the number of T's in the three tosses. Let $h_{X,Y}$ denote the joint distribution of X and Y. $h_{X,Y}(1,2)$ equals

(a) 5/8 (b) 4/8 (c) 3/8 (d) 2/8 (e) 1/8

19. Which of the following is equal to $\text{Cov}(X + Y, X - Y)$, where X and Y are random variables on a sample space S?

(a) $\text{Var}(X) - \text{Var}(Y)$

(b) $\text{Var}(X^2) - \text{Var}(Y^2)$

(c) $\text{Var}(X^2) + 2\text{Cov}(X,Y) + \text{Var}(Y^2)$

(d) $\text{Var}(X^2) - 2\text{Cov}(X,Y) + \text{Var}(Y^2)$

(e) $(\text{Var}(X))^2 - (\text{Var}(Y))^2$

20. Which of the following is equal to $\text{Var}(2X - 3Y)$, where X and Y are random variables on S?

(a) $4\text{Var}(X) + 12\text{Cov}(X,Y) + 9\text{Var}(Y)$

(b) $2\text{Var}(X) - 3\text{Var}(Y)$

(c) $2\text{Var}(X) + 6\text{Cov}(X,Y) + 3\text{Var}(Y)$

(d) $4\text{Var}(X) - 12\text{Cov}(X,Y) + 9\text{Var}(Y)$

(e) $2\text{Var}(X) - 6\text{Cov}(X,Y) + 3\text{Var}(Y)$

21. The strictly decreasing functions in $\underline{100}^3$ are listed in lex order. How many are there before the function (9,5,4)?

(a) 18 (b) 23 (c) 65 (d) 98 (e) 180

22. All but one of the following have the same answer. Which one is different?

(a) The number of multisets of size 20 whose elements lie in $\underline{5}$.

(b) The number of strictly increasing functions from $\underline{20}$ to $\underline{24}$.

(c) The number of subsets of size 20 whose elements lie in $\underline{24}$.

(d) The number of weakly decreasing 4-lists made from $\underline{21}$.

(e) The number of strictly decreasing functions from $\underline{5}$ to $\underline{24}$.

23. Let X be a random variable with Poisson distribution $p(k; \lambda)$ Let $Y = (X+2)(X+1)$. What is the value of $E(Y)$?

(a) $\lambda^2 + 3\lambda + 1$

(b) $\lambda^2 + 3\lambda + 2$

(c) $\lambda^2 + 4\lambda + 2$

(d) $3\lambda^2 + 3\lambda + 2$

(e) $4\lambda^2 + 4\lambda + 2$

Answers: 1 (c), 2 (d), 3 (a), 4 (b), 5 (a), 6 (c), 7 (c), 8 (c), 9 (a), 10 (c), 11 (a), 12 (c), 13 (b), 14 (d), 15 (a), 16 (c), 17 (b), 18 (d), 19 (a), 20 (d), 21 (c), 22 (e), 23 (c).

Unit DT

Decision Trees and Recursion

In many situations one needs to make a series of decisions. This leads naturally to a structure called a "decision tree." Decision trees provide a geometrical framework for organizing the decisions. The important aspect is the decisions that are made. Everything we do in this unit could be rewritten to avoid the use of trees; however, trees

- give us a powerful intuitive basis for viewing the problems of this chapter,
- provide a language for discussing the material,
- allow us to view the collection of all decisions in an organized manner.

We'll begin with elementary examples of decision trees. We then show how decision trees can be used to study recursive algorithms. Next we shall look at decision trees and "Bayesian methods" in probability theory. Finally we relate decision trees to induction and recursive equations.

Section 1: Basic Concepts of Decision Trees

One area of application for decision trees is systematically listing a variety of functions. The simplest general class of functions to list is the entire set $\underline{n}^{\underline{k}}$. We can create a typical element in the list by choosing an element of $\underline{n}$ and writing it down, choosing another element (possibly the same as before) of $\underline{n}$ and writing it down next, and so on until we have made k decisions. This generates a function in one line form sequentially: First $f(1)$ is chosen, then $f(2)$ is chosen and so on. We can represent all possible decisions pictorially by writing down the decisions made so far and then some downward "edges" indicating the possible choices for the next decision.

We begin this section by discussing the picture of a decision tree, illustrating this with a variety of examples. Then we study how a tree is traversed, which is a way computers deal with the trees.

Decision Trees and Recursion

What is a Decision Tree?

Example 1 (Decision tree for $\underline{2}^3$) Here is an example of a decision tree for the functions $\underline{2}^3$. We've omitted the commas; for example, 121 stands for the function 1,2,1 in one-line form.

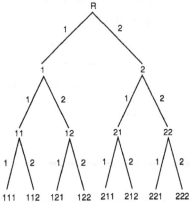

The set
$$V = \{R, 1, 2, 11, 12, 21, 22, 111, 112, 121, 122, 211, 212, 221, 222\}$$
is called the set of *vertices* of the decision tree. The vertex set for a decision tree can be any set, but must be specified in describing the tree. You can see from the picture of the decision tree that the places where the straight line segments (called *edges*) of the tree end is where the vertices appear in the picture. Each vertex should appear exactly once in the picture. The symbol R stands for the *root* of the decision tree. Various choices other than R can be used as the symbol for the root.

The *edges* of a decision tree such as this one are specified by giving the pair of vertices at the two ends of the edge, top vertex first, as follows: (R, 1), (21, 212), etc. The vertices at the ends of an edge are said to be "incident" on that edge. The complete set of edges of this decision tree is the set $E = \{$ (R,1), (R,2), (1,11), (1,12), (2,21),(2,22), (11,111), (11,112), (12,121), (12,122), (21,211), (21,212), (22,221), (22,222)$\}$. In addition to the edges, there are "labels," either a "1" or a "2," shown on the line segments representing the edges in the picture. This labeling of the edges can be thought of as a function from the set of edges, E, to the set $\{1, 2\}$.

If $e = (v, w)$ is an edge, the vertex w, is called a *child* of v, and v is the *parent* or w. The children of 22 are 221 and 222. The parent of 22 is 2.

The *degree* of a vertex v is the number of edges incident on that vertex. The *down degree* of v is the number of edges $e = (v, w)$ incident on that vertex (and below it); in other words, it is the number of children of v. The degree of 22 is 3 (counting edges $(2, 22), (22, 221), (22, 222)$). The down degree of 22 is 2 (counting edges $(22, 221), (22, 222)$). Vertices with down degree 0 are called *leaves*. All other vertices are called *internal* vertices.

Section 1: Basic Concepts of Decision Trees

For any vertex v in a decision tree there is a unique list of edges $(x_1, x_2), (x_2, x_3), \ldots,$ (x_k, x_{k+1}) from the root x_1 to $v = x_{k+1}$. This sequence is called the *path to a vertex* in the decision tree. The number of edges, k, is the length of this path and is called the *height* or *distance* of vertex v to from the root. The path to 22 is $(R, 2), (2, 22)$. The height of 22 is 2. □

The decision tree of the previous example illustrates the various ways of generating a function in $\underline{2}^{\underline{3}}$ sequentially. It's called a *decision tree for generating the functions in* $\underline{2}^{\underline{3}}$. Each edge in the decision tree is labeled with the choice of function value to which it corresponds. Note that the labeling does not completely describe the corresponding decision — we should have used something like "Choose 1 for the value of $f(2)$" instead of simply "1" on the line from 1 to 11.

In this terminology, a vertex v represents the partial function constructed so far, when the vertex v has been reached by starting at the root, following the edges to v, and making the decisions that label the edges on that unique path from the root to v. The edges leading out of a vertex are labeled with all possible decisions that can be made next, given the partial function at the vertex. We labeled the edges so that the labels on edges out of each vertex are in order, 1,2, when read left to right. The leaves are the finished functions. Notice that the leaves are in lexicographic order. In general, if we agree to label the edges from each vertex in order, then any set of functions generated sequentially by specifying $f(i)$ at the ith step will be in lex order.

To create a single function we start at the root and choose downward edges (i.e., make decisions) until we reach a leaf. This creates a *path* from the root to a leaf. We may describe a path in any of the following ways:

- the sequence of vertices $v_0, v_1, \ldots, v_m$ on the path from the root v_0 to the leaf v_m;

- the sequence of edges $e_1, e_2, \ldots, e_m$, where $e_i = (v_{i-1}, v_i)$, $i = 1, \ldots, m$;

- the sequence of decisions $D_1, D_1, \ldots, D_m$, where e_i is labeled with decision D_i.

We illustrate with three descriptions of the path from the root R to the leaf 212 in Example 1:

- the vertex sequence is R, 2, 21, 212;

- the edge sequence is $(R, 2), (2, 21), (21, 212)$;

- the decision sequence is 2, 1, 2.

Decision trees are a part of a more general subject in discrete mathematics called "graph theory," which is studied in another unit.

It is now time to look at some more challenging examples so that we can put decision trees to work for us. The next example involves counting words where the decisions are based on patterns of consonants and vowels.

Decision Trees and Recursion

Example 2 (Counting words) Using the 26 letters of the alphabet and considering the letters AEIOUY to be vowels how many five letter "words" (i.e. five long lists of letters) are there, subject to the following rules?

(a) No vowels are ever adjacent.

(b) There are never three consonants adjacent.

(c) Adjacent consonants are always different.

To start with, it would be useful to have a list of all the possible patterns of consonants and vowels; e.g., CCVCV (with C for consonant and V for vowel) is possible but CVVCV and CCCVC violate conditions (a) and (b) respectively and so are not possible. We'll use a decision tree to generate these patterns in lex order. Of course, a pattern CVCCV can be thought of as a function f where $f(1) = C$, $f(2) = V$, ..., $f(5) = V$.

We could simply try to list the patterns (functions) directly without using a decision tree. The decision tree approach is preferable because we are less likely to overlook something. The resulting tree can be pictured as follows:

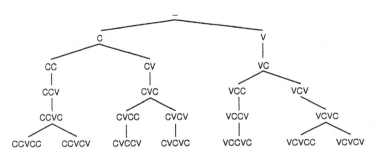

At each vertex there are potentially two choices, but at some vertices only one is possible because of rules (a) and (b) above. Thus there are one or two decisions at each vertex. You should verify that this tree lists all possibilities systematically.

We have used the dash "–" as the symbol for the root. This stands for the empty word on the letters C and V. The set of labels for the vertices of this decision tree T is a set of words of length 0 through 5. The vertex set is determined by the rules (or "syntax") associated with the problem (rules (a), (b), and (c) above).

Using the rules of construction, (a), (b), and (c), we can now easily count the number of words associated with each leaf. The total number of words is the sum of these individual counts. For CCVCC we obtain $(20 \times 19)^2 \times 6$; for CCVCV, CVCCV, VCCVC, and VCVCC we obtain $(20 \times 19) \times 20 \times 6^2$; for CVCVC we obtain $20^3 \times 6^2$; for $VCVCV$ we obtain $20^2 \times 6^3$. □

Definition 1 (Rank of an element of a list) *The rank of an element in a list is the number of elements that appear before it in the list. The rank of a leaf of a decision tree is the number of leaves that are to the left of it in the picture of the tree. The rank is denoted by the function RANK.*

In Example 1, RANK(212) = 5 and RANK(111) = 0.

Section 1: Basic Concepts of Decision Trees

Rank can be used to store data in a computer. For example, suppose we want to store information for each of the $10! \approx 3.6 \times 10^6$ permutations of $\underline{10}$. The naive way to do this is to have a $10 \times 10 \times \cdots \times 10$ array and store information about the permutation f in location $f(1), \ldots, f(10)$. This requires storage of size 10^{10}. If we store information about permutations in a one dimensional array with information about f stored at RANK(f), we only need $10!$ storage locations, which is much less. We'll discuss ranking permutations soon.

The inverse of the rank function is also useful. Suppose we want to generate objects at random from a set of n objects. Let RANK be a rank function for them. Generate a number k between 0 and $n-1$ inclusive at random. Then RANK$^{-1}(k)$ is a random object.

Example 3 (Permutations in lexicographic order) Recall that we can think of a permutation on $\underline{3}$ as a bijection $f : \underline{3} \to \underline{3}$. Its one-line form is $f(1), f(2), f(3)$. Here is an example of a decision trees for this situation (omitting commas):

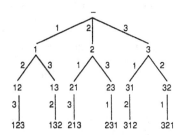

Because we first chose $f(1)$, listing its values in increasing order, then did the same with $f(2)$ and finally with $f(3)$, the leaves are listed lexicographically, that is, in "alphabetical" order like a dictionary only with numbers instead of letters.

We could have abbreviated this decision tree a bit by shrinking the edges coming from vertices with only one decision and omitting labels on nonleaf vertices. As you can see, there is no "correct" way to label a decision tree. The intermediate labels are simply a tool to help you correctly list the desired objects (functions in this case) at the leaves. Sometimes one may even omit the function at the leaf and simply read it off the tree by looking at the labels on the edges or vertices associated with the decisions that lead from the root to the leaf. In this tree, the labels on an edge going down from vertex v tell us what values to add to the end of v to get a "partial permutation" that is one longer than v. ☐

Decision Trees and Recursion

Permutations

We now look at two ways of ranking permutations. The method for generating random permutations in Example 23 of Unit Fn can provide another method for ranking permutations. If you'd like a challenge, you can think about how to do that. We won't discuss it.

Example 4 (Permutations in direct insertion order) Another way to create a permutation is by *direct insertion*. (Often this is simply called "insertion.") Suppose that we have an ordered list of k items into which we want to insert a new item. It can be placed in any of $k+1$ places; namely, at the end of the list or immediately before the ith item where $1 \leq i \leq k$. By starting out with 1, choosing one of the two places to insert 2 in this list, choosing one of the three places to insert 3 in the new list and, finally, choosing one of the four places to insert 4 in the newest list, we will have produced a permutation of $\underline{4}$. To do this, we need to have some convention as to how the places for insertion are numbered when the list is written from left to right. The obvious choice is from left to right; however, right to left is often preferred. We'll use right to left. One reason for this choice is that the leftmost leaf is $12\ldots n$ as it is for lex order.

If there are $k+1$ possible positions to insert something, we number the positions $0, 1, \ldots, k$, starting with the rightmost position as number 0. We'll use the notation $(\ )_i$ to stand for position i so that we can keep track of the positions when we write a list into which something is to be inserted.

Here's the derivation of the permutation of $\underline{4}$ associated with the insertions 1, 1 and 2.

- Start with $(\ )_1 1 (\ )_0$.

- Choose the insertion of the symbol 2 into position 1 (designated by $(\ )_1$) to get 21. With positions of possible insertions indicated, this is $(\ )_2 2 (\ )_1 1 (\ )_0$.

- Now insert symbol 3 into position 1 (designated by $(\ )_1$) to get 231 or, with possible insertions indicated $(\ )_3 2 (\ )_2 3 (\ )_1 1 (\ )_0$.

- Finally, insert symbol 4 into position 2 to get 2431.

Here is the decision tree for permutations of $\underline{4}$ in direct insertion order. We've turned

Section 1: Basic Concepts of Decision Trees

the vertices sideways so that rightmost becomes topmost in the insertion positions.

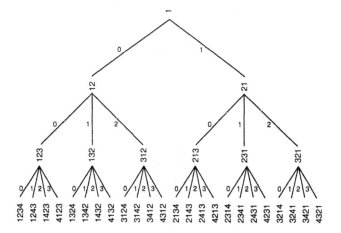

The labels on the vertices are, of course, the partial permutations, with the full permutations appearing on the leaves. The decision labels on the edges are the positions in which to insert the next number. Notice that the labels on the leaves are no longer in lex order because we constructed the permutations differently. Had we labeled the vertices with the positions used for insertion, the leaves would then be labeled in lex order. For example, 2413 becomes 1,0,2, which is gotten by reading edge labels on the path from the root to the leaf labeled 2413. Similarly 4213 becomes, 1,0,3.

Like the method of lex order generation, the method of direct insertion generation can be used for other things besides permutations. However, direct insertion cannot be applied as widely as lex order. Lex order generation works with anything that can be thought of as an (ordered) list, but direct insertion requires more structure. Note that the RANK(3412) = 10 and RANK(4321) = 23. What would RANK(35412) be for the permutations on $\underline{5}$ in direct insertion order? □

Traversing Decision Trees

We conclude this section with an important class of search algorithms called *backtracking algorithms*. In many computer algorithms it is necessary either to systematically inspect all the vertices of a decision tree or to find the leaves of the tree. An algorithm that systematically inspects all the vertices (and so also finds the leaves) is called a *traversal of the tree*. How can we create such an algorithm? To understand how to do this, we first look at how a tree can be "traversed."

Decision Trees and Recursion

Example 5 (Traversals of a decision tree) Here is a sample decision tree T with edges labeled A through J and vertices labeled 1 through 11.

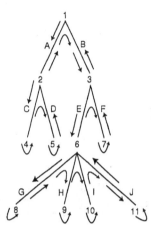

The arrows are not part of the decision tree, but will be helpful to us in describing certain ideas about linear orderings of vertices and edges that are commonly associated with decision trees. Imagine going around ("traversing") the decision tree following arrows. Start at the root, 1, go down edge A to vertex 2, etc. Here is the sequence of vertices as encountered in this process: 1, 2, 4, 2, 5, 2, 1, 3, 6, 8, 6, 9, 6, 10, 6, 11, 6, 3, 7, 3, 1. This sequence of vertices is called the *depth first vertex sequence*, DFV(T), of the decision tree T. The number of times each vertex appears in DFV(T) is one plus the down degree of that vertex. For edges, the corresponding sequence is A, C, C, D, D, A, B, E, G, G, H, H, I, I, J, J, E, F, F, B. This sequence is the *depth first edge sequence*, DFE(T), of the tree. Every edge appears exactly twice in DFE(T). If the vertices of the tree are read left to right, top to bottom, we obtain the sequence 1, 2, 3, 4, 5, 6, 7, 8, 9, 10, 11. This is called the *breadth first vertex sequence*, BFV(T). Similarly, the *breadth first edge sequence*, BFE(T), is A, B, C, D, E, F, G, H, I, J.

The sequences BFV(T) and BFE(T) are linear orderings of the vertices and edges of the tree T (i.e., each vertex or edge appears exactly once in the sequence). We also associate two linear orderings with DFV(T):

- PREV(T), called the *preorder sequence of vertices* of T, is the sequence of *first* occurrences of the vertices of T in DFV(T).

- POSV(T), called the *postorder sequence of vertices* of T, is the sequence of *last* occurrences of the vertices of T in DFV(T).

For the present tree

$$\text{PREV}(T) = 1,2,4,5,3,6,8,9,10,11,7 \quad \text{and} \quad \text{POSV}(T) = 4,5,2,8,9,10,11,6,7,3,1.$$

With a little practice, you can quickly construct PREV(T) and POSV(T) directly from the picture of the tree. For PREV(T), follow the arrows and list each vertex the first time you encounter it (and only then). For POSV(T), follow the arrows and list each vertex the last time you encounter it (and only then). Notice that the order in which the leaves of T appear, 4, 5, 8, 9, 10, 11, is the same in both PREV(T) and POSV(T). □

Section 1: Basic Concepts of Decision Trees

We now return to the problem of creating a traversal algorithm. One way to imagine doing this is to generate the depth first sequence of vertices and/or edges of the tree as done in the preceding example. We can describe our traversal more precisely by giving an algorithm. Here is one which traverses a tree whose leaves are associated with functions and lists the functions in the order of PREV(T).

Theorem 1 (Systematic Traversal Algorithm) *The following procedure systematically visits the vertices of a tree T in depth-first order, DFV(T), listing the leaves as they occur in the list DFV(T).*

1. **Start:** *Mark all edges as unused and position yourself at the root.*

2. **Leaf:** *If you are at a leaf, list the function.*

3. **Decide case:** *If there are no unused edges leading out from the vertex, go to Step 4; otherwise, go to Step 5.*

4. **Backtrack:** *If you are at the root, STOP; otherwise, return to the vertex just above this one and go to Step 3.*

5. **Decision:** *Select the leftmost unused edge out of this vertex, mark it used, follow it to a new vertex and go to Step 2.*

Step 4 is labeled **Backtrack**. What does this mean? If you follow the arrows in the tree pictured in Example 5, backtracking corresponds to going toward the root on an edge that has already been traversed in the opposite direction. In other words, backtracking refers to the process of moving along an edge *back* toward the root of the tree. Thinking in terms of the decision sequence, backtracking corresponds to undoing (i.e., backtracking on) a decision previously made. Notice that the algorithm only needs to keep track of the decisions from the root to the present vertex — when it backtracks, it can "forget" the decision it is backtracking from. You should take the time now to apply the algorithm to Example 1, noting which decisions you need to remember at each time.

So far in our use of decision trees, it has always been clear what decisions are reasonable; i.e., are on a path to a solution. (In this case every leaf is a solution.) This is because we've looked only at simple problems such as listing *all* permutations of $\underline{n}$ or listing *all* functions in $\underline{n}^{\underline{k}}$. We now look at a situation where that is not the case.

Example 6 (Restricted permutations) Consider the following problem.

List all permutations f of $\underline{n}$ such that
$|f(i) - f(i+1)| \leq 3$ for $1 \leq i < n$.

It's not at all obvious what decisions are reasonable in this case. For instance, when $n = 9$, the partially specified one line function 124586 cannot be completed to a permutation.

There is a simple cure for this problem: We will allow ourselves to make decisions which lead to "dead ends," situations where we cannot continue on to a solution. With this expanded notion of a decision tree, there are often many possible decision trees that appear reasonable for doing something. Suppose that we're generating things in lex order and we've

Decision Trees and Recursion

reached the vertex 12458. What do we do now? We'll simply continue to generate more of the permutation, making sure that $|f(i) - f(i+1)| \leq 3$ is satisfied for that portion of the permutation we have generated so far. The resulting portion of the tree that starts with the partial permutation 12458 is represented by the following decision tree. Because the names the of the vertices are so long, we've omitted all of the name, except for the function value just added, for all but the root. Thus the rightmost circled vertex is 124589763.

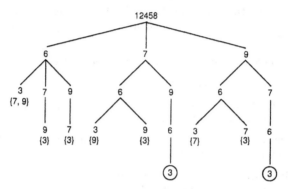

Each vertex is labeled with an additional symbol of the permutation after the symbols 12458. The circled leaves represent solutions. Each solution is obtained by starting with 12458 and adding all symbols on vertices of the path from the root, 12458, to the circled leaf. Thus the two solutions are 124587963 and 124589763. The leaves that are not circled are places where there is no way to extend the partial permutation corresponding to that leaf such that $|f(i) - f(i+1)| \leq 3$ is satisfied. Below each leaf that is not a solution, you can see the set of values available for completing the partial permutation at that leaf. It is clear in those cases that no completion satisfying $|f(i) - f(i+1)| \leq 3$ is possible.

Had we been smarter, we might have come up with a simple test that would have told us that 124586 could not be extended to a solution. This would have lead to a different decision tree in which the vertex corresponding to the partial permutation 124586 would have been a leaf.

You should note here that the numbers 3, 6, 7, 9 are *labels* on the vertices of this tree. A vertex of this tree is the partial permutation gotten by concatenating (attaching) the labels on the path from the root to that vertex to the permutation 12458. Thus, 124586 is a vertex with label 6. The path from the root to that vertex is 12458, 124586 (corresponding to the edge (12458, 124586)). The vertices are not explicitly shown, but can be figured out from the rules just mentioned. The labels on the vertices correspond to the decisions to be made (how to extend the partial permutation created thus far).

Our tree traversal algorithm, Theorem 1, requires a slight modification to cover this type of extended decision tree concept where a leaf need not be a solution: Change Step 2 to

2'. **Leaf:** If you are at a leaf, take appropriate action.

For the tree rooted at 12458 in this example, seven of the leaves are not solutions and two are. For the two that are, the "appropriate action" is to print them out, jump and shout, or, in some way, proclaim success. For the leaves that are not solutions, the appropriate action is to note failure in some way (or just remain silent) and **Backtrack** (Step (4)

Section 1: Basic Concepts of Decision Trees

of Theorem 1). This backtracking on leaves that are not solutions is where this type of algorithm gets its name: "backtracking algorithm." □

How can there be more than one decision tree for generating solutions in a specified order? Suppose someone who was not very clever wanted to generate all permutations of $\underline{n}$ in lex order. He might program a computer to generate all functions in $\underline{n}^{\underline{n}}$ in lex order and to then discard those functions which are not permutations. This leads to a much bigger tree because n^n is much bigger than $n!$, even when n is as small as 3. A somewhat cleverer friend might suggest that he have the program check to see that $f(k) \neq f(k-1)$ for each $k > 1$. This won't slow down the program very much and will lead to only $n(n-1)^{n-1}$ functions. Thus the program should run faster. Someone else might suggest that the programmer check at each step to see that the function produced so far is an injection. If this is done, nothing but permutations will be produced, but the program may be much slower.

The lesson to be learned from the previous paragraph is that there is often a trade off between the size of the decision tree and the time that must be spent at each vertex determining what decisions to allow. Because of this, different people may develop different decision trees for the same problem. The differences between computer run times for different decision trees can be truly enormous. By carefully defining the criteria that allow one to decide that a vertex is a leaf, people have changed problems that were too long to run on a supercomputer into problems that could be easily run on a personal computer. We'll conclude this section with two examples of backtracking of the type just discussed.

Example 7 (Domino coverings) We are going to consider the problem of covering a m by n board (for example, $m = n = 8$ gives a chess board) with 1 by 2 rectangles (called "dominoes"). A domino can be placed either horizontally or vertically so that it covers two squares and does not overlap another domino. Here is a picture of the situation for $m = 3$, $n = 4$. (The sequences of h's and v's under eleven covered boards will be explained below.)

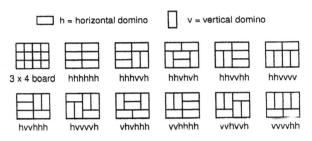

If the squares of the board are numbered systematically, left to right, top to bottom, from 1 to 12, we can describe any placement of dominoes by a sequence of 6 h's and v's: Each of the domino placements in the above picture has such a description just below it. Take as an example, hhvhvh (the third domino covering in the picture). We begin with no

Decision Trees and Recursion

dominoes on the board. None of the squares, numbered 1 to 12 are covered. The list of "unoccupied squares" is as follows:

```
 1   2   3   4
 5   6   7   8
 9  10  11  12
```

Thus, the smallest unoccupied square is 1. The first symbol in hhvhvh is the h. That means that we take a horizontal domino and cover the square 1 with it. That forces us to cover square 2 also. The list of unoccupied squares is as follows:

```
         3   4
 5   6   7   8
 9  10  11  12
```

Now the smallest unoccupied square is 3. The second symbol in hhvhhv is also an h. Cover square 3 with a horizontal domino, forcing us to cover square 4 also. The list of unoccupied squares is as follows:

```
 5   6   7   8
 9  10  11  12
```

At this point, the first row of the board is covered with two horizontal dominoes (check the picture). Now the smallest unoccupied square is 5 (the first square in the second row). The third symbol in hhvhvh is v. Thus we cover square 5 with a vertical domino, forcing us to cover square 9 also. The list of unoccupied squares is as follows:

```
     6   7   8
    10  11  12
```

We leave it to you to continue this process to the bitter end and obtain the domino covering shown in the picture.

Here is the general description of the process. Place dominoes sequentially as follows. If the first unused element in the sequence is h, place a horizontal domino on the first unoccupied square and the square to its right. If the first unused element in the sequence is v, place a vertical domino on the first unoccupied square and the square just below it. Not all sequences correspond to legal placements of dominoes (try hhhhhv). For a 2 × 2 board, the only legal sequences are hh and vv For a 2 × 3 board, the legal sequences are hvh, vhh and vvv. For a 3 × 4 board, there are eleven legal sequences as shown in the picture at the start of this example.

To find these sequences in lex order we used a decision tree for generating sequences of h's and v's in lex order. Each decision is required to lead to a domino that lies entirely

DT-12 [100]

Section 1: Basic Concepts of Decision Trees

on the board and does not overlap another domino. Here is our decision tree:

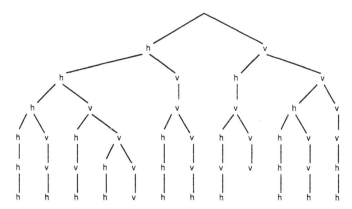

Note that in this tree, the decision (label) that led to a vertex is placed at the vertex rather than on the edge. The actual vertices, not explicitly labeled, are the sequences of choices from the root to that vertex (e.g., the vertex hvv has label v). The leaf vhvvv associated with the path v, h, v, v, v does not correspond to a covering. It has been abandoned (i.e., declared a leaf but not a solution) because there is no way to place a domino on the lower left square of the board, which is the first free square. Draw a picture of the board to see what is happening. Our criterion for deciding if a vertex is a leaf is to check if that vertex corresponds to a solution or to a placement that does not permit another domino to be placed on the board. It is not hard to come up with a criterion that produces a smaller decision tree. For example, vhvv leaves the lower left corner of the board isolated. That means that vhvv cannot be extended to a solution, even though more dominoes can be placed on the board. But, checking this more restrictive criterion is more time consuming.

Exercises for Section 1

1.1. List the nonroot vertices of the decision tree in Example 2 in PREV, POSV and BFV orders.

1.2. Let $RANK_L$ denote the rank in lex order and let $RANK_I$ denote the rank in insertion order on permutations of n. Answer the following questions and give reasons for your answers:

(a) For $n = 3$ and $n = 4$ which permutations σ have $RANK_L(\sigma) = RANK_I(\sigma)$?

(b) What is $RANK_L(2314)$? $RANK_L(45321)$?

(c) What is $RANK_I(2314)$? $RANK_I(45321)$?

(d) What permutation σ of $\underline{4}$ has $RANK_L(\sigma) = 15$?

(e) What permutation σ of $\underline{4}$ has $RANK_I(\sigma) = 15$?

Decision Trees and Recursion

(f) What permutation σ of $\underline{5}$ has $\text{RANK}_L(\sigma) = 15$?

1.3. Draw the decision tree to list all sequences of length six of A's and B's that satisfy the following conditions:

- There are no two adjacent A's.
- There are never three B's adjacent.
- If each leaf is thought of as a word, the leaves are in alphabetical order.

1.4. Draw a decision tree for $D(\underline{6}^{\underline{4}})$, the strictly decreasing functions from $\underline{4}$ to $\underline{6}$. You should choose a decision tree so that the leaves are in lex order when read from left to right

(a) What is the rank of 5431? of 6531?

(b) What function has rank 0? rank 7?

(c) Your decision tree should contain the decision tree for $D(\underline{5}^{\underline{4}})$. Indicate it and use it to list those functions in lex order.

(d) Indicate how all of the parts of this exercise can be interpreted in terms of subsets of a set.

1.5. Modify Theorem 1 to list all vertices in PREV order. Do the same for POSV order.

1.6. The president of Hardy Hibachi Corporation decided to design a series of different grills for his square-topped hibachis. They were to be collectibles. He hoped his customers would want one of each different design (and spend big bucks to get them). Having studied combinatorics in college, his undergrad summer intern suggested that these grills be modeled after the patterns associated with domino arrangements on 4 × 4 boards. Their favorite grill was in the design which has the code vvhvvhhh. The student, looking at some old class notes, suggested seven other designs: vvhhvhvh, hhvvhvvh, vhvhhvvh, hvvhvhvh, hvvvvhhh, vhvhvvhh, hhhvvvh. These eight grills were fabricated out of sturdy steel rods, put in a box, and shipped to the boss. When he opened up the box, much to his disgust, he found that all of the grills were the same. What went wrong? How should the collection of different grills be designed? (This is called an *isomorph rejection* problem.)

The favorite grill: vvhvvhhh =

Section 2: Recursive Algorithms

A *recursive algorithm* is an algorithm that refers to itself when it is executing. As with any recursive situation, when an algorithm refers to itself, it must be with "simpler" parameters so that it eventually reaches one of the "simplest" cases, which is then done without recursion. Let's look at a couple of examples before we try to formalize this idea.

Example 8 (A recursive algorithm for 0-1 sequences) Suppose you are interested in listing all sequences of length eight, consisting of four zeroes and four ones. Suppose that you have a friend who does this sort of thing, but will only make such lists if the length of the sequence is seven or less. "Nope," he says, "I can't do it — the sequence is too long." There is a way to trick your friend into doing it. First give him the problem of listing all sequences of length seven with three ones. He doesn't mind, and gives you the list 1110000, 1011000, 0101100, etc. that he has made. You thank him politely, sneak off, and put a "1" in front of every sequence in the list he has given you to obtain 11110000, 11011000, 10101100, etc. Now, you return to him with the problem of listing all strings of length seven with four ones. He returns with the list 1111000, 0110110, 0011101, etc. Now you thank him and sneak off and put a "0" in front of every sequence in the list he has given you to obtain 01111000, 00110110, 00011101, etc. Putting these two lists together, you have obtained the list you originally wanted.

How did your friend produce these lists that he gave you? Perhaps he had a friend that would only do lists of length 6 or less, and he tricked this friend in the same way you tricked him! Perhaps the "6 or less" friend had a "5 or less friend" that he tricked, etc. If you are sure that your friend gave you a correct list, it doesn't really matter how he got it. ▫

Next we consider an example from sorting theory. We imagine we are given a set of objects which have a linear order described on them (perhaps, but not necessarily, lexicographic order of some sort). As a concrete example, we could imagine that we are given a set of integers S, perhaps a large number of them. They are not in order as presented to us, be we want to list them in order, smallest to largest. That problem of putting the set S in order is called *sorting* S. On the other hand, if we are given two ordered lists, like $(25, 235, 2333, 4321)$ and $(21, 222, 2378, 3421, 5432)$, and want to put the combined list in order, in this case $(21, 25, 222, 235, 2333, 2378, 3421, 4321, 5432)$, this process is called *merging* the two lists. Our next example considers the relationship between sorting and merging.

Example 9 (Sorting by recursive merging) Sorting by recursive merging, called *merge sorting*, can be described as follows.

- The lists containing just one item are the simplest and they are already sorted.
- Given a list of $n > 1$ items, choose k with $1 \leq k < n$, sort the first k items, sort the last $n - k$ items and merge the two sorted lists.

Decision Trees and Recursion

This algorithm builds up a way to sort an n-list out of procedures for sorting shorter lists. Note that we have not specified how the first k or last $n - k$ items are to be sorted, we simply assume that it has been done. Of course, an obvious way to do this is to simply apply our merge sorting algorithm to each of these sublists.

Let's implement the algorithm using people rather than a computer. Imagine training a large number of obedient people to carry out two tasks: (a) splitting a list for other people to sort and (b) merging two lists. We give one person the unsorted list and tell him to sort it using the algorithm and return the result to us. What happens?

- Anyone who has a list with only one item returns it unchanged to the person he received it from.

- Anyone with a list having more than one item splits it and gives each piece to a person who has not received a list, telling each person to sort it and return the result. When the results have been returned, this person merges the two lists and returns the result to whoever gave him the list.

If there are enough obedient people around, we'll eventually get our answer back.

Notice that *no one needs to pay any attention to what anyone else is doing* to a list. This is makes a *local description* possible; that is, we tell each person what to do and they do not need to concern themselves with what other people are doing. This can also be seen in the pseudocode for merge sorting a list L:

    Sort(L)
        If length is 1, return L
        Else
            Split L into two lists L1 and L2
            S1 = Sort(L1)
            S2 = Sort(L2)
            S = Merge(L1, L2)
            Return S
        End if
    End

The procedure is not concerned with what goes on when it calls itself recursively. This is very much like proof by induction. (We discuss proof by induction in the last section of this unit.) To see that, let's prove that the algorithm sorts correctly. We assume that splitting and merging have been shown to be correct — that's a separate problem. We induct on the length n of the list. The base case, $n = 1$ is handled correctly by the program since it returns the list unchanged. Now for induction. Splitting L results in shorter lists and so, by the induction hypothesis, S1 and S2 are sorted. Since merging is done correctly, S is also sorted.

This algorithm is another case of divide and conquer since it splits the sorting problem into two smaller sorting problems whose answers are combined (merged) to obtain the solution to the original sorting problem. □

Let's summarize some of the above observations with two definitions.

Section 2: Recursive Algorithms

Definition 2 (Recursive approach) *A recursive approach to a problem consists of two parts:*

1. *The problem is reduced to one or more problems of the same kind which are simpler in some sense.*

2. *There is a set of simplest problems to which all others are reduced after one or more steps. Solutions to these simplest problems are given.*

The preceding definition focuses on tearing down (reduction to simpler cases). Sometimes it may be easier or better to think in terms of building up (construction of bigger cases):

Definition 3 (Recursive solution) *We have a recursive solution to the problem (proof, algorithm, data structure, etc.) if the following two conditions hold.*

1. *The set of simplest problems can be dealt with (proved, calculated, sorted, etc.).*

2. *The solution to any other problem can be built from solutions to simpler problems, and this process eventually leads back to the original problem.*

The recursion $C(n,k) = C(n-1,k-1) + C(n-1,k)$ for computing binomial coefficients can be viewed as a recursive algorithm. Such algorithms for computing can be turned into algorithms for constructing the things we are counting. To do this, it helps to have a more systematic way to think about recursive algorithms. In the next example we introduce a tree to represent the local description of a recursive algorithm.

Example 10 (Permutations in lex order) The following figure represents the *local description* of a decision tree for listing the permutations of an ordered set

$$S = \{s_1, s_2, \ldots, s_n\} \quad \text{with} \quad s_1 < s_2 < \cdots < s_n.$$

The permutations in the figure are listed in one-line form. The vertices of this decision tree are of the form L(X) where X is some set. The simplest case, shown below, is where the tree has one edge. The labels on the edges are of the form (t), where t is an element of the set X associated with the uppermost vertex L(X) incident on that edge.

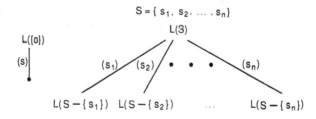

The leaves of the recursive tree tell us to construct permutations of the set S with the already chosen element removed from the set. (This is because permutations are injections.)

Decision Trees and Recursion

One way to think of the local description is to regard it as a rule for recursively constructing an entire decision tree, once the set S is specified. Here this construction has been carried out for $S = \{1, 2, 3, 4\}$.

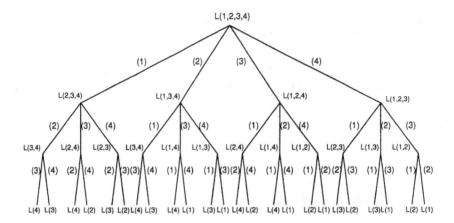

To obtain a permutation of $\underline{4}$, read the labels (t) on the edges from the root to a particular leaf. For example the if this is done for the preorder first leaf, one obtains $(1)(2)(3)L(4)$. $L(4)$ is a "simplest case" and has the label (4), giving the permutation 1234 in one line notation. Repeating this process for the leaves from left to right gives the list of permutations of $\underline{4}$ in lex order. For example, the tenth leaf gives the permutation 2341.

We'll use induction to prove that this is the correct tree. When $n = 1$, it is clear. Suppose it is true for all S with cardinality less than n. The permutations of S in lex order are those beginning with s_1 followed by those beginning with s_2 and so on. If s_k is removed from those permutations of S beginning with s_k, what remains is the set of permutations of $S - \{s_k\}$ listed in lex order. By the induction hypothesis, these are given by $L(S - \{s_k\})$. Note that the validity of our proof does not depend on how they are given by $L(S - \{s_k\})$. □

No discussion of recursion would be complete without the entertaining example of the Towers of Hanoi puzzle. We shall explore additional aspects of this problem in the exercises. Our approach will be the same as the previous example. We shall give a local description of the recursion. Having done so, we construct the trees for some examples and try to gain insight into the sequence of moves associated with the general Towers of Hanoi problem.

Example 11 (Towers of Hanoi) The *Towers of Hanoi* puzzle consists of n different sized washers (i.e., discs with holes in their centers) and three poles. Initially the washers

Section 2: Recursive Algorithms

are stacked on one pole as shown below.

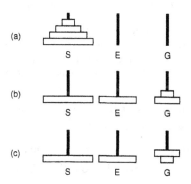

The object is to switch all of the washers from the pole S to G using pole E as a place for temporary placement of discs. A legal move consists of taking the top washer from a pole and placing on top of the pile on another pole, provided it is not placed on a smaller washer. Configuration (a), above, is the starting configuration, (b) is an intermediate stage, and (c) is illegal.

We want an algorithm H(n, S, E, G) that takes washers numbered $1, 2, \ldots, n$ that are stacked on the pole called S and moves them to the pole called G. The pole called E is also available. A call of this procedure to move 7 washers might be H(7, "start", "extra", "goal").

Here is a recursive description of how to solve the Towers of Hanoi. To move the largest washer, we must move the other $n - 1$ to the spare peg. After moving the largest, we can then move the other $n - 1$ on top of it. Let the washers be numbered 1 to n from smallest to largest. When we are moving any of the washers 1 through k, we can ignore the presence of all larger washers beneath them. Thus, moving washers 1 through $n - 1$ from one peg to another when washer n is present uses the same moves as moving them when washer n is not present. Since the problem of moving washers 1 through $n - 1$ is simpler, we practically have a recursive description of a solution. All that's missing is the observation that the simplest case, $n = 1$, is trivial. The following diagram gives the local description of a decision tree that represents this recursive algorithm.

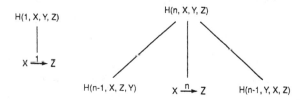

The "simplest case," n equal 1 is shown on the left. The case for general n is designated by H(n, X, Y, Z). You can think of the symbol H(n, X, Y, Z) as designating a vertex of a decision tree. The local description tells you how to construct the rest of the decision tree, down to and including the simplest cases. There is a simple rule for deciding how to rearrange X, Y and Z:

Decision Trees and Recursion

- for the left child: X is fixed, Y and Z are switched;
- for the right child: Z is fixed, X and Y are switched.

All leaves of the decision tree are designated by symbols of the form "$U \xrightarrow{k} V$." This symbol has the interpretation "move washer number k from pole U to pole V." These leaves in preorder (left to right order in the tree) give the sequence of moves needed to solve the Towers of Hanoi puzzle. The local description tells us that, in order to list the leaves of H(n, S, E, G), we

- list the leaves of H(n-1, S, G, E), moving the top n-1 washers from S to E using G
- move the largest washer from S to G
- list the leaves of H(n-1, E, S, G), moving the top n-1 washers from E to G using S

For example, the leaves of the tree with root H(2,S,E,G) are, in order,

$$S \xrightarrow{1} E, \quad S \xrightarrow{2} G, \quad E \xrightarrow{1} G.$$

The leaves of H(3,S,E,G) are gotten by concatenating (piecing together) the leaves of the subtree rooted at H(2,S,G,E) with $S \xrightarrow{3} G$ and the leaves of the subtree rooted at H(2,E,S,G). This gives

$$S \xrightarrow{1} G, S \xrightarrow{2} E, G \xrightarrow{1} E, S \xrightarrow{3} G, E \xrightarrow{1} S, E \xrightarrow{2} G, S \xrightarrow{1} G. \quad \square$$

Example 12 (The Towers of Hanoi decision tree for $n=4$) Starting with the local description of the general decision tree for the Towers of Hanoi and applying the rules of construction specified by it, we obtain the decision tree for the Towers of Hanoi puzzle with $n = 4$. For example, we start with n = 4, X = S, Y = E and Z = G at the root of the tree. To match the H(n,X,Y,Z) pattern when we expand the rightmost son of the root (namely H(3,E,S,G)), we have n = 3, X = E, Y = S and Z = G.

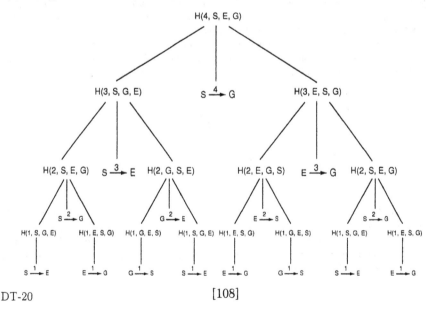

Section 2: Recursive Algorithms

There are fifteen leaves. You should apply sequentially the moves specified by these leaves to the starting configuration ((a) of Example 11) to see that the rules are followed and the washers are all transferred to G.

There are some observations we can make from this example. There are $2^4 - 1 = 15$ leaves or, equivalently, "moves" in transferring all washers from S to G. If h_n is the number of moves required for n washers, then the local description of the decision tree implies that, in general, $h_n = 2h_{n-1} + 1$. Computing some numbers for the h_n gives 1, 3, 7, 15, 31, etc. It appears that $h_n = 2^n - 1$. This fact can be proved easily by induction.

Note that the washer number 1, the smallest washer moves every other time. It moves in a consistent pattern. It starts on S, then E, then G, then S, then E, then G, etc. For H(3, S, E, G), the pattern is S, G, E, S, G, E, etc. In fact, for n odd, the pattern is always S, G, E, S, G, E, etc. For n even, the pattern is always S, E, G, S, E, G, etc. This means that if someone shows you a configuration of washers on discs for the H(n, S, E, G) and says to you, "It's the smallest washer's turn to move," then you should be able to make the move. If they tell you it is not the smallest washer's turn, then you should also be able to make the move. Why? Only one move *not* involving the smallest washer is legal! □

Example 13 (The Towers of Hanoi, recursion, and stacks) One way to generate the moves of H(n, S, E, G) is to use the local description to generate the depth first vertex sequence (Example 5).

- The depth first vertex list for $n = 4$ would start as follows:

    H(4, S, E, G), H(3, S, G, E), H(2, S, G, E), H(1, S, G, E), S $\xrightarrow{1}$ E

    At this point we have gotten to the first leaf. It should be printed out.

- The next vertex in the depth first vertex sequence is H(1, S, G, E) again. We represent this by removing S $\xrightarrow{1}$ E to get

    H(4, S, E, G), H(3, S, G, E), H(2, S, G, E), H(1, S, G, E).

- Next we remove H(1, S, G E) to get

    H(4, S, E, G), H(3, S, G, E), H(2, S, G, E).

- The next vertex in depth first order is S $\xrightarrow{2}$ G. We add this to our list to get

    H(4, S, E, G), H(3, S, G, E), H(2, S, G, E), S $\xrightarrow{2}$ G.

Continuing in this manner we generate, for each vertex in the decision tree, the path from the root to that vertex. The vertices occur in depth first order. Computer scientists refer to the path from the root to a vertex v as the *stack* of v. Adding a vertex to the stack is called *pushing* the vertex on the stack. Removing a vertex is *popping* the vertex from the stack. Stack operations of this sort reflect how most computers carry out recursion. This "one dimensional" view of recursion is computer friendly, but the geometric picture provided by the local tree is more people friendly. □

Decision Trees and Recursion

Example 14 (The Towers of Hanoi configuration analysis) In the figure below, we show the starting configuration for H(6, S, E, G) and a path,

$$H(6, S, E, G), \ H(5, E, S, G), \ H(4, E, G, S), \ H(3, G, E, S), \ G \xrightarrow{3} S.$$

This path goes from the root H(6, S, E, G) to the leaf $G \xrightarrow{3} S$. Given this path, we want to construct the configuration of washers corresponding to that path, assuming that the move $G \xrightarrow{3} S$ has just been carried out. This is also shown in the figure and we now explain how we obtained it.

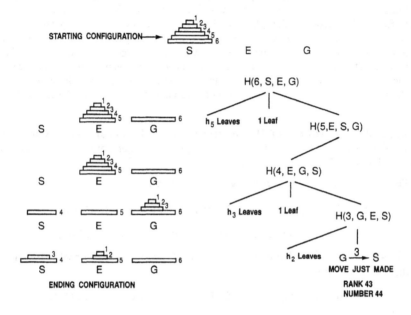

The first part of the path shows what happens when we use the local tree for H(6,S,E,G). Since we are going to H(5,E,S,G), the first edge of the path "slopes down to the right." At this point, the left edge, which led to H(5,S,G,E) moved washers 1 through 5 to pole E using $h_5 = 2^5 - 1 = 31$ moves and move $S \xrightarrow{6} G$ has moved washer 6. This is all shown in the figure and it has taken $31 + 1 = 32$ moves.

Next, one replaces H(5,E,S,G) with the local tree (being careful with the S, E, G labels!). This time the path "slopes to the left". Continuing in this manner we complete the entire path and have the configuration that is reached.

We can compute the rank of this configuration by noticing how many moves were made to reach it. Each move, except our final one $G \xrightarrow{3} S$, is a leaf to the left of the leaf corresponding to the move $G \xrightarrow{3} S$ at the end of the path. We can see from the figure that there were

$$(h_5 + 1) + 0 + (h_3 + 1) + h_2 = (31 + 1) + (7 + 1) + 3 = 43$$

such moves and so the rank of this configuration is 43.

Section 2: Recursive Algorithms

You should study this example carefully. It represents a very basic way of studying recursions. In particular

(a) You should be able to do the same analysis for a different path.

(b) You should be able to start with an ending configuration and reconstruct the path.

(c) You should be able to start with a configuration and, by attempting to reconstruct the path (and failing), be able to show that the configuration is illegal (can never arise).

We've already discussed (a). When you know how to do (b), you should be able to do (c). How can we reconstruct the path from the ending configuration? Look at the final configuration in the previous figure and note where the largest washer is located. Since it is on G, it must have been moved from S to G. This can only happen on the middle edge leading out from the root. Hence we must take the rightmost branch and are on the path that starts H(5,E,S,G). We are now dealing with a configuration where washers 1–5 start out stacked on E, washer 6 is on G and washer 6 will never move again. This takes care of washer 6 and we ignore it from now on. We are now faced with a new problem: There are 5 washers starting on E. In the final configuration shown in the figure, washer 5 is still on E, so it has not moved in this new problem. Therefore, we must take leftmost branch from H(5,E,S,G). This is H(4,E,G,S). Again, we have a new problem with 4 washers starting out on E. Since washer 4 must end up on S, we take the rightmost branch out of the vertex H(4,E,G,S). We continue in this manner until we reach H(1,...). If washer 1 must move, that is the last move. Otherwise, the last move is the leaf to the left of the last right-pointing branch that we took. In our particular case, from H(4,E,G,S) we go right to H(3,G,E,S), right to H(2,E,G,S), left to H(1,E,S,G). Since washer 1 is on E, it has not yet moved in doing H(1,E,S,G). (Of course, it may have moved several times earlier.) The last right branch was H(2,E,G,S) from H(3,G,E,S) so that the last move was washer 3 from G to S.

How could the previous process ever fail? Not all configurations arise. For example, if washer 5 were on S, we would have decided to move it in H(5,E,S,G) since it is not on E. But the only time H(5,E,S,G) moves washer 5 is from E to G so it cannot end up on S. □

We conclude this section with another "canonical" example of recursive algorithm and decision trees. We want to look at all subsets of $\underline{n}$. It will be more convenient to work with the representation of subsets by functions with domain $\underline{n}$ and range $\{0,1\}$. For a subset S of $\underline{n}$, define

$$\chi_S(i) = \begin{cases} 1 & \text{if } i \in S, \\ 0 & \text{if } i \notin S. \end{cases}$$

This function is called the *characteristic function* of S. We have a characteristic function for every subset S of $\underline{n}$. In one line notation, these functions become n-strings of zeroes and ones: The string $a_1 \ldots a_n$ corresponds to the subset T where $i \in T$ if and only if $a_i = 1$. Thus the all zeroes string corresponds to the empty set and the all ones string to $\underline{n}$. This correspondence is called the *characteristic function* interpretation of subsets of $\underline{n}$.

Our goal is to make a list of all subsets of $\underline{n}$ such that subsets adjacent to each other in the list are "close" to each other. Before we can begin to look for such a *Gray code*, we must say what it means for two subsets (or, equivalently, two strings) to be close. Two strings will be considered close if they differ in exactly one position. In set terms, this means one

Decision Trees and Recursion

of the sets can be obtained from the other by removing or adding a single element. With this notion of closeness, a Gray code for all subsets when $n = 1$ is 0, 1. A Gray code for all subsets when $n = 2$ is 00, 01, 11, 10.

How can we produce a Gray code for all subsets for arbitrary n? There is a simple recursive procedure. The following construction of the Gray code for $n = 3$ illustrates it.

$$\begin{array}{cc} 0\ 00 & 1\ 10 \\ 0\ 01 & 1\ 11 \\ 0\ 11 & 1\ 01 \\ 0\ 10 & 1\ 00 \end{array}$$

You should read down the first column and then down the second. Notice that the sequences in the first column begin with 0 and those in the second with 1. The rest of the first column is simply the Gray code for $n = 2$ while the second column is the Gray code for $n = 2$, read from the last sequence to the first.

We now prove that this two column procedure for building a Gray code for subsets of an n-set from the Gray code for subsets of an $(n-1)$-set always works. Our proof will be by induction. For $n = 1$, we have already exhibited a Gray code. Suppose that $n > 1$ and that we have a Gray code for $n - 1$. (This is the induction assumption.)

- Between the bottom of the first column and the top of the second, the only change is in the first position since the remaining $n - 1$ positions are the last element of our Gray code for $n - 1$.

- Within a column, there is never any change in the first position and there is only a single change from line to line in the remaining positions because they are a Gray code by the induction assumption.

This completes the proof.

As an extra benefit, we note that the last element of our Gray code differs in only one position from the first element (Prove it!), so we can cycle around from the last element to the first by a single change.

Example 15 (Decision tree for the subset Gray code) Here is another notation for describing our subset Gray code. Let $\overrightarrow{\text{GRAY}}(1) = 0, 1$, and let $\overleftarrow{\text{GRAY}}(1) = 1, 0$. As the arrows indicate, $\overrightarrow{\text{GRAY}}(1)$ is the Gray code for $n = 1$ listed from first to last element, while $\overleftarrow{\text{GRAY}}(1)$ is this Gray code in reverse order. In general, if $\overrightarrow{\text{GRAY}}(n)$ is the Gray code for n-bit words, then $\overleftarrow{\text{GRAY}}(n)$ is defined to be that list in reverse order.

We define
$$\overrightarrow{\text{GRAY}}(2) = 0\overrightarrow{\text{GRAY}}(1), 1\overleftarrow{\text{GRAY}}(1).$$

The meaning of $0\overrightarrow{\text{GRAY}}(1)$ is that 0 is put at the front of every string in $\overrightarrow{\text{GRAY}}(1)$. Juxtaposing the two lists (or "concatenation") means just listing the second list after the first. Thus, $0\overrightarrow{\text{GRAY}}(1) = 00, 01$, and $1\overleftarrow{\text{GRAY}}(1) = 11, 10$. Hence,

$$\overrightarrow{\text{GRAY}}(2) = 0\overrightarrow{\text{GRAY}}(1), 1\overleftarrow{\text{GRAY}}(1) = 00, 01, 10, 11.$$

Section 2: Recursive Algorithms

If we read $\overrightarrow{\mathrm{GRAY}}(2)$ in reverse order, we obtain $\overleftarrow{\mathrm{GRAY}}(2)$. You should verify the following equality.
$$\overleftarrow{\mathrm{GRAY}}(2) = 1\overrightarrow{\mathrm{GRAY}}(1), 0\overleftarrow{\mathrm{GRAY}}(1).$$

What we did for $n = 2$ works in general: The following diagram gives the local description of a decision tree for constructing subset Gray codes:

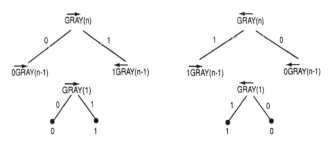

The left side of the figure is a *definition* for $\overrightarrow{\mathrm{GRAY}}(n)$. We must verify two things:

- This definition gives us a Gray code.
- Given the left figure and the fact that $\overleftarrow{\mathrm{GRAY}}(n)$ is the reversal of $\overrightarrow{\mathrm{GRAY}}(n)$, the right figure is correct.

The first part was already done because the figure simply describes the construction we gave before we started this example. The second part is easy when we understand what the tree means. Reading the $\overrightarrow{\mathrm{GRAY}}(n)$ tree from the right, we start with the reversal of $1\overleftarrow{\mathrm{GRAY}}(n-1)$. Since $\overleftarrow{\mathrm{GRAY}}$ and $\overrightarrow{\mathrm{GRAY}}$ are defined to be reversals of each other, we get $1\overrightarrow{\mathrm{GRAY}}(n-1)$. Similarly, reversing $0\overrightarrow{\mathrm{GRAY}}(n-1)$ gives $0\overleftarrow{\mathrm{GRAY}}(n-1)$.

If we apply the local description to the case $n = 3$, we obtain the following decision tree:

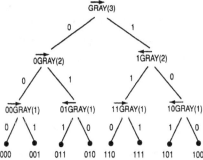

In the above decision tree for $\overrightarrow{\mathrm{GRAY}}(3)$, the elements of the Gray code for $n = 3$ are obtained by listing the labels on the edges for each path that ends in a leaf. These paths are listed in preorder of their corresponding leaves (left to right in the picture). This gives 000, 001, 011, 010, 110, 111, 101, 100. You should practice doing the configuration analysis for this recursion, analogous to Example 14. In particular, given a sequence, 10011101 say,

Decision Trees and Recursion

construct its path in the decision tree. What is the RANK of 10011101 in $\overrightarrow{\text{GRAY}}(8)$? What is the element in $\overrightarrow{\text{GRAY}}(8)$ just before 10011101; just after 10011101?

Note in the above decision tree for $\overrightarrow{\text{GRAY}}(3)$ that every time an edge with label 1 is encountered (after the first such edge), that edge changes direction from the edge just prior to it in the path. By "changing direction," we mean that if an edge is sloping downward to the right (downward to the left) and the previous edge in the path sloped downward to the left (downward to the right), then a change of direction has occurred. Conversely, every time an edge with label 0 is encountered (after the first such edge), that edge does not change direction from the edge just prior to it in the path. This is a general rule that can be proved by induction. □

Exercises for Section 2

2.1. Suppose the permutations on $\underline{8}$ are listed in lexicographic order.

  (a) What is the RANK in the list of all such permutations of 87612345?

  (b) What permutation has RANK 20,160?

2.2. Consider the Towers of Hanoi puzzle, H(8, S, E, G). Suppose that pole S has washers 6, 5, 2, 1; pole E has no washers; pole G has washers 8, 7, 4, 3. Call this the *basic configuration*.

  (a) What is the path in the decision tree that corresponds to the basic configuration?

  (b) What was the move that produced the basic configuration and what was the configuration from which that move was made?

  (c) What was the move just prior to the one that produced the basic configuration and what was the configuration from which that move was made?

  (d) What will be the move just after the one that produced the basic configuration?

  (e) What is the RANK, in the list of all moves of H(8, S, E, G), of the move that produced the basic configuration?

2.3. Consider $\overrightarrow{\text{GRAY}}(9)$.

  (a) What is the element just before 110010000? just after 110010000?

  (b) What is the first element of the second half of the list?

  (c) What is the RANK of 111111111?

  (d) What is the element of RANK 372?

*2.4. Consider the Towers of Hanoi puzzle with four poles and n washers. The rules are the same, except that there are two "extra" poles E and F. The problem is to

Section 3: Decision Trees and Conditional Probability

transfer all of the n washers from S to G using the extra poles E and F as temporary storage. Let h'_n denote the optimal number of moves needed to solve the three pole problem. Let f_n denote the optimal number of moves needed to solve the four pole problem with n washers.

(a) Recall that $h_n = 2^n - 1$ is the number of moves in the recursive algorithm H(n, S, E, G). Prove by induction that $h'_n = h_n$.

(b) Compute f_n for $n = 1, 2, 3$, describing, in the process, optimal sequences of moves.

Let's adopt a specific strategy for doing four poles and n washers. Choose integers $p \geq 0$ and $q > 0$ so that $p + q = n$. We now describe strategy G(p, q, S, E, F, G). To execute G(p, q, S, E, F, G), proceed as follows:

(i) If $p = 0$, then $q = n$. Use H(n, S, E, G) to move washers $1, \ldots, n$ to G.

(ii) If $p > 0$, choose integers $i \geq 0$ and $j > 0$ such that $i + j = p$. Use G(i, j, S, E, G, F) to move washers $1, 2, \ldots, p$ to pole F (the washers are numbered in order of size). Next, use H(q, S, E, G) to move washers $q, \ldots, n$ to G. Finally, use G(i, j, F, S, E, G) to move $1, 2, \ldots, p$ to pole G, completing the transfer. For all possible choices of i and j, choose the one that minimizes the number of moves.

Finally, to move the n washers, choose that G(p, q, S, E, F, G) with $n = p + q$ which has the minimum number of moves. Call this number s_n.

(c) What are the simplest cases in this recursive algorithm? How can you compute the values of i and j to minimize the number of moves? Use your method to solve the problem for $n \leq 6$.

(d) What is the recursion for s_n?

(e) Prove that $f_n \leq 2\min(f_{n-q} + h_q)$, where the minimum is over $q > 0$ and $f_0 = 0$. recursion.

Section 3: Decision Trees and Conditional Probability

We conclude our discussion of decision trees by giving examples of the use of decision trees in elementary probability theory. In particular, we focus on what are called *conditional probabilities* and *Bayesian methods* in probability.

Definition 4 (Conditional probability) Let U be a sample space with probability function P. If $A \subseteq U$ and $B \subseteq U$ are events (subsets) of U then the conditional probability of B given A, denoted by $P(B|A)$, is

$$P(B|A) = \begin{cases} P(A \cap B)/P(A), & \text{if } P(A) \neq 0, \\ \text{undefined}, & \text{if } P(A) = 0. \end{cases}$$

Decision Trees and Recursion

How should we interpret $P(B|A)$? If an experiment is performed n times and b of those times B occurs, then b/n is nearly $P(B)$. Furthermore, as n increases, the ratio b/n almost surely approaches $P(B)$ as a limit.* Now suppose an experiment is performed n times but we are only interested in those times when A occurs. Furthermore, suppose we would like to know the chances that B occurs, given that A has occurred. Let the count for A be a and that for $A \cap B$ be c. Since we are interested only in the cases when A occurs, only a of the experiments matter. In these a experiments, B occurred c times. Hence the probability that B occurs given that A has occurred is approximately $c/a = (c/n)/(a/n)$, which is approximately $P(A \cap B)/P(A)$, which is the definition of $P(B|A)$. As n increases, the approximations almost surely approach $P(B|A)$. Hence

> $P(B|A)$ should be thought of as the probability that B occurred, given that we know A occurred.

Another way you can think of this is that we are changing to a new sample space A. To define a probability function P_A on this sample space, we rescale P so that $\sum_{a \in A} P_A(a) = 1$. Since $\sum_{a \in A} P(a) = P(A)$, we must set $P_A(a) = P(a)/P(A)$. Then, the probability that B occurs is the sum of $P_A(a)$ over all $a \in B$ that are in our new sample space A. Thus

$$P_A(B) = \sum_{a \in A \cap B} P_A(a) = \sum_{a \in A \cap B} P(a)/P(A) = P(A \cap B)/P(A),$$

which is our definition of $P(B|A)$.

The following theorem contains some simple but important properties of conditional probability.

Theorem 2 (Properties of conditional probability) Let (U, P) be a probability space. All events in the following statements are subsets of U and the conditional probabilities are assumed to be defined. (Recall that $P(C|D)$ is undefined when $P(D) = 0$.)

(a) $P(B|U) = P(B)$ and $P(B|A) = P(A \cap B \mid A)$.

(b) A and B are independent events if and only if $P(B|A) = P(B)$.

(c) (Bayes' Theorem) $P(A|B) = P(B|A)P(A)/P(B)$.

(d) $P(A_1 \cap \cdots \cap A_n) = P(A_1)\, P(A_2 \mid A_1)\, P(A_3 \mid A_1 \cap A_2) \cdots P(A_n \mid A_1 \cap \cdots \cap A_{n-1})$.

You can think of (b) as a justification for the terminology "independent" since $P(B|A) = P(B)$ says that the probability of B having occurred is unchanged even if we know that A occurred; in other words, A does not influence B's chances. We will encounter other forms of Bayes' Theorem. All of them involve reversing the order in conditional probabilities. (Here, $A|B$ and $B|A$.)

Proof: All the proofs are simple applications of the definition of conditional probability, so we prove just (b) and (d) and leave (a) and (c) as exercises.

---

* For example, we might toss fair coin 100 times and obtain 55 heads, so $a/n; = 55/100$ is nearly $1/2 = P(\text{head})$. With 10,000 tosses, we might obtain 4,930 heads and $4{,}930/10{,}000$ is even closer to $1/2$ than $55/100$. (This is the sort of accuracy one might realistically expect.)

Section 3: Decision Trees and Conditional Probability

We prove (b). Suppose A and B are independent. By the definition of independence, this means that $P(A \cap B) = P(A)P(B)$. Dividing both sides by $P(A)$ and using the definition of conditional probability, we obtain $P(B|A) = P(B)$. For the converse, suppose $P(B|A) = P(B)$. Using the definition of conditional probability and multiplying by $P(A)$, we obtain $P(A \cap B) = P(A)P(B)$, which is the definition of independence.

We prove (d) simply by using the definition of conditional probability and doing a lot of cancellation of adjacent numerators and denominators:

$$P(A_1)\, P(A_2 \mid A_1)\, P(A_3 \mid A_1 \cap A_2) \cdots P(A_n \mid A_1 \cap \cdots \cap A_{n-1})$$
$$= P(A_1)\, \frac{P(A_2 \cap A_1)}{P(A_1)}\, \frac{P(A_1 \cap A_2 \cap A_3)}{P(A_1 \cap A_2)} \cdots \frac{P(A_1 \cap \cdots \cap A_n)}{P(A_1 \cap \cdots \cap A_{n-1})}$$
$$= P(A_1 \cap \cdots \cap A_n).$$

This completes the proof.

An alternative proof of (d) can be given by induction on n. For $n = 1$, (d) becomes $P(A_1) = P(A_1)$, which is obviously true. For $n > 1$ we have

$$P(A_1)\, P(A_2 \mid A_1) \cdots P(A_{n-1} \mid A_1 \cap \cdots \cap A_{n-2})\, P(A_n \mid A_1 \cap \cdots \cap A_{n-1})$$
$$= \Big(P(A_1)\, P(A_2 \mid A_1) \cdots P(A_{n-1} \mid A_1 \cap \cdots \cap A_{n-2})\Big) P(A_n \mid A_1 \cap \cdots \cap A_{n-1})$$
$$= P(A_1 \cap \cdots \cap A_{n-1}) P(A_n \mid A_1 \cap \cdots \cap A_{n-1}) \qquad \text{by induction}$$
$$= P(A_1 \cap \cdots \cap A_n) \qquad \text{Definition 4.}$$

This completes the proof. □

Example 16 (Diagnosis and Bayes' Theorem) Suppose we are developing a test to see if a person has a disease, say the dreaded wurfles. It's known that 1 person in about 500 has the wurfles. To measure the effectiveness of the test, we tried it on a lot of people. Of the 87 people with wurfles, the test always detected it, so we decide it is 100% effective at detection. We also tried the test on a large number of people who do not have wurfles and found that the test incorrectly told us that they have wurfles 3% of the time. (These are called "false positives.")

If the test is released for general use, what is the probability that a person who tests positive actually has wurfles?

Let's represent our information mathematically. Our probability space will be the general population with the uniform distribution. The event W will correspond to having wurfles and the event T will correspond to the test being positive. Our information can be written

$$P(W) = 1/500 = 0.002 \qquad P(T|W) = 1 \qquad P(T|W^c) = 0.03,$$

and we are asked to find $P(W|T)$. Bayes' formula (Theorem 2(c)) tells us

$$P(W|T) = \frac{P(T|W)\, P(W)}{P(T)}.$$

Decision Trees and Recursion

Everything on the right is known except $P(T)$. How can we compute it? The idea is to partition T using W and then convert to known conditional probabilities:

$$P(T) = P(T \cap W) + P(T \cap W^c) \qquad \text{partition } T$$
$$= P(T|W)P(W) + P(T|W^c)P(W^c) \qquad \text{convert to conditional}$$
$$= 1 \times 0.002 + 0.03 \times (1 - 0.002) \approx 0.032,$$

where we have rounded off. Thus $P(W|T) \approx 1 \times 0.002/0.032 \approx 6\%$. In other words, even if the test is positive, you only have a 6% chance of having wurfles. This shows how misleading a rather accurate test can be when it is used to detect a rare condition. ☐

Example 17 (Decision trees and conditional probability) We can picture the previous example using a decision tree. We start out with the sample space U at the root. Since we have information about how the test behaves when wurfles are present and when they are absent, the first decision partitions U into W (has wurfles) and W^c (does not have wurfles). Each of these is then partitioned according to the test result, T (test positive) and T^c (test negative). Each edge has the form (A, B) and is labeled with the conditional probability $P(B|A)$. The labels $P(W)$ and $P(W^c)$ are equal to $P(W|U)$ and $P(W^c|U)$ respectively (by Theorem 2(a)). Here is the decision tree for our wurfles test.

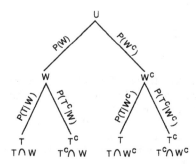

If you follow the path (U, W, T), your choices were first W (has wurfles) then T (tests positive). In terms of sets, these choices correspond to the event (i.e., set) $T \cap W$ of all people who both test positive and have wurfles. Accordingly, the leaf that is at the end of this path is labeled with the event $T \cap W$. Similar "event" labels are placed at the other leaves.

Using the definition of conditional probability, you should be able to see that the probability of the event label at a vertex is simply the product of the probabilities on the edges along the path from the root to the vertex. For example, to compute $P(T \cap W^c)$ we multiply $P(W^c)$ and $P(T|W^c) = P(T \cap W^c)/P(W^c)$. Numerically this is $0.998 \times 0.03 \approx 0.03$. To compute $P(T \cap W)$ we multiply $P(W)$ and $P(T|W)$. Numerically this is $0.002 \times 1.0 = 0.002$. Here is the tree with the various numerical values of the probabilities

Section 3: Decision Trees and Conditional Probability

shown.

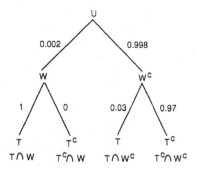

Using the above tree and the computational rules described in the previous paragraph, we can compute $P(W|T) = P(W \cap T)/P(T) = P(T \cap W)/P(T)$ as follows.

1. Compute $P(T)$ by adding up the probabilities of the event labels of *all leaves* that are associated with the decision T. (These are the event labels $T \cap W$ and $T \cap W^c$.) Thus, $P(T) = P(T \cap W) + P(T \cap W^c)$. Using the actual probabilities on the edges of the decision tree we get $P(T) = 0.002 \times 1.0 + 0.998 \times 0.03 \approx 0.032$.

2. Compute $P(W|T)$ using $P(W|T) = P(T \cap W)/P(T)$. Using the computation in step (1), we get $P(W|T) = P(T \cap W)/P(T) = (0.002 \times 1.0)/0.032 \approx 0.06$ These are the same calculations we did in the previous example, so why go to the extra trouble of drawing the tree? The tree gives us a systematic method for recording data and carrying out the calculations. ☐

In the previous example, each vertex was specifically labeled with the event, such as $W \cap T^c$, associated with it. In the next example, we simply keep track of the information we need to compute our answer.

Example 18 (Another decision tree with probabilities) We are given an urn with one red ball and one white ball. A fair die is thrown. If the number is 1, then 1 red ball and 2 white balls are added to the urn. If the number is 2 or 3, then 2 red balls and 3 white balls are added to the urn. If the number is 4, 5, or 6, then 3 red balls and 4 white balls are added to the urn. A ball is then selected uniformly at random from the urn. We represent the situation in the following decision tree.

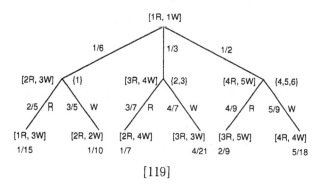

[119]

DT-31

Decision Trees and Recursion

The root is represented by the initial composition of the urn. The children of the root $[1R, 1W]$ are $[2R, 3W]$, $[3R, 4W]$, and $[4R, 5W]$. Beside each of these children of the root is the outcome set of the roll of the die that produces that urn composition: $\{1\}$, $\{2,3\}$, $\{4,5,6\}$. The probabilities on the edges incident on the root are the probabilities of the outcome sets of the die. The probabilities on the edges incident on the leaves are the conditional probabilities as discussed in Example 17. Thus, 3/7 is the conditional probability that the final outcome is R, given that the outcome of the die was in the set $\{2,3\}$.

Here is a typical sort of question asked about this type of probabilistic decision tree: "Given that the ball drawn was red, what is the probability that the outcome of the die was in the set $\{2,3\}$." We could write this mathematically as $P(\{2,3\} \mid R)$, where $\{2,3\}$ represents the result of rolling the die and R represents the result of the draw. Note in this process that the basic data given are conditional probabilities of the form $P(\text{drawing is R} \mid \text{die in S})$. We are computing conditional probabilities of the form $P(\text{die roll in S} \mid \text{drawing is R})$. This is exactly the same situation as in Example 17. Thus our question is answered by carrying out the two steps in Example 17:

1. Add up the probabilities of *all* leaves resulting from the drawing of a red ball to obtain $P(R) = 1/15 + 1/7 + 2/9 = 136/315$. (The probabilities of the leaves were computed by multiplying along the paths from the root. The results for all leaves are shown in the picture of the decision tree.)

2. Compute the conditional probability $P(\{2,3\} \mid R)$ by dividing $P(\{2,3\} \cap R) = 1/7$ by $P(R)$. Divide this by the answer from part (1). In this case, we get $(1/7)/(136/315) = 0.331$.

If you wish, you can think of this problem in terms of a new sample space. The elements of the sample space are the leaves. Step 1 (multiplying probabilities along paths) computes the probability function for this sample space. Since an event is a subset of the sample space, an event is a set of leaves and its probability is the sum of the probabilities of the leaves it contains. Can we interpret the nonleaf vertices? Yes. Each such vertex represents an event that consists of the set of leaves below it. Many people prefer this alternative way of thinking about the decision tree. ∎

The procedure we used to compute conditional probabilities in Steps 1 to 3 of two previous examples can be stated as a formula, which is another form of Bayes' Theorem:

Theorem 3 (Bayes' Theorem) Let (U, P) be a probability space, let $\{A_i : i = 1, 2, \ldots, n\}$ be a partition of U, and let $B \subset U$. Then

$$P(A_i|B) = \frac{P(A_i)P(B|A_i)}{\sum_{t=1}^{n} P(A_t)P(B|A_t)}.$$

Most students find decision trees much easier to work with than trying to apply the formal statement of Bayes' theorem. Our proof will closely follow the terminology of Example 17.

Proof: We can draw a decision tree like the ones in the previous examples, but now there are n edges of the decision tree coming down from the root and 2 edges coming down from

Section 3: Decision Trees and Conditional Probability

each child of the root. Here is a decision tree for this generalization:

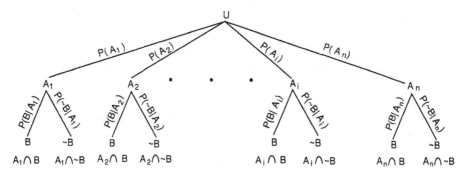

We follow the two step process of Example 17. In doing this, we need to compute, for $1 \leq t \leq n$, the products of the probabilities along the path passing through the vertex A_t and leading to the leaves labeled by the events $A_t \cap B = B \cap A_t$.

1. Add up the probabilities of *all leaves* contained in B, i.e., add up $P(A_t)P(B|A_t)$ over $1 \leq t \leq n$ to obtain $P(B)$.

2. Compute $P(A_i|B) = P(A_i \cap B)/P(B)$. Since $P(A_i \cap B) = P(A_i)P(B|A_i)$, this quotient is the formula in the theorem.

This process gives the formula in the theorem. $\square$

All of our probabilistic decision trees discussed thus far have had height two. However, probabilistic decision trees can have leaves at any distance from the root and different leaves may be at different distances. The two step procedure in Example 17 contains no assumptions about the height of leaves and, in fact, will work for all trees. The next example illustrates this.

Example 19 (Tossing coins) Suppose you have two coins. One has heads on both sides and the other is a normal coin. You select a coin randomly and toss it. If the result is heads, you switch coins; otherwise you keep the coin you just tossed. Now toss the coin you're holding. What is the probability that the result of the toss is heads? Here is the decision tree.

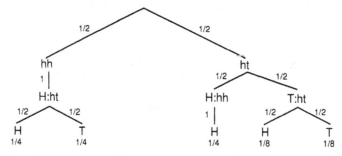

The labels hh and ht indicate which coin you're holding — two headed or normal. The labels H and T indicate the result of the toss. A label like H:ht means the toss was H and

[121]

DT-33

Decision Trees and Recursion

so I am now holding the ht coin. The conditional probabilities are on the edges and the leaf probabilities were computed by multiplying the probabilities along the paths, as required by Step 1. Adding up, we find that the probability of heads is $1/4 + 1/4 + 1/8 = 5/8$.

Given that the final toss is heads, what is the probability that you're holding the double-headed coin? The leaf where you're holding the double headed coin and tossed a head is the middle leaf, which has probability $1/4$, so the answer is $(1/4)/(5/8) = 2/5$.

Given that the final toss is heads, what is the probability that the coin you picked up at the start was not double headed? This is a bit different than what we've done before because there are two leaves associated with this event. Since the formula for conditional probability is

$$\frac{P((\text{chose ht}) \cap (\text{second toss was H}))}{P(\text{second toss was H})},$$

we simply add up the probability of those two leaves to get the numerator and so our answer is $(1/4 + 1/8)/(5/8) = 3/5$.

In the last paragraph we introduced a generalization of our two step procedure: If an event corresponds to more than one leaf, we add up the probability of those leaves. □

---

*Generating Objects at Random

To test complicated algorithms, we may want to run the algorithm on a lot of random problems. Even if we know the algorithm works, we may want to do this to study the speed of the algorithm. Computer languages include routines for generating random numbers. What can we do if we want something more complicated?

In Section 4 of Unit Fn, we gave an algorithm for generating random permutations. Here we show how to generate random objects using a decision tree.

Let (U, P) be a probability space. Suppose we want to choose elements of U at random according to the probability function P. In other words, $u \in U$ will have a probability $P(u)$ of being chosen each time we choose an element. This is easy to do if we have a decision tree whose leaves correspond to the elements of U. The process is best understood by looking at an example

Example 20 (Generating random words) In Example 2 we looked at the problem of counting certain types of "words." Go back and review that example before continuing.

• • •

We want to generate those words at random.

We'll use a two step approach. First, we'll select a CV-pattern corresponding to one of the leaves in the tree from Example 2. (We've reproduced the figure below. For the

Section 3: Decision Trees and Conditional Probability

present, ignore the numbers in the figure.) Second, we'll generate a word at random that fits the pattern.

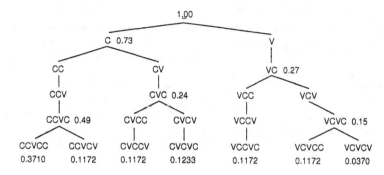

Generating a random word to fit the pattern is simple. To illustrate, suppose the pattern is CCVCV. Since there are 20 choices for the first C, use the computer software to generate a random number between 1 and 21 to decide what consonant to choose for C. The second C has 19 choices since adjacent consonants must be different and so on. Here's the result of some random choices

position & type	number of choices	random number	letter chosen	comments
1 C	20	5	G	5th among consonants (BCDFG...)
2 C	19	11	P	11th among consonants except G
3 V	6	2	E	2nd among vowels (AEIOUY)
4 C	20	11	N	11th among consonants (BCDFG...)
5 V	6	3	I	3rd among vowels (AEIOUY)

How should we choose a pattern? We discovered in Example 2 that some patterns fit more words than other patterns fit. Each pattern should be chosen in proportion to the number of words it fits so that each word will have an equal chance. Using the counts in Example 2, we computed the probabilities of the leaves in the preceding figure. Thus

$$P(\text{leaf}) = \frac{\text{number of words with leaf pattern}}{\text{total number of words}}$$

You should compute those values yourself. We have constructed a probability space where U is the set of leaves in the tree and P has the values shown at the leaves. Each vertex in the tree corresponds to an event, namely the set of leaves that are below it in the tree. Thus we can compute the probability of each vertex in the tree by adding up the probabilities of the leaves below that vertex. Many of those probabilities are shown in the previous figure.

How do we generate a leaf at random using the probabilities we've computed? We start at the root of the tree and choose a path randomly as follows. If we are at a vertex v that has edges (v, w), (v, x) and (v, y), we simply choose among w, x and y by using the conditional probabilities $P(w|v) = P(w)/P(v)$, $P(x|v) = P(x)/P(v)$ and $P(y|v) = P(y)/P(v)$. In other words, choose w with probability $P(w|v)$ and so on. (This can be done using random number generators on computers.)

Decision Trees and Recursion

Someone might say:

> All that work with the tree is not necessary since we can use the following "direct" method: Using the leaf probabilities, generate a random pattern.

That approach is not always feasible. For example, suppose we wanted to generate a random strictly decreasing function from 200 to 100. We learned in Section 3 of Unit Fn that there are $\binom{200}{100}$ of these functions. This number is about 3×10^{58}. Many random number generators cannot reliably generate random integers between 1 and a number this large. Thus we need a different method. One way is to use a decision tree that lists the functions. It's a much bigger tree than we've looked at, but we don't need to construct the tree. All we need to know is how to compute the conditional probabilities so that each leaf will have probability $1/\binom{200}{100}$. It turns out that this can be done rather easily. In summary, the tree method can be used when the "direct" method is not practical. □

---

*The First Moment Method and the SAT Problem

We now review briefly the concept of conjunctive normal form. Suppose $p, q, r, \ldots$ are Boolean variables (that is, variables that can be 0 or 1). The operations $\sim$, $\vee$, and $\wedge$ stand for "negation", "or", and "and", respectively. A *disjunctive clause* is a list of Boolean variables and negations of Boolean variables joined by $\vee$. Here are four examples of disjunctive clauses:

$$q \vee r, \qquad p \vee (\sim q) \vee r, \qquad (\sim p) \vee (\sim q) \vee (\sim r), \qquad (\sim r) \vee q.$$

Conjunctive normal form is a statement form consisting of disjunctive clauses joined by $\wedge$; for example

$$(q \vee r) \wedge (p \vee (\sim q) \vee (\sim r)) \wedge ((\sim p) \vee (\sim q) \vee r) \wedge ((\sim r) \vee q).$$

(*Disjunctive normal form* is the same as conjunctive normal form except that $\wedge$ and $\vee$ are switched.) The *satisfiability problem* is the following. Given a statement in conjunctive normal form, is there some choice of values for the Boolean variables that make the statement equal to 1? One may also want to know what choice of variables does this. The satisfiability problem is also called the *SAT problem*. The SAT problem is known to be hard in general. (The technical term is "NP-complete".)

Given a statement in conjunctive normal form, how might we try to solve the satisfiability problem? One way is with the following backtracking algorithm for a statement involving $p_1, p_2, \ldots, p_n$

Step 1. Set $k = 1$.

Step 2. Set $p_k = 0$.

Step 3. (Test) Check to see if any of the clauses that contain only $p_1, \ldots, p_k$ are 0. If so, go to Step 4_0; if not, go to Step 4_1.

Step 4_0. (Failure) If $p_k = 0$, set $p_k = 1$ and go to Step 3. If $p_k = 1$, replace k with $k-1$ and go to Step 4_0.

DT-36 [124]

Section 3: Decision Trees and Conditional Probability

Step 4_1. (Partial success) If $k = n$, stop because the current values of the variables make the statement 1. If $k < n$, replace k with $k + 1$ and go to Step 2.

You should use the algorithm on the conjunctive normal form statement given earlier.

The following example shows that we can sometimes guarantee that the algorithm will succeed if there are not too many clauses. However, it does not give us values of the variables that will make the statement 1. In the example after that, we will see how to use the idea from the next example to find those values without backtracking.

Example 21 (SAT with just a few clauses) Suppose we have a conjunctive normal form statement $S = C_1 \wedge C_2 \wedge \cdots \wedge C_k$, where the C_i are clauses in the Boolean variables $p_1, \ldots, p_n$. Make the set $\times^n \{0, 1\}$, the possible values for $p_1, \ldots, p_n$, into a probability space by letting each n-tuple have probability $1/2^n$.

Let X_i be a random variable whose value is 1 if C_i has the value 0, and let $X_i = 0$ if C_i has the value 1. (Be careful: note the reversal — X_i is the opposite of C_i.) The number of clauses which are 0 is $X = X_1 + \cdots + X_k$. If we can show that $P(X = 0) = 0$, we will have shown that there is some choice of $p_1, \ldots, p_n$ for which all clauses are 1 and so S will be 1 as well. How can we do this? Here is one tool:

Theorem 4 (First Moment Method) *Suppose that X is an integer-valued random variable and $E(X) < m + 1$, then $P(X \leq m)$ is greater than 0.*

This is easy to prove:

$$m + 1 > E(X) = \sum_k kP(X = k) \geq \sum_{k \geq m+1}(m+1)P(X = k) = (m+1)P(X \geq m+1).$$

Thus $P(X \geq m + 1) < 1$ and so $P(X \leq m) = 1 - P(X \geq m + 1) > 0$.

To apply this, we need to compute $E(X) = E(X_1) + \cdots + E(X_k)$. Let v_i be the number of variables and their negations appearing in C_i. We claim that $E(X_i) = 2^{-v_i}$. Why is this? Note that $E(X_i)$ equals the probability that C_i has the value 0. To make C_i have the value 0, each variable in C_i must be chosen correctly: 0 if it appears without being negated and 1 if it appears negated. The variables not appearing in C_i can have any values whatsoever.

We have shown that $E(X) = 2^{-v_1} + \cdots + 2^{-v_k}$. By the First Moment Method, we are done if this is less than 1. We have proved:

Theorem 5 (SAT for few clauses) *Suppose we have a conjunctive normal form statement $S = C_1 \wedge C_2 \wedge \cdots \wedge C_k$, where the C_i are clauses in the Boolean variables $p_1, \ldots, p_n$. Let v_i be the number of variables (and negations of variables) that appear in C_i. If $2^{-v_1} + \cdots + 2^{-v_k} < 1$, then there is a choice of values for $p_1, \ldots, p_n$ which gives S the value 1.*

Let's apply the theorem to

$$S = (q \vee r) \wedge (p \vee (\sim q) \vee (\sim r)) \wedge ((\sim p) \vee (\sim q) \vee r) \wedge ((\sim r) \vee q).$$

Decision Trees and Recursion

We have $v_1 = 2$, $v_2 = 3$, $v_3 = 3$, and $v_4 = 2$. Thus
$$E(X) = 2^{-2} + 2^{-3} + 2^{-3} + 2^{-2} = 3/4 < 1.$$
Thus there is a choice of variables that give S the value 1. If you carried out the backtracking algorithm as you were asked to earlier, you found such an assignment. Of course, you may find the assignment rather easily without backtracking. However, the theorem tells us a lot more: It doesn't look at the structure of the clauses, so you could change p to $\sim p$ and so on in any of the clauses you wish and the statement would still be satisfiable. $\square$

Example 22 (Satisfiability without backtracking) Suppose the situation in the preceding example holds; that is, $E(X) < 1$. We want to find values for $p_1, \ldots, p_n$ that satisfy S (give it the value 1). We have
$$\begin{aligned} E(X) &= P(p_n = 0)\, E(X \mid p_n = 0) + P(p_n = 1)\, E(X \mid p_n = 1) \\ &= \tfrac{1}{2} E(X \mid p_n = 0) + \tfrac{1}{2} E(X \mid p_n = 1). \end{aligned}$$
Since $E(X) < 1$ at least one of $E(X \mid p_n = 0)$ and $E(X \mid p_n = 1)$ must be less than 1. Suppose that $E(X \mid p_n = 0) < 1$. Set $p_n = 0$ and simplify S to get a new statement S' in $p_1, \ldots, p_{n-1}$. To get this new statement S' from S when $p_n = 0$:

- any clause not containing p_n or $\sim p_n$ is unchanged;
- any clause containing $\sim p_n$ will have the value 1 regardless of the remaining variables and so is dropped;
- any clause containing p_n depends on the remaining variables for its value and so is kept, with p_n removed.

When $p_n = 1$, the last two cases are reversed to produce S'. This method will be illustrated soon.

Let X' be for S' what X is for S. You should show that
$$E(X') = E(X \mid p_n = 0) < 1.$$
We can now repeat the above procedure for S', which will give us a value for p_{n-1}. Continuing in this way, we find values for $p_n, p_{n-1}, \ldots, p_1$.

Let's apply this to
$$S = (q \vee r) \wedge (p \vee (\sim q) \vee (\sim r)) \wedge ((\sim p) \vee (\sim q) \vee r) \wedge ((\sim r) \vee q).$$
When $p = 0$, this reduces to
$$(q \vee r) \wedge ((\sim q) \vee (\sim r)) \wedge ((\sim r) \vee q),$$
and so $E(X \mid p = 0) = 2^{-2} + 2^{-2} + 2^{-2} < 1$. Thus we can take the previous statement to be S'. Suppose we try $q = 0$. Then
$$(q \vee r) \wedge ((\sim q) \vee (\sim r)) \wedge ((\sim r) \vee q)$$
reduces to $r \wedge (\sim r)$ because the middle clause disappears. The expectation is $2^{-1} + 2^{-1} = 1$, so this is a bad choice. (Of course this is obviously a bad choice, but we're applying the method blindly like a computer program would.) Thus we must choose $q = 1$. The statement reduces to $\sim r$, and we choose $r = 0$. $\square$

Section 3: Decision Trees and Conditional Probability

Exercises for Section 3

3.1. A box contains 3 white and 4 green balls.

(a) Two balls are sampled *with* replacement, what is the probability that the second is white if the first is green? If the first is white?

(b) Two balls are sampled *without* replacement, what is the probability that the second is white if the first is green? If the first is white?

3.2. Two dice are rolled and the total is six.

(a) What is the probability that at least one die is three?

(b) What is the probability that at least one die is four?

(c) What is the probability that at least one die is odd?

3.3. In a certain college, 10 percent of the students are physical science majors, 40 percent are engineering majors, 20 percent are biology majors and 30 percent are humanities majors. Of the physical science majors, 10 percent have read Hamlet, of the engineering majors, 50 percent have read Hamlet, of the biology majors, 30 percent have read Hamlet, and of the humanities majors, 20 percent have read Hamlet.

(a) Given that a student selected at random has read Hamlet, what is the probability that that student is a humanities major?

(b) Given that a student selected at random has not read Hamlet, what is the probability that that student is an engineering or physical science major?

3.4. We are given an urn that has one red ball and one white ball. A fair die is thrown. If the number is a 1 or 2, one red ball is added to the urn. Otherwise three red balls are added to the urn. A ball is then drawn at random from the urn.

(a) Given that a red ball was drawn, what is the probability that a 1 or 2 appeared when the die was thrown?

(b) Given that the final composition of the urn contained more than one red ball, what is the probability that a 1 or 2 appeared when the die was thrown?

3.5. A man starts with one dollar in a pot. A "play" consists of flipping a fair coin and, if heads occurs, a dollar is added to the pot, if tails occurs, a dollar is removed from the pot. The game ends if the man has zero dollars or if he has played four times. Let X denote the random variable which, for each outcome of the game, specifies the maximum amount of money that was ever in the pot, from (and including) the

Decision Trees and Recursion

start of the game to (and including) that final outcome. What is the expected value $E(X)$?

3.6. The probability of team A winning any game is $1/3$, of B winning $2/3$ (no ties in game play). Team A plays team B in a tournament. If either team wins two games in a row, that team is declared the winner. At most four games are played in the tournament and, if no team has won the tournament at the end of four games, the tournament is declared a draw. What is the expected number of games in the tournament?

3.7. The platoon commander knows:

- If the air strike is successful, there is a 60% probability that the ground forces *will not* encounter enemy fire.

- If the air strike is not successful, there is a 80% probability that the ground forces *will* encounter enemy fire.

- There is a 70% probability that the air strike will be successful.

Answer the following questions.

(a) What is the **probability** that the ground forces *will not* encounter enemy fire?

(b) The ground forces did not encounter enemy fire. What is the **probability** that the air strike was successful?

Section 4: Inductive Proofs and Recursive Equations

Proof by induction, familiar from prior courses and used occasionally in earlier sections, is central to the study of recursive equations. We'll begin by reviewing proof by induction. Then we'll look at recursions (another name for recursive equations). The two subjects are related since induction proofs use smaller cases to prove larger cases and recursions use previous values in a sequence to compute later values. A "solution" to a recursion is a formula that tells us how to compute any term in the sequence without first computing the previous terms. We will find that it is usually easy to verify a solution to a recursion if someone gives it to us; however, it can be quite difficult to find the solution on our own — in fact there may not even be a simple solution even when the recursion looks simple.

Section 4: Inductive Proofs and Recursive Equations

Induction

Suppose $\mathcal{A}(n)$ is an assertion that depends on n. We use *induction* to prove that $\mathcal{A}(n)$ is true when we show that

- it's true for the smallest value of n and
- if it's true for everything less than n, then it's true for n.

Closely related to proof by induction is the notion of a recursion. A *recursion* describes how to calculate a value from previously calculated values. For example, $n!$ can be calculated by using $n! = 1$ if $n = 0$, $n! = n(n-1)!$ if $n > 0$.

Notice the similarity between the two ideas: There is something to get us started and then each new thing depends on similar previous things. Because of this similarity, recursions often appear in inductively proved theorems. We'll study inductive proofs and recursive equations in this section.

Inductive proofs and recursive equations are special cases of the general concept of a recursive approach to a problem. Thinking recursively is often fairly easy when one has mastered it. Unfortunately, people are sometimes defeated before reaching this level. In Section 3 we look at some concepts related to recursive thinking and recursive algorithms.

We recall the theorem on induction and some related definitions:

Theorem 6 (Induction) Let $\mathcal{A}(m)$ be an assertion, the nature of which is dependent on the integer m. Suppose that $n_0 \leq n_1$. If we have proved the two statements

(a) "$\mathcal{A}(n)$ is true for $n_0 \leq n \leq n_1$" and

(b) "If $n > n_1$ and $\mathcal{A}(k)$ is true for all k such that $n_0 \leq k < n$, then $\mathcal{A}(n)$ is true."

Then $\mathcal{A}(m)$ is true for all $m \geq n_0$.

Let's look at a common special case: $n_0 = n_1$ and, in proving (b) we use only $\mathcal{A}(n-1)$. Then the theorem becomes

Let $\mathcal{A}(m)$ be an assertion, the nature of which is dependent on the integer m. If we have proved the two statements

(a) "$\mathcal{A}(n_0)$ is true" and

(b) "If $n > n_0$ and $\mathcal{A}(n-1)$ is true, then $\mathcal{A}(n)$ is true."

Then $\mathcal{A}(m)$ is true for all $m \geq n_0$.

Some people use terms like "weak induction", "simple induction" and "strong induction" to distinguish the various types of induction.

Decision Trees and Recursion

Definition 5 (Induction hypothesis) *The statement "$A(k)$ is true for all k such that $n_0 \leq k < n$" is called the induction assumption or induction hypothesis and proving that this implies $A(n)$ is called the inductive step. $A(n_0), \ldots, A(n_1)$ are called the base cases or simplest cases.*

Proof: We now prove the theorem. Suppose that $A(n)$ is false for some $n \geq n_0$. Let m be the least such n. We cannot have $m \leq n_1$ because (a) says that $A(n)$ is true for $n_0 \leq n \leq n_1$. Thus $m > n_1$.

Since m is as small as possible, $A(k)$ is true for $n_0 \leq k < m$. By (b), the inductive step, $A(m)$ is also true. This contradicts our assumption that $A(n)$ is false for some $n \geq n_0$. Hence the assumption is false; in other words, $A(n)$ is never false for $n \geq n_0$. This completes the proof. ☐

Example 23 (Every integer is a product of primes) A positive integer $n > 1$ is called a *prime* if its only divisors are 1 and n. The first few primes are 2, 3, 5, 7, 11, 13, 17, 19, 23. If a number is not a prime, such as 12, it can be written as a product of primes (*prime factorization*: $12 = 2 \times 2 \times 3$). We adopt the terminology that a single prime p is a product of one prime, itself. We shall prove $A(n)$ that "every integer $n \geq 2$ is a product of primes." Our proof will be by induction. We start with $n_0 = n_1 = 2$, which is a prime and hence a product of primes. The induction hypothesis is the following:

"Suppose that for some $n > 2$, $A(k)$ is true for all k such that $2 \leq k < n$."

Assume the induction hypothesis and consider n. If n is a prime, then it is a product of primes (itself). Otherwise, $n = st$ where $1 < s < n$ and $1 < t < n$. By the induction hypothesis, s and t are each a product of primes, hence $n = st$ is a product of primes. ☐

In the example just given, we needed the induction hypothesis "for all k such that $2 \leq k < n$." In the next example we have the more common situation where we only need to assume "for $k = n-1$." We can still make the stronger assumption and the proof is valid, but the stronger assumption is not used; in fact, we are using the simpler form of induction described after the theorem.

Example 24 (Sum of first n integers) We would like a formula for the sum of the first n integers. Let us write $S(n) = 1 + 2 + \ldots + n$ for the value of the sum. By a little calculation,

$$S(1) = 1, \quad S(2) = 3, \quad S(3) = 6, \quad S(4) = 10, \quad S(5) = 15, \quad S(6) = 21.$$

What is the general pattern? It turns out that $S(n) = \frac{n(n+1)}{2}$ is correct for $1 \leq n \leq 6$. Is it true in general? This is a perfect candidate for an induction proof with

$$n_0 = n_1 = 1 \quad \text{and} \quad A(n): \quad \text{"}S(n) = \frac{n(n+1)}{2}\text{."}$$

Section 4: Inductive Proofs and Recursive Equations

Let's prove it. We have shown that $\mathcal{A}(1)$ is true. In this case we need only the restricted induction hypothesis; that is, we will prove the formula for $S(n)$ by assuming the formula for $k = n - 1$. Thus, we assume only $\mathcal{A}(n-1)$ is true. Here it is (the inductive step):

$$\begin{aligned}
S(n) &= 1 + 2 + \cdots + n & &\text{by the definition of } S(n)\\
&= \Big(1 + 2 + \cdots + (n-1)\Big) + n\\
&= S(n-1) + n & &\text{by definition of } S(n-1),\\
&= \frac{(n-1)\big((n-1)+1\big)}{2} + n & &\text{by } \mathcal{A}(n-1),\\
&= \frac{n(n+1)}{2} & &\text{by algebra.}
\end{aligned}$$

This completes the proof. We call your attention to the fact that, in the third line we proved $S(n) = S(n-1) + n$. □

Recursive Equations

The equation $S(n) = S(n-1) + n$ (for $n > 1$) that arose in the inductive proof in the preceding example is called a *recurrence relation*, *recursion*, or *recursive equation*. A recursion is *not complete* unless there is information on how to get started. In this case the information was $S(1) = 1$. This information is called the *initial condition* or, if there is more than one, *initial conditions*. Many examples of such recurrence relations occur in computer science and mathematics. We discussed recurrence relations in Section 3 of Unit CL (Basic Counting and Listing) for binomial coefficients $C(n,k)$ and Stirling numbers $S(n,k)$.

In the preceding example, we found that $S(n) = n(n+1)/2$. This is a *solution* to the recursion because it tells us how to compute $S(n)$ without having to compute $S(k)$ for any other values of k. If we had used the recursion $S(n) = S(n-1) + n$, we would have had to compute $S(n-1)$, which requires $S(n-2)$, and so on all the way back to $S(1)$.

A recursion tells us how to compute values in a sequence a_n from earlier values $a_{n-1}, a_{n-2}, \ldots$ and n. We can denote this symbolically by writing $a_n = G(n, a_{n-1}, a_{n-2}, \ldots)$. For example, in the case of the sum of the first n integers, which we called $S(n)$, we would have

$$a_n = S(n) \quad \text{and} \quad G = a_{n-1} + n \quad \text{since} \quad S(n) = S(n-1) + n.$$

Induction proofs deduce the truth of $\mathcal{A}(n)$ from earlier statements. Thus it's natural to use induction to prove that a formula for the solution to a recursion is correct. That's what we did in the previous example. There's a way to avoid giving an inductive proof each time we have such a problem: It turns out that the induction proofs for solutions to recursions all have the same form. A general pattern often means there's a general theorem. If we can find and prove the theorem, then we could use it to avoid giving an inductive

Decision Trees and Recursion

proof in each special case. That's what the following theorem is about. (The a_n and $f(n)$ of the theorem are generalizations of S_n and $\frac{n(n+1)}{2}$ from the previous example.)

Theorem 7 (Verifying the solution of a recursion) *Suppose we have initial conditions that give a_n for $n_0 \leq n \leq n_1$ and a recursion that allows us to compute a_n when $n > n_1$. To verify that $a_n = f(n)$, it suffices to do two things:*

Step 1. Verify that f satisfies the initial conditions.

Step 2. Verify that f satisfies the recursion.

Proof: The goal of this theorem is to take care of the inductive part of proving that a formula is the solution to a recursion. Thus we will have to prove it by induction. We must verify (a) and (b) in Theorem 6. Let $\mathcal{A}(n)$ be the assertion "$a_n = f(n)$." By Step 1, $\mathcal{A}(n)$ is true for $n_0 \leq n \leq n_1$, which proves (a). Suppose the recursion is $a_n = G(n, a_{n-1}, a_{n-2}, \ldots)$ for some formula G. We have

$$f(n) = G\Big(n, f(n-1), f(n-2), \ldots\Big) \quad \text{by Step 2,}$$
$$= G(n, a_{n-1}, a_{n-2}, \ldots) \quad \text{by } \mathcal{A}(k) \text{ for } k < n,$$
$$= a_n \quad \text{by the recursion for } a_n.$$

This proves (b) and so completes the proof. $\square$

Example 25 (Proving a formula for the solution of a recursion) Let $S(n)$ be the sum of the first n integers. The initial condition $S(1) = 1$ and the recursion $S(n) = n + S(n-1)$ allow us to compute $S(n)$ for all $n \geq 1$. It is claimed that $f(n) = \frac{n(n+1)}{2}$ equals $S(n)$.

The initial condition is for $n = 1$. Thus $n_0 = n_1 = 1$. Since $f(1) = 1$, f satisfies the initial condition. (This is Step 1.) For $n > 1$ we have

$$n + f(n-1) = n + \frac{n(n-1)}{2} = \frac{n(n+1)}{2} = f(n)$$

and so f satisfies the recursion. (This is Step 2.)

We now consider a different problem. Suppose we are given that

$$a_0 = 2, \quad a_1 = 7, \quad \text{and} \quad a_n = 3a_{n-1} - 2a_{n-2} \text{ when } n > 1$$

and we are asked to prove that $a_n = 5 \times 2^n - 3$ for $n \geq 0$.

Let's verify that the formula is correct for $n = 0$ and $n = 1$ (the initial conditions — Step 1 in our theorem):

$$n = 0: \quad a_0 = 2 = 5^0 - 3 \qquad n = 1: \quad a_1 = 7 = 5^1 - 3.$$

Now for Step 2, the recursion. Let $f(x) = 5 \times 2^x - 3$ and assume that $n > 1$. We have

$$3f(n-1) - 2f(n-2) = 3(5 \times 2^{n-1} - 3) - 2(5 \times 2^{n-2} - 3)$$
$$= (3 \times 5 \times 2 - 2 \times 5)2^{n-2} - 3$$
$$= 5 \times 2^n - 3 = f(n).$$

Section 4: Inductive Proofs and Recursive Equations

This completes the proof.

As a final example, suppose $b_0 = b_1 = 1$ and $b_{n+1} = n(b_n + b_{n-1})$ for $n \geq 1$. We want to prove that $b_n = n!$. Since our theorem stated the recursion for a_n, let's rewrite our recursion to avoid confusion. Let $n+1 = k$ in the recursion to get $b_k = (k-1)(b_{k-1}+b_{k-2})$. The initial conditions are $b_0 = 1 = 0!$ and $b_1 = 1 = 1!$, so we've done Step 1. Now for Step 2:

$$\text{Is} \quad k! = (k-1)((k-1)! + (k-2)!) \quad \text{true?}$$

Yes because $(k-1)! = (k-1) \times (k-2)!$ and so $(k-1)! + (k-2)! = ((k-1)+1)(k-2)! = k \times (k-2)!$. We could have done this without changing the subscripts in the recursion: Just check that $(n+1)! = n(n! + (n-1)!)$. We'll let you do that. □

So far we have a method for checking the solution to a recursion, which we just used in the previous example. How can we find a solution in the first place? If we're lucky, someone will tell us. If we're unlucky, we need a clever guess or some tools. Let's look at how we might guess.

Example 26 (Guessing solutions to recurrence relations)

(1) Let $r_k = -r_{k-1}/k$ for $k \geq 1$, with $r_0 = 1$. Writing out the first few terms gives $1, -1, 1/2, -1/6, 1/24, \ldots$. Guessing, it looks like $r_k = (-1)^k/k!$ is a solution.

(2) Let $t_k = 2t_{k-1} + 1$ for $k > 0$, $t_0 = 0$. Writing out some terms gives $0, 1, 3, 7, 15, \ldots$. It looks like $t_k = 2^k - 1$, for $k \geq 0$.

(3) What is the solution to $a_0 = 0$, $a_1 = 1$ and $a_n = 4a_{n-1} - 4a_{n-2}$ for $n \geq 2$? Let's compute some values

n:	0	1	2	3	4	5	6	7
a_n:	0	1	4	12	32	80	192	448

These numbers factor nicely: $4 = 2^2$, $12 = 2^2 \times 3$, $32 = 2^5$, $80 = 2^4 \times 5$, $192 = 2^6 \times 3$, $448 = 2^6 \times 7$. Can we see a pattern here? We can pull out a factor of 2^{n-1} from a_n:

n:	0	1	2	3	4	5	6	7
a_n:	0	1	4	12	32	80	192	448
$a_n/2^{n-1}$:	0	1	2	3	4	5	6	7

Now the pattern is clear: $a_n = n2^{n-1}$. That was a lot of work, but we're not done yet -- this is just a guess. We have to prove it. You can use the theorem to do that. We'll do it a different way in a little while.

(4) Let $b_n = b_1 b_{n-1} + b_2 b_{n-2} + \cdots + b_{n-1} b_1$ for $n \geq 2$, with $b_1 = 1$. Here are the first few terms: $1, 1, 2, 5, 14, 42, 132, 429, 1430, 4862, \ldots$. Each term is around 3 or 4 times the preceding one. Let's compute the ratio exactly

n	2	3	4	5	6	7	8	9	10
b_n/b_{n-1}	1	2	5/2	14/5	3	22/7	13/4	10/3	17/5

[133]

DT-45

Decision Trees and Recursion

These ratios have surprisingly small numerators and denominators. Can we find a pattern? The large primes 13 and 17 in the numerators for $n = 8$ and 10 suggest that maybe we should look for $2n-3$ in the numerator.[1] Let's adjust our ratios accordingly:

n	2	3	4	5	6	7	8	9	10
$b_n/(2n-3)b_{n-1}$	1	2/3	1/2	2/5	1/3	2/7	1/4	2/9	1/5

Aha! These numbers are just $2/n$. Our table leads us to guess $b_n = 2(2n-3)b_{n-1}/n$, a much simpler recursion than the one we started with.

This recursion is so simple we can "unroll" it:

$$b_n = \frac{2(2n-3)}{n} b_{n-1} = \frac{2(2n-3)}{n} \frac{2(2n-5)}{n-1} b_{n-2} = \cdots = \frac{2^{n-1}(2n-3)(2n-5)\cdots 1}{n!}.$$

This is a fairly simple formula. Of course, it is still only a conjecture and it is not easy to prove that it is the solution to the original recursion because the computations in Theorem 7 using this formula and the recursion $b_n = b_1 b_{n-1} + b_2 b_{n-2} + \cdots + b_{n-1} b_1$ would be very messy.

(5) Let $d_n = (n-1)d_{n-1} + (n-1)d_{n-2}$ for $n \geq 2$, with $d_0 = 1$ and $d_1 = 0$. In the previous example, we looked at a recursion that was almost like this: The only difference was that $d_1 = 1$. In that case we were told that the answer was $n!$, so maybe these numbers look like $n!$. If this were like $n!$, we'd expect nd_{n-1} to equal d_n. Here are the first few values of d_n together with nd_{n-1}:

n	0	1	2	3	4	5	6
d_n	1	0	1	2	9	44	265
nd_{n-1}	–	1	0	3	8	45	264

We're close! The values of d_n and nd_{n-1} only differ by 1. Thus we are led to guess that $d_n = nd_{n-1} + (-1)^n$. This is not a solution—it's another recursion. Nevertheless, we might prefer it because it's a bit simpler than the one we started with. □

As you can see from the previous example, guessing solutions to recursions can be difficult. Now we'll look at a couple of theorems that tell us the solutions without any guessing.

Theorem 8 (Solutions to Some Recursions) *Let $a_0, a_1, \ldots, a_n, \ldots$ be a sequence of numbers. Suppose there are constants b and c such that b is not 0 or 1 and $a_n = ba_{n-1} + c$ for $n \geq 1$. Then*

$$a_n = Ab^n + K \quad \text{where} \quad K = \frac{c}{1-b} \quad \text{and} \quad A = a_0 - K = a_0 - \frac{c}{1-b}.$$

---

[1] "Why look at large primes?" you ask. Because they are less likely to have come from a larger number that has lost a factor due to reduction of the fraction.

Section 4: Inductive Proofs and Recursive Equations

This gives us the solution to the recursion $t_k = 2t_{k-1} + 1$ (with $t_0 = 0$) of the previous example: $K = \frac{1}{1-2} = -1$ and $A = 0 - (-1) = 1$. That gives the solution $t_k = 2^k - 1$, no guessing needed!

Proof: (of Theorem 8) We'll use Theorem 7. The initial condition is simple:
$$Ab^0 + K = A + K = (a_0 - K) + K = a_0.$$
That's Step 1: For Step 2 we want to show that $a_n = Ab^n + K$ satisfies the recursion. We have
$$b(Ab^{n-1} + K) + c = Ab^n + bK + c = Ab^n + \frac{bc}{1-b} + c$$
$$= Ab^n + \frac{bc + (1-b)c}{1-b} = Ab^n + \frac{c}{1-b} = Ab^n + K.$$
We're done. □

Example 27 (I forgot the formulas for A and K!) If you remember the b^n, you can still solve the recursion even if the initial condition is not at a_0. Let's do the example
$$a_1 = 3 \quad \text{and} \quad a_n = 4a_{n-1} - 7 \quad \text{for} \quad n > 1.$$
The solution will be $a_n = A4^n + K$ for some A and K. If we know a_n for two values of n, then we can solve for A and K. We're given $a_1 = 3$ and we compute $a_2 = 4 \times 3 - 7 = 5$. Thus
$$a_1 \text{ gives us } \quad 3 = A4^1 + K \quad \text{and} \quad a_2 \text{ gives us } \quad 5 = A4^2 + K.$$
Subtracting the first equation from the second: $12A = 2$ so $A = 1/6$. From a_1, $3 = 4/6 + K$ and so $K = 7/3$. □

Now let's look at recursions where a_n depends on a_{n-1} and a_{n-2} in a simple way.

Theorem 9 (Solutions to Some Recursions) Let $a_0, a_1, \ldots, a_n, \ldots$ be a sequence of numbers. Suppose there are constants b and c such that $a_n = ba_{n-1} + ca_{n-2}$ for $n \geq 2$. Let r_1 and r_2 be the roots of the polynomial $x^2 - bx - c$.

- If $r_1 \neq r_2$, then $a_n = K_1 r_1^n + K_2 r_2^n$ for $n \geq 0$, where K_1 and K_2 are solutions to the equations
$$K_1 + K_2 = a_0 \quad \text{and} \quad r_1 K_1 + r_2 K_2 = a_1.$$

- If $r_1 = r_2$, then $a_n = K_1 r_1^n + K_2 n r_1^n$ for $n \geq 0$, where K_1 and K_2 are solutions to the equations
$$K_1 = u_0 \quad \text{and} \quad r_1 K_1 + r_2 K_2 = r_1 K_1 + r_1 K_2 = a_1.$$

The equation $x^2 - bx - c = 0$ is called the characteristic equation of the recursion.

Before proving the theorem, we give some examples. In all cases, the roots of $x^2 - bx - c$ can be found either by factoring it or by using the quadratic formula
$$r_1, r_2 = \frac{b \pm \sqrt{b^2 + 4c}}{2}.$$

Decision Trees and Recursion

Example 28 (Applying Theorem 9) Let's redo the recursion

$$a_0 = 2, \quad a_1 = 7, \quad \text{and} \quad a_n = 3a_{n-1} - 2a_{n-2} \text{ when } n > 1$$

from Example 25. We have $b = 3$ and $c = -2$. The characteristic equation is $x^2 - 3x + 2 = 0$. The roots of $x^2 - 3x + 2$ are $r_1 = 2$ and $r_2 = 1$, which you can get by using the quadratic formula or by factoring $x^2 - 3x + 2$ into $(x-2)(x-1)$. Since $r_1 \neq r_2$, we are in the first case in the theorem. Thus we have to solve

$$K_1 + K_2 = 2 \quad \text{and} \quad 2K_1 + K_2 = 7.$$

The solution is $K_1 = 5$ and $K_2 = -3$ and so $a_n = 5 \times 2^n - 3 \times 1^n = 5 \times 2^n - 3$.

- As another example, we'll solve the recursion $a_0 = 0$, $a_1 = 1$, and $a_n = 4a_{n-1} - 4a_{n-2}$ for $n \geq 2$. Applying the theorem, $r_1 = r_2 = 2$ and so $a_n = K_1 2^n + K_2 n 2^n$ where $K_1 = 0$ and $2K_1 + 2K_2 = 1$. Thus $K_1 = 0$, $K_2 = 1/2$, and $a_n = (1/2)n 2^n = n 2^{n-1}$.

- As a final example, consider the recursion

$$F_0 = F_1 = 1 \quad \text{and} \quad F_k = F_{k-1} + F_{k-2} \quad \text{when } k \geq 2.$$

This is called the *Fibonacci recursion*. We want to find an explicit formula for F_k.

The characteristic equation is $x^2 - x - 1 = 0$. By the quadratic formula, its roots are $r_1 = \frac{1+\sqrt{5}}{2}$ and $r_2 = \frac{1-\sqrt{5}}{2}$. Thus, we need to solve the equations

$$K_1 + K_2 = 1 \quad \text{and} \quad r_1 K_1 + r_2 K_2 = 1.$$

High school math gives

$$K_1 = \frac{1 - r_2}{r_1 - r_2} = \frac{1 + \sqrt{5}}{2\sqrt{5}} = \frac{r_1}{\sqrt{5}}$$

$$K_2 = \frac{1 - r_1}{r_2 - r_1} = \frac{1 - \sqrt{5}}{-2\sqrt{5}} = -\frac{r_2}{\sqrt{5}}.$$

Thus

$$F_n = \frac{1}{\sqrt{5}} \left(\frac{1+\sqrt{5}}{2} \right)^{n+1} - \frac{1}{\sqrt{5}} \left(\frac{1-\sqrt{5}}{2} \right)^{n+1}.$$

It would be difficult to guess this solution from a few values of F_n! □

Section 4: Inductive Proofs and Recursive Equations

Example 29 (A shifted index) Let's solve the recursion

$$a_1 = 0, \quad a_2 = 1 \quad \text{and} \quad a_n = 5a_{n-1} + 6a_{n-2} \quad \text{for } n \geq 3.$$

This doesn't quite fit the theorem since it starts with a_1 instead of a_0. What can we do? The same thing we did in Example 27: Use values of a_n to get two equations in the two unknowns K_1 and K_2.

Let's do this. The characteristic equation $x^2 - 5x - 6 = 0$ gives us $r_1 = 6$ and $r_2 = -1$ and so $a_n = K_1 6^n + K_2(-1)^n$. Using a_1 and a_2:

$$a_1 \text{ gives us} \quad 0 = 6K_1 - K_2 \quad \text{and} \quad a_2 \text{ gives us} \quad 1 = 6^2 K_1 + K_2.$$

Adding the two equations: $1 = 42K_1$. Thus $K_1 = 1/42$. a_1 gives us $0 = 6/42 - K_2$ and so $K_2 = 1/7$. Thus $a_n = (1/42)6^n + (1/7)(-1)^n$. $\square$

We conclude this section with a proof of Theorem 9.

Proof: (of Theorem 9) We apply Theorem 7 with $n_0 = 0$ and $n_1 = 1$.

We first assume that $r_1 \neq r_2$ and we set $f(n) = K_1 r_1^n + K_2 r_2^n$ where K_1 and K_2 are as given by the theorem. Step 1 is simple because the equations for K_1 and K_2 are simply the equations $f(0) = a_0$ and $f(1) = a_1$. Here's Step 2

$$\begin{aligned} bf(n-1) + cf(n-2) &= b(K_1 r_1^{n-1} + K_2 r_2^{n-1}) + c(K_1 r_1^{n-2} + K_2 r_2^{n-2}) \\ &= K_1 r_1^{n-2}(br_1 + c) + K_2 r_2^{n-2}(br_2 + c) \\ &= K_1 r_1^{n-2} r_1^2 + K_2 r_2^{n-2} r_2^2 \\ &= f(n). \end{aligned}$$

Wait! Something must be wrong — the theorem says $r_1 \neq r_2$ and we never use that fact! What happened? We *assumed* that the equations could be solved for K_1 and K_2. How do we know that they have a solution? One way is to actually solve them using high school algebra. We find that

$$K_1 = \frac{a_0 r_2 - a_1}{r_2 - r_1} \quad \text{and} \quad K_2 = \frac{a_0 r_1 - a_1}{r_1 - r_2}.$$

Now we can see where $r_1 \neq r_2$ is needed: The denominators in these formulas must be nonzero.

We now consider the case $r_1 = r_2$. This is similar to $r_1 \neq r_2$. We sketch the ideas and leave it to you to fill in the details of the proof. Here it's clear that we can solve the equations for K_1 and K_2. Step 1 in Theorem 7 is checked as it was for the $r_1 \neq r_2$ case. Step 2 requires algebra similar to that needed for $r_1 \neq r_2$. The only difference is that we end up needing to show that

$$K_2 r_2^{n-2}((n-1)br_2 + (n-2)c) = K_2 n r_2^n.$$

You should be able to see that this is the same as showing $-br_2 - 2c = 0$. This follows from the fact that the only way we can have $r_1 = r_2$ is to have $\sqrt{b^2 + 4c} = 0$. In this case $r_2 = b/2$. $\square$

[137]

DT-49

Decision Trees and Recursion

Exercises for Section 4

4.1. Compute a_0, a_1, a_3 and a_4 for the following recursions. (Recall that $\lfloor x \rfloor$ is the greatest integer not exceeding x. For example $\lfloor 5.4 \rfloor = 5$ and $\lfloor -5.4 \rfloor = -6$.)

(a) $a_0 = 1$, $a_n = 3a_{n-1} - 2$ for $n \geq 1$.

(b) $a_0 = 0$, $a_n = \lfloor n/2 \rfloor + a_{n-1}$ for $n > 0$.

(c) $a_0 = 1$, $a_n = n + a_{\lfloor n/2 \rfloor}$ for $n > 0$.

(d) $a_0 = 0$, $a_1 = 1$, $a_n = 1 + \min(a_1 a_{n-1}, \ldots, a_k a_{n-k}, \ldots, a_{n-1} a_1)$ for $n > 1$.

4.2. We computed the first few values of some sequences that were defined by recursions. A table of values is given below. Guess simple formulas for each sequence.

$n:$	0	1	2	3	4	5	...
$a_n:$	0	0	1	1	2	2	...
$b_n:$	1	-1	2	-2	3	-3	...
$c_n:$	1	2	5	10	17	26	...
$d_n:$	1	1	2	6	24	120	...

4.3. What are the characteristic equations for the recursions $a_n = 6a_{n-1} - 5a_{n-2}$, $a_n = a_{n-1} + 2a_{n-2}$ and $a_n = 5(a_{n-1} + a_{n-2})$? What are the roots of these equations?

4.4. Solve the recursion $a_0 = 0$, $a_1 = 3$ and $a_n = 6a_{n-1} - 9a_{n-2}$ for $n > 2$.

4.5. Solve the recursion $a_2 = 1$, $a_3 = 3$ and $a_n = 3a_{n-1} + 2a_{n-2}$ for $n > 3$.

4.6. Solve the recursion $a_k = 2a_{k-1} - a_{k-2}$, $k \geq 2$, $a_0 = 2$, $a_1 = 1$.

4.7. Suppose $A \neq 1$. Let $G(n) = 1 + A + A^2 + \ldots + A^{n-1}$ for $n \geq 1$.

(a) Using induction, prove that $G(n) = (1 - A^n)/(1 - A)$ for $n \geq 1$. (This is the formula for the sum of a *geometric series*.)

(b) Obtain a simple recursion for $G(n)$ from $G(n) = 1 + A + A^2 + \ldots + A^{n-1}$, including initial conditions.

(c) Use the recursion in (b) and Theorem 7 to prove that $G(n) = (1 - A^n)/(1 - A)$ for $n \geq 1$.

(d) By setting $A = y/x$ and doing some algebra, prove that

$$\frac{x^{k+1} - y^{k+1}}{x - y} = x^k y^0 + x^{k-1} y^1 + \cdots + x^0 y^{k-1} \quad \text{when } x \neq y.$$

Section 4: Inductive Proofs and Recursive Equations

4.8. In each of the following, find an explicit formula for a_k that satisfies the given recursion. Prove your formula.

(a) $a_k = a_{k-1}/(1 + a_{k-1})$ for $k \geq 1$, $a_0 = A > 0$.

(b) $a_k = Aa_{k-1} + B$, $k \geq 1$, $a_0 = C$.

4.9. Consider $a_k = a_{k-1} + Bk(k-1)$, $k \geq 1$, $a_0 = A$. Prove that $a_k = A + Bk(k^2-1)/3$, $k \geq 0$, is the solution to this recursion.

4.10. Consider $a_k = A2^k - a_{k-1}$, $k \geq 1$, $a_0 = C$.

(a) Prove that

$$a_k = A(2^k(-1)^0 + 2^{k-1}(-1)^1 + \cdots + 2^1(-1)^{k-1}) + (-1)^k C, \quad k \geq 1,$$

is the solution to this recursion.

(b) Write the formula for a_k more compactly using Exercise 4.7.

4.11. A gambler has $t \geq 0$ dollars to start with. He bets one dollar each time a fair coin is tossed. If he wins Q, $Q \geq t$, dollars, he quits, a happy man. If he loses all his money he quits, a sad man. What is the probability q_t that he wins Q dollars instead of losing all his money and quits a happy man? What is the probability p_t that he loses all his money and quits a sad man (i.e., ruined)? This problem is called the *Gambler's Ruin* problem.

Decision Trees and Recursion

Multiple Choice Questions for Review

1. In each case, two permutations on $\underline{6}$ are listed. In which case is the first permutation less than the second in direct insertion order?

   (a) $2, 3, 1, 4, 5, 6$ $1, 3, 2, 4, 5, 6$

   (b) $2, 3, 1, 4, 5, 6$ $2, 1, 3, 4, 5, 6$

   (c) $2, 3, 1, 4, 5, 6$ $4, 5, 6, 1, 3, 2$

   (d) $6, 1, 2, 3, 4, 5$ $2, 1, 3, 4, 5, 6$

   (e) $6, 2, 3, 1, 4, 5$ $2, 3, 1, 4, 5, 6$

2. What is the rank, in direct insertion order, of the permutation $5, 4, 6, 3, 2, 1$?

   (a) 3   (b) 4   (c) 715   (d) 716   (e) 717

3. What is the rank, in lex order, of the permutation $6, 1, 2, 3, 4, 5$?

   (a) 20   (b) 30   (c) 480   (d) 600   (e) 619

4. Consider the list of all sequences of length six of A's and B's that satisfy the following conditions:

   (i) There are no two adjacent A's.

   (ii) There are never three B's adjacent.

   What is the next sequence after ABBABB in lex order?

   (a) ABABAB

   (b) ABBABA

   (c) BABABA

   (d) BABBAB

   (e) BBABBA

5. Which of the following 4×4 domino covers represent two distinct hibachi grills?

   (a) hhhhhvvh and hvvhhhhh

   (b) hvvhvvhh and vvhhvvhh

   (c) vhvvvhvh and hhvhvhvv

   (d) vvhhvvhh and hhhhvvvv

   (e) vvvvvvvv and hhhhhhhh

6. Given that $a_0 = 1$, $a_n = n + (-1)^n a_{n-1}$ for $n \geq 2$ What is the value of a_4?

   (a) 1   (b) 4   (c) 5   (d) 8   (e) 11

7. Given that $a_k = a_{k-1}/(1 + a_{k-1})$ for $k \geq 1$, $a_0 = 1$. Which of the following gives an explicit formula for a_k?

   (a) $1/3^k$, $k = 0, 1, 2, 3, \ldots$

DT-52   [140]

Review Questions

(b) $1/2^k$, $k = 0, 1, 2, 3, \ldots$

(c) $1/(3^{k+1} - 2)$, $k = 0, 1, 2, 3, \ldots$

(d) $1/(k+1)$, $k = 0, 1, 2, 3, \ldots$

(e) $2/(k+2)$, $k = 0, 1, 2, 3, \ldots$

8. Consider the recurrence relation $a_k = -8a_{k-1} - 15a_{k-2}$ with initial conditions $a_0 = 0$ and $a_1 = 2$. Which of the following is an explicit solution to this recurrence relation?

   (a) $a_k = (-3)^k - (-5)^k$

   (b) $a_k = k(-3)^k - k(-5)^k$

   (c) $a_k = k(-3)^k - (-5)^k$

   (d) $a_k = (-5)^k - (-3)^k$

   (e) $a_k = k(-5)^k - k(-3)^k$

9. Consider the recurrence relation $a_k = 6a_{k-1} - 9a_{k-2}$ with initial conditions $a_0 = 0$ and $a_1 = 2$. Which of the following is an explicit solution to this recurrence relation, provided the constants A and B are chosen correctly?

   (a) $a_n = A3^n + B3^n$

   (b) $a_n = A3^n + B(-3)^n$

   (c) $a_n = A3^n + nB3^n$

   (d) $a_n = A(-3)^n + nB(-3)^n$

   (e) $a_n = nA3^n + nB3^n$

10. In the Towers of Hanoi puzzle H(8, S, E, G), the configuration is

    Pole S: 6, 5;   Pole E: 1;   Pole G: 8,7,4,3,2.

    What move was just made to create this configuration?

    (a) washer 1 from S to E

    (b) washer 1 from G to E

    (c) washer 2 from S to G

    (d) washer 2 from E to G

    (e) washer 5 from G to S

11. In the Towers of Hanoi puzzle H(8, S, E, G), the configuration is

    Pole S: 6, 5;   Pole E: empty;   Pole G: 8, 7, 4 ,3 ,2, 1 .

    What are the next two moves?

    (a) washer 1 from G to E followed by washer 2 from G to S

    (b) washer 1 from G to S followed by washer 2 from G to E

    (c) washer 5 from S to E followed by washer 1 from G to E

Decision Trees and Recursion

(d) washer 5 from S to E followed by washer 1 from G to S

(e) washer 5 from S to E followed by washer 2 from G to S

12. In the Towers of Hanoi puzzle H(8, S, E, G), the configuration is

    Pole S: 6, 5, 2; Pole E: 1; Pole G: 8, 7, 4 ,3.

    The next move is washer 2 from S to G. What is the RANK of this move in the list of all moves for H(8, S, E, G)?

    (a) 205   (b) 206   (c) 214   (d) 215   (e) 216

13. In the subset Gray code for $n = 6$, what is the next element after 111000?

    (a) 000111
    (b) 101000
    (c) 111001
    (d) 111100
    (e) 101100

14. In the subset Gray code for $n = 6$, what is the element just before 110000?

    (a) 010000
    (b) 100000
    (c) 110001
    (d) 110100
    (e) 111000

15. In the subset Gray code for $n = 6$, what is the RANK of 110000?

    (a) 8   (b) 16   (c) 32   (d) 48   (e) 63

16. In the subset Gray code for $n = 6$, what is the element of RANK 52?

    (a) 101011
    (b) 101110
    (c) 101101
    (d) 110000
    (e) 111000

17. The probability of team A winning any game is 1/3. Team A plays team B in a tournament. If either team wins two games in a row, that team is declared the winner. At most three games are played in the tournament and, if no team has won the tournament at the end of three games, the tournament is declared a draw. What is the expected number of games in the tournament?

    (a) 3   (b) 19/9   (c) 22/9   (d) 25/9   (e) 61/27

18. The probability of team A winning any game is 1/2. Team A plays team B in a tournament. If either team wins two games in a row, that team is declared the winner. At

most four games are played in the tournament and, if no team has won the tournament at the end of four games, the tournament is declared a draw. What is the expected number of games in the tournament?

(a) 4   (b) 11/4   (c) 13/4   (d) 19/4   (e) 21/8

19. A man starts with one dollar in a pot. A "play" consists of flipping a fair coin and,
 - if heads occurs, doubling the amount in the pot,
 - if tails occurs, losing one dollar from the pot. The game ends if the man has zero dollars or if he has played three times. Let Y denote the random variable which, for each outcome of the game, specifies the amount of money in the pot. What is the value of Var(Y)?

(a) 9/8   (b) 10/8   (c) 12/8   (d) 14/8   (e) 447/64

20. We are given an urn that has one red ball and one white ball. A fair die is thrown. If the number is a 1 or 2, one red ball is added to the urn. Otherwise two red balls are added to the urn. A ball is then drawn at random from the urn. Given that a red ball was drawn, what is the probability that a 1 or 2 appeared when the die was thrown?

(a) 4/13   (b) 5/13   (c) 6/13   (d) 7/13   (e) 8/13

21. In a certain college,
 - 10 percent of the students are science majors.
 - 10 percent are engineering majors.
 - 80 percent are humanities majors.
 - Of the science majors, 20 percent have read Newsweek.
 - Of the engineering majors, 10 percent have read Newsweek.
 - Of the humanities majors, 20 percent have read Newsweek.

Given that a student selected at random has read Newsweek, what is the probability that that student is an engineering major?

(a) 1/19   (b) 2/19   (c) 5/19   (d) 9/19   (e) 10/19

22. The probability of team A winning any game is 1/3. Team A plays team B in a tournament. If either team wins two games in a row, that team is declared the winner. At most *four* games are played and, if no team has won the tournament at the end of four games, a draw is declared. Given that the tournament lasts more than two games, what is the probability that A is the winner?

(a) 1/9   (b) 2/9   (c) 4/9   (d) 5/9   (e) 6/9

23. Ten percent of the students are science majors (S), 20 percent are engineering majors (E), and 70 percent are humanities majors (H). Of S, 10 percent have read 2 or more articles in Newsweek, 20 percent 1 article, 70 percent 0 articles. For E, the corresponding percents are 5, 15, 80. For H they are 20, 30, 50. Given that a student has read 0 articles in Newsweek, what is the probability that the student is S or E (i.e., not H)?

(a) 21/58   (b) 23/58   (c) 12/29   (d) 13/29   (e) 1/2

[143]

Decision Trees and Recursion

Answers: **1** (d), **2** (e), **3** (d), **4** (c), **5** (b), **6** (c), **7** (d), **8** (a), **9** (c), **10** (c), **11** (d), **12** (a), **13** (b), **14** (a), **15** (c), **16** (b), **17** (c), **18** (b), **19** (e), **20** (a), **21** (a), **22** (b), **23** (b).

Unit GT

Basic Concepts in Graph Theory

Section 1: What is a Graph?

There are various types of graphs, each with its own definition. Unfortunately, some people apply the term "graph" rather loosely, so you can't be sure what type of graph they're talking about unless you ask them. After you have finished this chapter, we expect you to use the terminology carefully, not loosely. To motivate the various definitions, we'll begin with some examples.

Example 1 (A computer network) Computers are often linked with one another so that they can interchange information. Given a collection of computers, we would like to describe this linkage in fairly clean terms so that we can answer questions such as "How can we send a message from computer A to computer B using the fewest possible intermediate computers?"

We could do this by making a list that consists of pairs of computers that are connected. Note that these pairs are unordered since, if computer C can communicate with computer D, then the reverse is also true. (There are sometimes exceptions to this, but they are rare and we will assume that our collection of computers does not have such an exception.) Also, note that we have implicitly assumed that the computers are distinguished from each other: It is insufficient to say that "A PC is connected to a Mac." We must specify which PC and which Mac. Thus, each computer has a unique identifying label of some sort.

For people who like pictures rather than lists, we can put dots on a piece of paper, one for each computer. We label each dot with a computer's identifying label and draw a curve connecting two dots if and only if the corresponding computers are connected. Note that the shape of the curve does not matter (it could be a straight line or something more complicated) because we are only interested in whether two computers are connected or not. Below are two such pictures of the same graph. Each computer has been labeled by the initials of its owner.

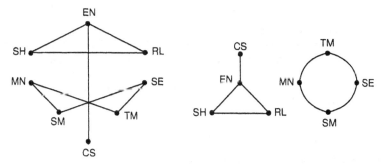

Computers (vertices) are indicated by dots (•) with labels. The connections (edges) are indicated by lines. When lines cross, they should be thought of as cables that lie on top of each other — not as cables that are joined. ☐

Basic Concepts in Graph Theory

The notation $\mathcal{P}_k(V)$ stands for the set of all k-element subsets of the set V. Based on the previous example we have

Definition 1 (Simple graph) *A simple graph G is a pair $G = (V, E)$ where*

- *V is a finite set, called the vertices of G, and*
- *E is a subset of $\mathcal{P}_2(V)$ (i.e., a set E of two-element subsets of V), called the edges of G.*

In our example, the vertices are the computers and a pair of computers is in E if and only if they are connected.

Example 2 (Routes between cities) Imagine four cities named, with characteristic mathematical charm, A, B, C and D. Between these cities there are various routes of travel, denoted by a, b, c, d, e, f and g. Here is picture of this situation:

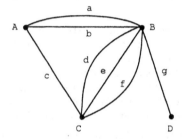

Looking at this picture, we see that there are three routes between cities B and C. These routes are named d, e and f. Our picture is intended to give us only information about the interconnections between cities. It leaves out many aspects of the situation that might be of interest to a traveler. For example, the nature of these routes (rough road, freeway, rail, etc.) is not portrayed. Furthermore, unlike a typical map, no claim is made that the picture represents in any way the distances between the cities or their geographical placement relative to each other. The object shown in this picture is called a *graph*.

Following our previous example, one is tempted to list the pairs of cities that are connected; in other words, to extract a simple graph from the information. Unfortunately, this does not describe the problem adequately because there can be more than one route connecting a pair of cities; e.g., d, e and f connecting cities B and C in the figure. How can we deal with this? Here is a precise definition of a graph of the type required to handle this type of problem. ☐

Definition 2 (Graph) *A graph is a triple $G = (V, E, \phi)$ where*

- *V is a finite set, called the vertices of G,*
- *E is a finite set, called the edges of G, and*
- *ϕ is a function with domain E and codomain $\mathcal{P}_2(V)$.*

Section 1: What is a Graph?

In the pictorial representation of the cities graph, $G = (V, E, \phi)$ where

$$V = \{A, B, C, D\}, \quad E = \{a, b, c, d, e, f, g\}$$

and

$$\phi = \begin{pmatrix} a & b & c & d & e & f & g \\ \{A,B\} & \{A,B\} & \{A,C\} & \{B,C\} & \{B,C\} & \{B,C\} & \{B,D\} \end{pmatrix}.$$

Definition 2 tells us that to specify a graph G it is necessary to specify the sets V and E and the function ϕ. We have just specified V and ϕ in *set theoretic* terms. The picture of the cities graph specifies the same V and ϕ in pictorial terms. The set V is represented clearly by dots ($\bullet$), each of which has a city name adjacent to it. Similarly, the set E is also represented clearly. The function ϕ is determined from the picture by comparing the name attached to a route with the two cities connected by that route. Thus, the route name d is attached to the route with endpoints B and C. This means that $\phi(d) = \{B, C\}$.

Note that, since part of the definition of a function includes its codomain and domain, ϕ determines $\mathcal{P}_2(V)$ and E. Also, V can be determined from $\mathcal{P}_2(V)$. Consequently, we could have said that a graph is a function ϕ whose domain is a finite set and whose codomain is $\mathcal{P}_2(V)$ for some finite set V. Instead, we choose to specify V and E explicitly because the vertices and edges play a fundamental role in thinking about a graph G.

The function ϕ is sometimes called the *incidence function* of the graph. The two elements of $\phi(x) = \{u, v\}$, for any $x \in E$, are called the vertices of the edge x, and we say u and v are *joined* by x. We also say that u and v are *adjacent vertices* and that u is *adjacent to v* or , equivalently, v is adjacent to u. For any $v \in V$, if v is a vertex of an edge x then we say x is *incident* on v. Likewise, we say v is a member of x, v is on x, or v is in x. Of course, v is a member of x actually means v is a member of $\phi(x)$.

Here are two additional pictures of the same cities graph given above:

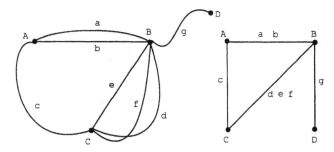

The drawings look very different but exactly the same set V and function ϕ are specified in each case. It is *very important* that you understand exactly what information is needed to completely specify the graph. When thinking in terms of cities and routes between them, you naturally want the pictorial representation of the cities to represent their geographical positioning also. If the pictorial representation does this, that's fine, but it is not a part of the information required to define a graph. Geographical location is extra information. The geometrical positioning of the vertices A, B, C and D is very different, in the first of the two pictorial representations above, than it was in our original representation of the cities. However, in each of these cases, the vertices on a given edge are the same and hence the

Basic Concepts in Graph Theory

graphs specified are the same. In the second of the two pictures above, a different method of specifying the graph is given. There, ϕ^{-1}, the inverse of ϕ, is given. For example, $\phi^{-1}(\{C, B\})$ is shown to be $\{d, e, f\}$. Knowing ϕ^{-1} determines ϕ and hence determines G since the vertices A, B, C and D are also specified.

Example 3 (Loops) A *loop* is an edge that connects a vertex to itself. Graphs and simple graphs as defined in Definitions 1 and 2 cannot have loops. Why? Suppose $e \in E$ is a loop in a graph that connects $v \in V$ to itself. Then $\phi(e) = \{v, v\} = \{v\}$ because repeated elements in the description of a set count only once — they're the same element. Since $\{v\} \notin \mathcal{P}_2(V)$, the range of ϕ, we cannot have $\phi(e) = \{v, v\}$. In other words, we cannot have a loop.

Thus, if we want to allow loops, we will have to change our definitions. For a graph, we expand the codomain of ϕ to be $\mathcal{P}_2(V) \cup \mathcal{P}_1(V)$. For a simple graph we need to change the set of allowed edges to include loops. This can be done by saying that E is a subset of $\mathcal{P}_2(V) \cup \mathcal{P}_1(V)$ instead of a subset of just $\mathcal{P}_2(V)$. For example, if $V = \{1, 2, 3\}$ and $E = \{\{1, 2\}, \{2\}, \{2, 3\}\}$, this simple graph has a loop at vertex 2 and vertex 2 is connected by edges to the other two vertices. When we want to allow loops, we speak of a graph with loops or a simple graph with loops.

Examples of graphs with loops appear in the exercises. □

We have two definitions, Definition 1 (simple graph) and Definition 2 (graph). How are they related? Let $G = (V, E)$ be a simple graph. Define $\phi: E \to E$ to be the identity map; i.e., $\phi(e) = e$ for all $e \in E$. The graph $G' = (V, E, \phi)$ is essentially the same as G. There is one subtle difference in the pictures: The edges of G are unlabeled but each edge of G' is labeled by a set consisting of the two vertices at its ends. But this extra information is contained already in the specification of G. Thus, simple graphs are a special case of graphs.

Definition 3 (Degrees of vertices) Let $G = (V, E, \phi)$ be a graph and $v \in V$ a vertex. Define the degree of v, $d(v)$ to be the number of $e \in E$ such that $v \in \phi(e)$; i.e., e is incident on v.
Suppose $|V| = n$. Let $d_1, d_2, \ldots, d_n$, where $d_1 \leq d_2 \leq \cdots \leq d_n$ be the sequence of degrees of the vertices of G, sorted by size. We refer to this sequence as the *degree sequence* of the graph G.

In the graph for routes between cities, $d(A) = 3$, $d(B) = 6$, $d(C) = 4$, and $d(D) = 1$. The degree sequence is 1,3,4,6.

Sometimes we are interested only in the "structure" or "form" of a graph and not in the names (labels) of the vertices and edges. In this case we are interested in what is called an unlabeled graph. A picture of an unlabeled graph can be obtained from a picture of a graph by erasing all of the names on the vertices and edges. This concept is simple enough, but is difficult to use mathematically because the idea of a picture is not very precise.

The concept of an *equivalence relation* on a set is an important concept in mathematics and computer science. We'll explore it here and will use it to develop an intuitive

Section 1: What is a Graph?

understanding of unlabeled graphs. Later we will use it to define connected components and biconnected components. Equivalence relations are discussed in more detail in *A Short Course in Discrete Mathematics*, the text for the course that precedes this course.

Definition 4 (Equivalence relation) *An equivalence relation on a set S is a partition of S. We say that $s, t \in S$ are equivalent if and only if they belong to the same block (called an equivalence class in this context) of the partition. If the symbol $\sim$ denotes the equivalence relation, then we write $s \sim t$ to indicate that s and t are equivalent.*

Example 4 (Equivalence relations) Let S be any set and let all the blocks of the partition have one element. Two elements of S are equivalent if and only if they are the same. This rather trivial equivalence relation is, of course, denoted by "=".

Now let the set be the integers $\mathbb{Z}$. Let's try to define an equivalence relation by saying that n and k are equivalent if and only if they differ by a multiple of 24. Is this an equivalence relation? If it is we should be able to find the blocks of the partition. There are 24 of them, which we could number $0, \ldots, 23$. Block j consists of all integers which equal j plus a multiple of 24; that is, they have a remainder of j when divided by 24. Since two numbers belong to the same block if and only if they both have the same remainder when divided by 24, it follows that they belong to the same block if and only if their difference gives a remainder of 0 when divided by 24, which is the same as saying their difference is a multiple of 24. Thus this partition does indeed give the desired equivalence relation.

Now let the set be $\mathbb{Z} \times \mathbb{Z}^*$, where $\mathbb{Z}^*$ is the set of all integers except 0. Write $(a, b) \sim (c, d)$ if and only if $ad = bc$. With a moment's reflection, you should see that this is a way to check if the two fractions a/b and c/d are equal. We can label each equivalence class with the fraction a/b that it represents. In an axiomatic development of the rationals from the integers, one defines a rational number to be just such an equivalence class and proves that it is possible to add, subtract, multiply and divide equivalence classes.

Suppose we consider all functions $S = \underline{m}^{\underline{n}}$. We can define a partition of S in a number of different ways. For example, we could partition the functions f into blocks where the sum of the integers in the Image(f) is constant, where the max of the integers in Image(f) is constant, or where the "type vector" of the function, namely, the number of 1's, 2's, etc. in Image(f), is constant. Each of these defines a partition of S. □

In the next theorem we provide necessary and sufficient conditions for an equivalence relation. Verifying the conditions is a useful way to prove that some particular situation is an equivalence relation. Recall that a *binary relation* on a set S is a subset R of $S \times S$.

Theorem 1 (Equivalence Relations) *Let S be a set and suppose that we have a binary relation $R \subseteq S \times S$. We write $s \sim t$ whenever $(s, t) \in R$. This is an equivalence relation if and only if the following three conditions hold.*

(i) *(Reflexive) For all $s \in S$ we have $s \sim s$.*

(ii) *(Symmetric) For all $s, t \in S$ such that $s \sim t$ we have $t \sim s$.*

(iii) *(Transitive) For all $r, s, t \in S$ such that $r \sim s$ and $s \sim t$ we have $r \sim t$.*

Basic Concepts in Graph Theory

Proof: We first prove that an equivalence relation satisfies (i)–(iii). Suppose that $\sim$ is an equivalence relation. Since s belongs to whatever block it is in, we have $s \sim s$. Since $s \sim t$ means that s and t belong to the same block, we have $s \sim t$ if and only if we have $t \sim s$. Now suppose that $r \sim s \sim t$. Then r and s are in the same block and s and t are in the same block. Thus r and t are in the same block and so $r \sim t$.

We now suppose that (i)–(iii) hold and prove that we have an equivalence relation. What would the blocks of the partition be? Everything equivalent to a given element should be in the same block. Thus, for each $s \in S$ let $B(s)$ be the set of all $t \in S$ such that $s \sim t$. We must show that the set of these sets form a partition of S.

In order to have a partition of S, we must have

(a) the $B(s)$ are nonempty and every $t \in S$ is in some $B(s)$ and

(b) for every $p, q \in S$, $B(p)$ and $B(q)$ are either equal or disjoint.

Since $\sim$ is reflexive, $s \in B(s)$, proving (a). Suppose $x \in B(p) \cap B(q)$ and $y \in B(p)$. We have, $p \sim x$, $q \sim x$ and $p \sim y$. Thus $q \sim x \sim p \sim y$ and so $y \in B(q)$, proving that $B(p) \subseteq B(q)$. Similarly $B(q) \subseteq B(p)$ and so $B(p) = B(q)$. This proves (b). □

Example 5 (Equivalent forms) Consider the following two graphs, represented by pictures:

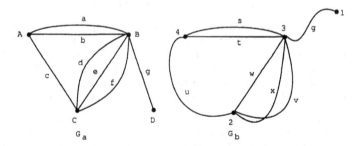

Now let's remove all symbols representing edges and vertices. What we have left are two "forms" on which the graphs were drawn. You can think of drawing a picture of a graph as a two step process: (1) draw the form; (2) add the labels. One student referred to these forms as "ghosts of departed graphs." Note that form F_a and form F_b have a certain eerie similarity (appropriate for ghosts).

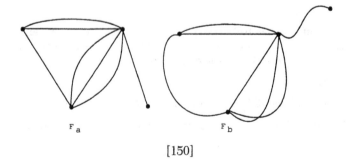

Section 1: What is a Graph?

If you use your imagination a bit you can see that form F_b can be transformed into form F_a by sliding vertices around and bending, stretching, and contracting edges as needed. The edges need not be detached from their vertices in the process and edges and vertices, while being moved, can pass through each other like shadows. Let's refer to the sliding, bending, stretching, and contracting process as "morphing" form F_a into F_b. Morphing is easily seen to define an equivalence relation $\sim$ on the set of all forms. Check out reflexive, symmetric, and transitive, for the morphing relation $\sim$. By Theorem 1, the morphing equivalence relation partitions the set of all forms of graphs into blocks or equivalence classes. This is a good example where it is easier to think of the relation $\sim$ than to think globally of the partition of the forms.

Now suppose we have any two graphs, $G_a = (V_a, E_a, \phi_a)$ and $G_b = (V_b, E_b, \phi_b)$. Think of these graphs not as pictures, but as specified in terms of sets and functions. Now choose forms F_a and F_b for G_a and G_b respectively, and draw their pictures. We leave it to your intuition to accept the fact that either $F_a \sim F_b$, no matter what you choose for F_a and F_b, or $F_a \not\sim F_b$ no matter what your choice is for the forms F_a and F_b. If $F_a \sim F_b$ we say that G_a and G_b are *isomorphic graphs* and write $G_a \approx G_b$. The fact that $\sim$ is an equivalence relation forces $\approx$ to be an equivalence relation also. In particular, two graphs G_a and G_b are isomorphic if and only if you can choose any form F_a for drawing G_a and use that same form for G_b. □

In general, deciding whether or not two graphs are isomorphic can be very difficult business. You can imagine how hard it would be to look at the forms of two graphs with thousands of vertices and edges and deciding whether or not those forms are equivalent. There are no general computer programs that do the task of deciding isomorphism well. For graphs known to have special features, isomorphism of graphs can sometimes be decided efficiently. In general, if someone presents you with two graphs and asks you if they are isomorphic, your best answer is "no." You will be right most of the time.

---

*Random Graphs

We now look briefly at a subject called *random graphs*. They often arise in the analysis of graphical algorithms and of systems which can be described graphically (such as the web). There are two basic ways to describe random graphs. One is to let the probability space be the set of all graphs with, for example, n vertices and q edges and use the uniform distribution. The other, which is often easier to study is described in the following definition. It is the one we study here.

Definition 5 (Random graph model) Let $\mathcal{G}(n,p)$ be the probability space obtained by letting the elementary events be the set of all n-vertex simple graphs with $V = \underline{n}$. If $G \in \mathcal{G}(n,p)$ has m edges, the $P(G) = p^m q^{N-m}$ where $q = 1-p$ and $N = \binom{n}{2}$.

We need to show that $\mathcal{G}(n,p)$ is a probability space. There is a nice way to see this by reinterpreting P. List the $N = \binom{n}{2}$ vertices $\mathcal{P}_2(V)$ in lex order. Let the sample space be $U = \times^N \{\text{choose, reject}\}$ with $P(a_1, \ldots, a_N) = P^*(a_1) \times \cdots \times P^*(a_N)$ where $P^*(\text{choose}) = p$

Basic Concepts in Graph Theory

and $P^*(\text{reject}) = 1 - p$. We've met this before in Unit Fn and seen that it is a probability space. To see that it is, note that $P \geq 0$ and

$$\sum_{a_1,\ldots,a_N} P(a_1,\ldots,a_N) = \sum_{a_1,\ldots,a_N} P^*(a_1) \times \cdots \times P^*(a_N)$$

$$= \left(\sum_{a_1} P^*(a_1)\right) \times \cdots \times \left(\sum_{a_N} P^*(a_N)\right)$$

$$= (p + (1-p)) \times \cdots \times (p + (1-p)) = 1 \times \cdots \times 1 = 1.$$

Why is this the same as the definition? Think of the chosen pairs as the edges of a graph chosen randomly from $\mathcal{G}(n,p)$. If G has m edges, then its probability should be $p^m(1-p)^{N-m}$ according to the definition. On the other hand, since G has m edges, exactly m of $a_1,\ldots,a_N$ equal "choose" and so, in the new space, $P(a_1,\ldots,a_N) = p^m(1-p)^{N-m}$ also. We say that we are choosing the edges of the random graph independently.

*Example 6 (The number of edges in random graph) Let X be a random variable that counts the number of edges in a random graph in $\mathcal{G}(n,p)$. What are the expected value and variance of X? In $U = \times^N \{\text{choose}, \text{reject}\}$, let

$$X_i(a_1,\ldots,a_N) = \begin{cases} 1 & \text{if } a_i = \text{choose}, \\ 0 & \text{if } a_i = \text{reject}. \end{cases}$$

You should be able to see that $X = X_1 + \cdots + X_N$ and that the X_i are independent random variables with $P(X_i = 1) = p$. This is just the binomial distribution (Unit Fn). We showed that the mean is Np and the variance is Npq, where $N = \binom{n}{2}$ and $q = 1 - p$. □

*Example 7 (Triangles in random graphs) How often can we find 3 vertices $\{u, v, w\}$ in a random graph so that $\{u,v\}$, $\{u,w\}$, and $\{v,w\}$ are all edges of the graph? In other words, how often can we find a "triangle"? How can we do this?

First, we need a sample space. It will be the random graph space introduced in Definition 5. Since we want to count something (triangles), we need a random variable. Let X be a random variable whose value is the number of triples of vertices such that the three possible edges connecting them are present in the random graph. In other words, X is defined for each graph, G, and its value, $X(G)$, is the number of triangles in the graph G. We want to compute $E(X)$. It would also be nice to compute $\text{Var}(X)$ since that gives us some idea of how much X tends to vary from graph to graph — large $\text{Var}(X)$ means there tends to be a lot of variation in the number of triangles from graph to graph and small $\text{Var}(X)$ means there tends to be little variation.

Let $X_{u,v,w}$ be a random variable which is 1 if the triangle with vertices $\{u,v,w\}$ is present and 0 otherwise. Then X is the sum of $X_{u,v,w}$ over all $\{u,v,w\} \in \mathcal{P}_3(V)$. Since expectation is linear, $E(X)$ is the sum of $E(X_{u,v,w})$ over all $\{u,v,w\} \in \mathcal{P}_3(V)$. Clearly $E(X_{u,v,w})$ does not depend on the particular triple. Since there are $\binom{n}{3}$ possibilities for $\{u,v,w\}$, $E(X) = \binom{n}{3} E(X_{1,2,3})$.

We want to compute $E(X_{1,2,3})$. It is given by

$$E(X_{1,2,3}) = 0 P(X_{1,2,3} = 0) + 1 P(X_{1,2,3} = 1) = P(X_{1,2,3} = 1).$$

Section 1: What is a Graph?

The only way $X_{1,2,3} = 1$ can happen is for the edges $\{1,2\}$, $\{1,3\}$, and $\{2,3\}$ to all be present in the graph. (We don't care about any of the other possible edges.) Since each of these events has probability p and the events are independent we have $P(X_{1,2,3} = 1) = p^3$. Thus $E(X_{1,2,3}) = p^3$ and so $E(X) = \binom{n}{3}p^3$. In other words, on average we see about $\binom{n}{3}p^3$ triangles. For example, if $p = 1/2$ all graphs are equally likely (You should show this.) and so the average number of triangles over all graphs with n vertices is $\binom{n}{3}/8$. When $n = 5$, this average is 1.25. Can you verify this by looking at all the 5-vertex graphs? How much work is involved?

What happens when n is very large? Then $\binom{n}{3} = \frac{n(n-1)(n-2)}{6}$ "behaves like" $n^3/6$. ("Behaves like" means that, as n goes to infinity, the limit of the ratio $\binom{n}{3}/(n^3/6)$ is 1.) Thus the expected number of triangles behaves like $(np)^3/6$.

What about the variance? We'll work it out in the next example. For now, we'll simply tell you that it behaves like $n^4p^3(1-p^2)/2$. What does this tell us for large n? The standard deviation behaves like $n^2p^{3/2}\sqrt{(1-p^2)/2}$. A more general version of the central limit theorem than we have discussed tells us the number of triangles tends to have a normal distribution with $\mu = (np)^3/6$ and $\sigma = n^2p^{3/2}\sqrt{(1-p^2)/2}$. If p is constant, σ will grow like a constant times n^2, which is much smaller than μ for large n. Thus the number of triangles in a random graph is almost surely close to $(np)^3/6$. $\square$

*Example 8 (Variance for triangles in random graphs) This is a continuation of the previous example. Since the various $X_{u,v,w}$ may not be independent, this is harder. Since $\mathrm{Var}(X) = E(X^2) - E(X)^2$, we will compute $E(X^2)$. Since X is a sum of terms of the form $X_{r,s,t}$, X^2 is a sum of terms of the form $X_{u,v,w}X_{a,b,c}$. Using linearity of expectation, we need to compute $E(X_{u,v,w}X_{a,b,c})$ for each possibility and add them up.

Now for the tricky part: This expectation depends on how many vertices $\{u,v,w\}$ and $\{a,b,c\}$ have in common.

- If $\{u,v,w\} = \{a,b,c\}$, then $X_{u,v,w}X_{a,b,c} = X_{u,v,w}$ and its expectation is p^3 by the previous example.

- If $\{u,v,w\}$ and $\{a,b,c\}$ have two vertices in common, then the two triangles have only 5 edges total because they have a common edge. Note that $X_{u,v,w}X_{a,b,c}$ is 1 if all five edges are present and is 0 otherwise. Reasoning as in the previous example, the expectation is p^5.

- If $\{u,v,w\}$ and $\{a,b,c\}$ have less than two vertices in common, we are concerned about six edges and obtain p^6 for the expectation.

To add up the results in the previous paragraph, we need to know how often each occurs in

$$X^2 = \left(\sum_{\{u,v,w\}\in\mathcal{P}_3(V)} X_{u,v,w}\right)\left(\sum_{\{a,b,c\}\in\mathcal{P}_3(V)} X_{a,b,c}\right) = \sum_{\substack{\{u,v,w\}\in\mathcal{P}_3(V)\\ \{a,b,c\}\in\mathcal{P}_3(V)}} X_{u,v,w}X_{a,b,c}.$$

- When $\{u,v,w\} = \{a,b,c\}$, we are only free to choose $\{u,v,w\}$ and this can be done in $\binom{n}{3}$ ways so we have $\binom{n}{3}p^3$ contributed to $E(X^2)$.

Basic Concepts in Graph Theory

- Suppose $\{u, v, w\}$ and $\{a, b, c\}$ have two vertices in common. How many ways can this happen? We can first choose $\{u, v, w\}$. Then choose two of u, v, w to be in $\{a, b, c\}$ and then choose the third vertex in $\{a, b, c\}$ to be different from $u, v,$ and w. This can be done in

$$\binom{n}{3} \times \binom{3}{2} \times (n-3) = 3(n-3)\binom{n}{3} = 12\binom{n}{4}$$

ways. Multiplying this by p^5 gives its contribution to $E(X^2)$.

- The remaining case, one vertex or no vertices in common, can be done in a similar fashion. Alternatively, we can simply subtract the above counts from all possible ways of choosing $\{u, v, w\}$ and $\{a, b, c\}$. This gives us

$$\binom{n}{3} \times \binom{n}{3} - \binom{n}{3} - 12\binom{n}{4}$$

for the third case. Multiplying this by p^6 gives its contribution to $E(X^2)$.

Since $E(X)^2 = \binom{n}{3}^2 p^6$, we have that

$$\mathrm{Var}(X) = E(X)^2 - E(X^2) = \binom{n}{3}(p^3 - p^6) + 12\binom{n}{4}(p^3 - p^5),$$

after a bit of algebra using the results in the preceding paragraph. Whew! $\square$

The previous material would be more difficult if we had used the model for random graphs that was suggested before Definition 5. Why is this? The model we are using lets us ignore possible edges that we don't care about. The other model does not because we must be sure that the total number of edges is correct.

Exercises for Section 1

1.1. We are interested in the number of simple graphs with $V = \underline{n}$.

  (a) Prove that there are $2^{\binom{n}{2}}$ (2 to the power $\binom{n}{2}$) such simple graphs.

  (b) How many of them have exactly q edges?

1.2. Let (V, E, ϕ) be a graph and let $d(v)$ be the degree of the vertex $v \in V$. Prove that $\sum_{v \in V} d(v) = 2|E|$, an even number. Conclude that the number of vertices v for which $d(v)$ is odd is even.

1.3. Let $Q = (V, E, \phi)$ be a graph where

$$V = \{A, B, C, D, E, F, G, H\}, \quad E = \{a, b, c, d, e, f, g, h, i, j, k\}$$

GT-10 [154]

Section 1: What is a Graph?

and
$$\phi = \begin{pmatrix} a & b & c & d & e & f & g & h & i & j & k \\ A & A & D & E & A & E & B & F & G & C & A \\ B & D & E & B & B & G & F & G & C & C & A \end{pmatrix}.$$

In this representation of ϕ, the first row specifies the edges and the two vertices below each edge specify the vertices incident on that edge. Here is a pictorial representation $P(Q)$ of this graph.

$P(Q)$: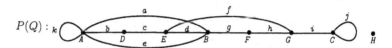

Note that $\phi(k) = \{A, A\} = \{A\}$. Such an edge is called a *loop*. (See Example 3.) Adding a loop to a vertex increases its degree by two. The vertex H, which does not belong to $\phi(x)$ for any edge x (i.e., has no edge incident upon it), is called an *isolated* vertex. The degree of an isolated vertex is zero. Edges, such as a and e of Q, with the property that $\phi(a) = \phi(e)$ are called *parallel* edges. If all edge and vertex labels are removed from $P(Q)$ then we get the following picture $P'(Q)$:

$P'(Q)$:

The picture $P'(Q)$ represents the "form" of the graph just described and is sometimes referred to as a pictorial representation of the "unlabeled" graph associated with Q. (See Example 5.) For each of the following graphs R, where $R = (V, E, \phi)$, $V = \{A, B, C, D, E, F, G, H\}$, draw a pictorial representation of R by starting with $P'(Q)$ removing and/or adding as few edges as possible and then labeling the resulting picture with the edges and vertices of R. A graph R which require no additions or removals of edges is said to be "of the same form as" or "isomorphic to" the graph Q (Example 5).

(a) Let
$$E = \{a, b, c, d, e, f, g, h, i, j, k\}$$
be the set of edges of R and
$$\phi = \begin{pmatrix} a & b & c & d & e & f & g & h & i & j & k \\ C & C & F & A & H & E & E & A & D & A & A \\ C & G & G & H & H & H & F & H & G & D & F \end{pmatrix}.$$

(b) Let
$$E = \{1, 2, 3, 4, 5, 6, 7, 8, 9, 10, 11\}$$
be the set of edges of R and
$$\phi = \begin{pmatrix} 1 & 2 & 3 & 4 & 5 & 6 & 7 & 8 & 9 & 10 & 11 \\ A & E & E & E & F & G & H & B & C & D & E \\ G & H & E & F & G & H & B & C & D & D & H \end{pmatrix}.$$

[155]

GT-11

Basic Concepts in Graph Theory

1.4. Let $Q = (V, E, \phi)$ be the graph where

$$V = \{A, B, C, D, E, F, G, H\}, \quad E = \{a, b, c, d, e, f, g, h, i, j, k, l\}$$

and

$$\phi = \begin{pmatrix} a & b & c & d & e & f & g & h & i & j & k & l \\ A & A & D & E & A & E & B & F & G & C & A & E \\ B & D & E & B & B & G & F & G & C & C & A & G \end{pmatrix}.$$

(a) What is the degree sequence of Q?

Consider the following unlabeled pictorial representation of Q

$P'(Q)$:

(a) Create a pictorial representation of Q by labeling $P'(Q)$ with the edges and vertices of Q.

(b) A necessary condition that a pictorial representation of a graph R can be created by labeling $P'(Q)$ with the vertices and edges of R is that the degree sequence of R be $(0, 2, 2, 3, 4, 4, 4, 5)$. True of false? Explain.

(c) A sufficient condition that a pictorial representation of a graph R can be created by labeling $P'(Q)$ with the vertices and edges of R is that the degree sequence of R be $(0, 2, 2, 3, 4, 4, 4, 5)$. True or false? Explain.

1.5. In each of the following problems information about the degree sequence of a graph is given. In each case, decide if a graph satisfying the specified conditions exists or not. Give reasons in each case.

(a) A graph Q with degree sequence $(1, 1, 2, 3, 3, 5)$?

(b) A graph Q with degree sequence $(1, 2, 2, 3, 3, 5)$, loops and parallel edges allowed?

(c) A graph Q with degree sequence $(1, 2, 2, 3, 3, 5)$, no loops but parallel edges allowed?

(d) A graph Q with degree sequence $(1, 2, 2, 3, 3, 5)$, no loops or parallel edges allowed?

(e) A simple graph Q with degree sequence $(3, 3, 3, 3)$?

(f) A graph Q with degree sequence $(3, 3, 3, 3)$, no loops or parallel edges allowed?

(g) A graph Q with degree sequence $(3, 3, 3, 5)$, no loops or parallel edges allowed?

(h) A graph Q with degree sequence $(4, 4, 4, 4, 4)$, no loops or parallel edges allowed?

(i) A graph Q with degree sequence $(4, 4, 4, 4, 6)$, no loops or parallel edges allowed?

1.6. Divide the following graphs into isomorphism equivalence classes and justify your answer; i.e., explain why you have the classes that you do. In all cases $V = \underline{4}$.

(a) $\phi = \begin{pmatrix} a & b & c & d & e & f \\ \{1,2\} & \{1,2\} & \{2,3\} & \{3,4\} & \{1,4\} & \{2,4\} \end{pmatrix}$

(b) $\phi = \begin{pmatrix} A & B & C & D & E & F \\ \{1,2\} & \{1,4\} & \{1,4\} & \{1,2\} & \{2,3\} & \{3,4\} \end{pmatrix}$

(c) $\phi = \begin{pmatrix} u & v & w & x & y & z \\ \{2,3\} & \{1,3\} & \{3,4\} & \{1,4\} & \{1,2\} & \{1,2\} \end{pmatrix}$

(d) $\phi = \begin{pmatrix} P & Q & R & S & T & U \\ \{3,4\} & \{2,4\} & \{1,3\} & \{3,4\} & \{1,2\} & \{1,2\} \end{pmatrix}$

*1.7. In Example 7, suppose that p is a function of n, say $p = p(n)$.

(a) Show that the expected number of triangles is behaves like 1 for large n if $p(n) = 6^{1/3}/n$.

(b) Suppose the expected number of triangles behaves like 1. How does the expected number of edges behave?

*1.8. Instead of looking for triangles as in Example 7, let's look for quadrilaterals having both diagonals. In other words, we'll look for sets of four vertices such that all of the $\binom{4}{2} = 6$ possible edges between them are present.

(a) Show that the expected number of such quadrilaterals is $\binom{n}{4}p^6$.

(b) Suppose n is large and p is a function of n so that we expect to see 1 quadrilateral on average. About how many edges do we expect to see?

(c) Generalize this problem from sets of 4 vertices to sets of k vertices.

*1.9. Show that the variance of X, the number of triangles in a random graph as computed in Example 8 satisfies

$$\text{Var}(X) = \binom{n}{3}p^3\left((1-p^3) + 3(n-3)(1-p^2)\right) < 3n\binom{n}{3}p^3(1-p^2).$$

Hint: $1 - p^3 < 1 - p^2 < 1$.

Section 2: Digraphs, Paths, and Subgraphs

In this section we introduce the notion of a directed graph and give precise definitions of some very important special substructures of both graphs and directed graphs.

Basic Concepts in Graph Theory

Example 9 (Flow of commodities) Look again at Example 2. Imagine now that the symbols a, b, c, d, e, f and g, instead of standing for route names, stand for commodities (applesauce, bread, computers, etc.) that are produced in one town and shipped to another town. In order to get a picture of the flow of commodities, we need to know the directions in which they are shipped. This information is provided by picture below:

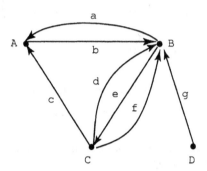

In set-theoretic terms, the information needed to construct the above picture can be specified by giving a pair $D = (V, E, \phi)$ where ϕ is a function. The domain of the function ϕ is $E = \{a, b, c, d, e, f, g\}$ and the codomain is $V \times V$. Specifically,

$$\phi = \begin{pmatrix} a & b & c & d & e & f & g \\ (B,A) & (A,B) & (C,A) & (C,B) & (B,C) & (C,B) & (D,B) \end{pmatrix}.$$

The structure specified by this information is an example of a *directed graph*, which we now define. ∎

Definition 6 (Directed graph) *A directed graph (or digraph) is a triple $D = (V, E, \phi)$ where V and E are finite sets and ϕ is a function with domain E and codomain $V \times V$. We call E the set of edges of the digraph D and call V the set of vertices of D.*

Just as with graphs, we can define a notion of a *simple digraph*. A simple digraph is a pair $D = (V, E)$, where V is a set, the vertex set, and $E \subseteq V \times V$ is the edge set. Just as with simple graphs and graphs, simple digraphs are a special case of digraphs in which ϕ is the identity function on E; that is, $\phi(e) = e$ for all $e \in E$.

There is a correspondence between simple graphs and simple digraphs that is fairly common in applications of graph theory. To interpret simple graphs in terms of simple digraphs, it is best to consider simple graphs with loops (see Example 3 and Exercises for Section 1). Thus consider $G = (V, E)$ where $E \subseteq \mathcal{P}_2(V) \cup \mathcal{P}_1(V)$. We can identify $\{u, v\} \in \mathcal{P}_2(V) \cup \mathcal{P}_1(V)$ with $(u, v) \in V \times V$ and with $(v, u) \in V \times V$. In the case were we have a loop, $u = v$, then we identify $\{u\}$ with (u, u). Here is a picture of a simple graph and its corresponding digraph:

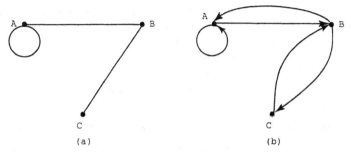

Each, edge that is not a loop in the simple graph is replaced by two edges "in opposite directions" in the corresponding simple digraph. A *loop* is replaced by a *directed loop* (e.g., $\{A\}$ is replaced by (A, A)).

Simple digraphs appear in mathematics under another important guise: *binary relations*. A binary relation on a set V is simply a subset of $V \times V$. Often the name of the relation and the subset are the same. Thus we speak of the binary relation $E \subseteq V \times V$. If you have absorbed all the terminology, you should be able to see immediately that (V, E) is a simple digraph and that any simple digraph (V', E') corresponds to a binary relation $E' \subseteq V' \times V'$.

Recall that a binary relation R is called *symmetric* if $(u, v) \in R$ implies $(v, u) \in R$. Thus simple graphs with loops correspond to symmetric binary relations.

An equivalence relation on a set S is a particular type of binary relation $R \subseteq S \times S$. For an equivalence relation, we have $(x, y) \in R$ if and only if x and y are equivalent (i.e., belong to the same equivalence class or block). Note that this is a symmetric relationship, so we may regard the associated simple digraph as a simple graph. Which simple graphs (with loops allowed) correspond to equivalence relations? As an example, take $S = \underline{7}$ and take the equivalence class partition to be $\{\{1, 2, 3, 4\}, \{5, 6, 7\}\}$. Since everything in each block is related to everything else, there are $\binom{4}{2} = 6$ non-loops and $\binom{4}{1} = 4$ loops associated with the block $\{1, 2, 3, 4\}$ for a total of ten edges. With the block $\{5, 6, 7\}$ there are three loops and three non-loops for a total of six edges Here is the graph of this equivalence relation:

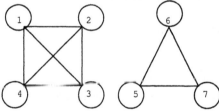

A *complete simple graph* $G=(V,E)$ *with loops* is a graph with every possible edge. That is, $E = \mathcal{P}_2(V) \cup \mathcal{P}_1(V)$. In the above graph, each block of the equivalence relation is replaced by the complete simple graph with loops on that block. This is the general rule.

A basic method for studying graphs and digraphs is to study substructures of these objects and their properties. One of the most important of these substructures is called a *path*.

Basic Concepts in Graph Theory

Definition 7 (Path, trail, walk and vertex sequence) Let $G = (V, E, \phi)$ be a graph.

Let $e_1, e_2, \ldots, e_{n-1}$ be a sequence of elements of E (edges of G) for which there is a sequence $a_1, a_2, \ldots, a_n$ of *distinct* elements of V (vertices of G) such that $\phi(e_i) = \{a_i, a_{i+1}\}$ for $i = 1, 2, \ldots, n-1$. The sequence of edges $e_1, e_2, \ldots, e_{n-1}$ is called a *path* in G. The sequence of vertices $a_1, a_2, \ldots, a_n$ is called the *vertex sequence* of the path. (Note that since the vertices are distinct, so are the edges.)

If we require that $e_1, \ldots, e_{n-1}$ be distinct, but not that $a_1, \ldots, a_n$ be distinct, the sequence of edges is called a *trail*.

If we do not even require that the edges be distinct, it is called a *walk*.

If $G = (V, E, \phi)$ is a directed graph, then $\phi(e_i) = \{a_i, a_{i+1}\}$ is replaced by $\phi(e_i) = (a_i, a_{i+1})$ in the above definition to obtain a *directed path, trail, and walk* respectively.

Note that the definition of a path requires that it not intersect itself (i.e., have repeated vertices), while a trail may intersect itself. Although a trail may intersect itself, it may not have repeated edges, but a walk may. If $P = (e_1, \ldots, e_{n-1})$ is a path in $G = (V, E, \phi)$ with vertex sequence $a_1, \ldots, a_n$ then we say that P is a *path from a_1 to a_n*. Similarly for a trail or a walk.

In the graph of Example 2, the sequence c, d, g is a path with vertex sequence A, C, B, D. If the graph is of the form $G = (V, E)$ with $E \subseteq \mathcal{P}_2(V)$, then the vertex sequence alone specifies the sequence of edges and hence the path. Thus, Example 1, the vertex sequence MN, SM, SE, TM specifies the path {MN, SM}, {SM, SE}, {SE, TM}. Similarly for digraphs. Consider the graph of Example 9. The edge sequence $P = (g, e, c)$ is a directed path with vertex sequence (D, B, C, A). The edge sequence $P = (g, e, c, b, a)$ is a directed trail, but not a directed path. The edge sequence $P = (d, e, d)$ is a directed walk, but not a directed trail.

Note that every path is a trail and every trail is a walk, but not conversely. However, we can show that, if there is a walk between two vertices, then there is a path. This rather obvious result can be useful in proving theorems, so we state it as a theorem.

Theorem 2 (Walk implies path) Suppose $u \neq v$ are vertices in the graph $G = (V, E, \phi)$. The following are equivalent:

(a) There is a walk from u to v.

(b) There is a trail from u to v.

(c) There is a path from u to v.

Furthermore, given a walk from u to v, there is a path from u to v all of whose edges are in the walk.

Proof: Since every path is a trail, (c) implies (b). Since every trail is a walk, (b) implies (a). Thus it suffices to prove that (a) implies (c). Let $e_1, e_2, \ldots, e_k$ be a walk from u to v. We use induction on n, the number of repeated vertices in a walk. If the walk has no repeated vertices, it is a path. This starts the induction at $n = 0$. Suppose $n > 0$. Let r be a repeated vertex. Suppose it first appears in edge e_i and last appears in edge e_j.

Section 2: Digraphs, Paths, and Subgraphs

If $r = u$, then $e_j, \ldots, e_k$ is a walk from u to v in which r is not a repeated vertex. If $r = v$, then $e_1, \ldots, e_i$ is a walk from u to v in which r is not a repeated vertex. Otherwise, $e_1, \ldots, e_i, e_j, \ldots, e_k$ is a walk from u to v in which r is not a repeated vertex. Hence there are less than n repeated vertices in this walk from u to v and so there is a path by induction. Since we constructed the path by removing edges from the walk, the last statement in the theorem follows. □

Note that the theorem and proof are valid if graph is replaced by digraph and walk, trail, and path are replaced by directed walk, trail, and path.

Another basic notion is that of a subgraph of $G = (V, E, \phi)$, which we will soon define. First we need some terminology about functions. By a *restriction* ϕ' of ϕ to $E' \subseteq E$, we mean the function ϕ' with domain E' and satisfying $\phi'(x) = \phi(x)$ for all $x \in E'$. (When forming a restriction, we may change the codomain. Of course, the new codomain must contain Image(ϕ') = $\phi(E)$. In the following definition, the codomain of ϕ' must be $\mathcal{P}_2(V')$ since G' is required to be a graph.)

Definition 8 (Subgraph) Let $G = (V, E, \phi)$ be a graph. A graph $G' = (V', E', \phi')$ is a *subgraph of G* if $V' \subseteq V$, $E' \subseteq E$, and ϕ' is the restriction of ϕ to E'.

As we have noted, the fact that G' is itself a graph means that $\phi(x) \in \mathcal{P}_2(V')$ for each $x \in E'$ and, in fact, the codomain of ϕ' must be $\mathcal{P}_2(V')$. If G is a graph with loops, the codomain of ϕ' must be $\mathcal{P}_2(V') \cup \mathcal{P}_1(V')$. This definition works equally well if G is a digraph. In that case, the codomain of ϕ' must be $V \times V$.

Example 10 (Subgraph — key information) For the graph $G = (V, E, \phi)$ below, let $G' = (V', E', \phi')$ be defined by $V' = \{A, B, C\}$, $E' = \{a, b, c, f\}$, and by ϕ' being the restriction of ϕ to E' with codomain $\mathcal{P}_2(V')$. Notice that ϕ' is determined completely from knowing V', E' and ϕ. Thus, to specify a subgraph G', the key information is V' and E'.

As another example from the same graph, we let $V' = V$ and $E' = \{a, b, c, f\}$. In this case, the vertex D is not a member of any edge of the subgraph. Such a vertex is called an *isolated vertex* of G'. (See also Exercises for Section 1.)

One way of specifying a subgraph is to give a set of edges $E' \subseteq E$ and take V' to be the set of all vertices on some edge of E'. In other words, V' is the union of the sets $\phi(x)$ over all $x \in E'$. Such a subgraph is called the *subgraph induced by the edge set E'* or the *edge induced subgraph* of E'. The first subgraph of this example is the subgraph induced by $E' = \{a, b, c, f\}$.

Likewise, given a set $V' \subseteq V$, we can take E' to be the set of all edges $x \in E$ such that $\phi(x) \subseteq V'$. The resulting subgraph is called the *subgraph induced by V'* or the *vertex induced subgraph* of V'. Referring to the picture again, the edges of the subgraph induced

Basic Concepts in Graph Theory

by $V' = \{C, B\}$, are $E' = \{d, e, f\}$.

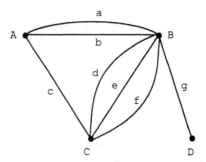

Look again at the above graph. In particular, consider the path c, a with vertex sequence C, A, B. Notice that the edge d has $\phi(d) = \{C, B\}$. The subgraph $G' = (V', E', \phi')$, where $V' = \{C, A, B\}$ and $E' = \{c, a, d\}$ is called a *cycle* of G. In general, whenever there is a path in G, say $e_1, \ldots, e_{n-1}$ with vertex sequence $a_1, \ldots, a_n$, and an edge x with $\phi(x) = \{a_1, a_n\}$, then the subgraph induced by the edges $e_1, \ldots, e_{n-1}, x$ is called a *cycle* of G. Parallel edges like a and b in the preceding figure induce a cycle. A loop also induces a cycle. □

The formal definition of a cycle is:

Definition 9 (Circuit and Cycle) Let $G = (V, E, \phi)$ be a graph and let $e_1, \ldots, e_n$ be a trail with vertex sequence $a_1, \ldots, a_n, a_1$. (It returns to its starting point.) The subgraph G' of G induced by the set of edges $\{e_1, \ldots, e_n\}$ is called a *circuit* of G. The *length* of the circuit is n.

- If the only repeated vertices on the trail are a_1 (the start and end), then the circuit is called a *simple circuit* or *cycle*.

- If "trail" is replaced by directed trail, we obtain a *directed circuit* and a *directed cycle*.

In our definitions, a path is a *sequence* of edges but a cycle is a *subgraph* of G. In actual practice, people often think of a cycle as a path, except that it starts and ends at the same vertex. This sloppiness rarely causes trouble, but can lead to problems in formal proofs. Cycles are closely related to the existence of multiple paths between vertices:

Theorem 3 (Cycles and multiple paths) Two vertices $u \neq v$ are on a cycle of G if and only if there are at least two paths from u to v that have no vertices in common except the endpoints u and v.

Proof: Suppose u and v are on a cycle. Follow the cycle from u to v to obtain one path. Then follow the cycle from v to u to obtain another. Since a cycle has no repeated vertices, the only vertices that lie in both paths are u and v. On the other hand, a path from u to v followed by a path from v to u is a cycle if the paths have no vertices in common other than u and v. □

Section 2: Digraphs, Paths, and Subgraphs

One important feature of a graph is whether or not any pair of vertices can be connected by a path. You can probably imagine, without much difficulty, applications of graph theory where this sort of "connectivity" is important. Not the least of such examples would be communication networks. Here is a formal definition of *connected* graphs.

Definition 10 (Connected graph) Let $G = (V, E, \phi)$ be a graph. If for any two distinct elements u and v of V there is a path P from u to v then G is a connected graph. If $|V| = 1$, then G is connected.

We make two observations about the definition.

- Because of Theorem 2, we can replace "path" in the definition by "walk" or "trail" if we wish. (This observation is used in the next example.)
- The last sentence in the definition is not really needed. To see this, suppose $|V| = 1$. Now G is connected if, for any two *distinct* elements u and v of V there is a path from u to v. This is trivially satisfied since we cannot find two distinct elements in the one element set V.

The graph of Example 1 has two distinct "pieces." It is not a connected graph. There is, for example, no path from $u = TM$ to $v = CS$. Note that one piece of this graph consists of the vertex induced subgraph of the vertex set $\{CS, EN, SH, RL\}$ and the other piece consists of the vertex induced subgraph of $\{TM, SE, MN, SM\}$. These pieces are called *connected components* of the graph. This is the case in general for a graph $G = (V, E, \phi)$: The vertex set is partitioned into subsets $V_1, V_2, \ldots, V_m$ such that if u and v are in the same subset then there is a path from u to v and if they are in different subsets there is no such path. The subgraphs $G_1 = (V_1, E_1, \phi_1), \ldots, G_m = (V_m, E_m, \phi_m)$ induced by the sets $V_1, \ldots, V_m$ are called the *connected components* of G. Every edge of G appears in one of the connected components. To see this, suppose that $\{u, v\}$ is an edge and note that the edge is a path from u to v and so u and v are in the same induced subgraph, G_i. By the definition of induced subgraph, $\{u, v\}$ is in G_i.

Example 11 (Connected components as an equivalence relation) You may have noticed that the "definition" that we have given of connected components is a bit sloppy: We need to know that the partitioning into such subsets can actually occur. To see that this is not trivially obvious, define two integers to be "connected" if they have a common factor. Thus 2 and 6 are connected and 3 and 6 are connected, but 2 and 3 are not connected and so we cannot partition the set $V = \{2, 3, 6\}$ into "connected components". We must use some property of the definition of graphs and paths to show that the partitioning of vertices is possible. One way to do this is to construct an equivalence relation.

For $u, v \in V$, write $u \sim v$ if and only if either $u = v$ or there is a walk from u to v. It is clear that $\sim$ is reflexive and symmetric. We now prove that it is transitive. Let $u \sim v \sim w$. The walk from u to v followed by the walk from v to w is a walk from u to w. This completes the proof that $u \sim v$ is an equivalence relation. The relation partitions V into subsets $V_1, \ldots, V_m$. By Theorem 2, the vertex induced subgraphs of the V_i satisfy Definition 10. □

When talking about connectivity, graphs and digraphs are different. In a digraph, the fact that there is a directed walk from u to v does not, in general, imply that there is a

Basic Concepts in Graph Theory

directed walk from v to u. Thus, the "directed walk relation", unlike the "walk relation" is not symmetric. This complicates the theory of connectivity for digraphs.

Example 12 (Eulerian graphs) We are going to describe a process for constructing a graph $G = (V, E, \phi)$ (with loops allowed). Start with $V = \{v_1\}$ consisting of a single vertex and with $E = \emptyset$. Add an edge e_1, with $\phi(e_1) = \{v_1, v_2\}$, to E. If $v_1 = v_2$, we have a graph with one vertex and one edge (a loop), else we have a graph with two vertices and one edge. Keep track of the vertices and edges in the order added. Here (v_1, v_2) is the sequence of vertices in the order added and (e_1) is the sequence of edges in order added. Suppose we continue this process to construct a sequence of vertices (not necessarily distinct) and sequence of *distinct* edges. At the point where k distinct edges have been added, if v is the last vertex added, then we add a new edge e_{k+1}, different from all previous edges, with $\phi(e_{k+1}) = \{v, v'\}$ where either v' is a vertex already added or a new vertex. Here is a picture of this process carried out with the edges numbered in the order added

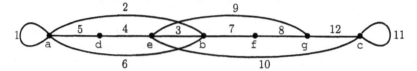

where the vertex sequence is

$$S = (a, a, b, e, d, a, b, f, g, e, c, c, g).$$

Such a graph is called a graph with an *Eulerian trail*. The edges, in the order added, are the Eulerian trail and S is the vertex sequence of the trail

By construction, if G is a graph with an *Eulerian trail*, then there is a trail in G that includes every edge in G. If there is a circuit in G that includes every edge of G then G is called an *Eulerian circuit graph* or graph with an *Eulerian circuit*. Thinking about the above example, if a graph has an Eulerian trail but no Eulerian circuit, then all vertices of the graph have even degree except the start vertex (a in our example with degree 5) and end vertex (g in our example with degree 3). If a graph has an Eulerian circuit then all vertices have even degree. The converses in each case are also true (but take a little work to show): If G is a connected graph in which every vertex has even degree then G has an Eulerian circuit. If G is a connected graph with all vertices but two of even degree, then G has an Eulerian trail joining the two vertices of odd degree. ☐

Here is a precise definition of Eulerian trail and circuit.

Definition 11 (Eulerian trail, circuit) Let $G = (V, E, \phi)$ be a connected graph. If there is a trail with edge sequence $(e_1, e_2, \ldots, e_k)$ in G which uses each edge in E, then $(e_1, e_2, \ldots, e_k)$ is called an Eulerian trail. If there is a circuit $C = (V', E', \phi')$ in G with $E' = E$, then C is called an Eulerian circuit.

The ideas of a directed Eulerian circuit and directed Eulerian trail for directed graphs are defined in exactly the same manner.

Section 2: Digraphs, Paths, and Subgraphs

An Eulerian circuit in a graph contains every edge of that graph. What about a cycle that contains every vertex but not necessarily every edge? Our next example discusses that issue.

Example 13 (Hamiltonian cycle) Start with a graph $G' = (V, E', \phi')$ that is a cycle and then add additional edges, without adding any new vertices, to obtain a graph $G = (V, E, \phi)$. As an example, consider

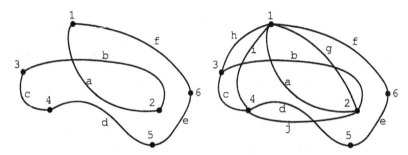

where the first graph $G' = (V, E', \phi')$ is the cycle induced by the edges $\{a, b, c, d, e, f\}$. The second graph $G = (V, E, \phi)$ is obtained from G' by adding edges g, h, i and j. A graph that can be constructed from such a two-step process is called a *Hamiltonian graph*. The cycle G' is called a *Hamiltonian cycle for G*.

Definition 12 (Hamiltonian cycle, Hamiltonian graph) A cycle in a graph $G = (V, E, \phi)$ is a Hamiltonian cycle for G if every element of V is a vertex of the cycle. A graph $G = (V, E, \phi)$ is Hamiltonian if it has a subgraph that is a Hamiltonian cycle for G.

Notice that an Eulerian circuit uses every edge exactly once and a Hamiltonian cycle uses every vertex exactly once. We gave a very simple characterization of when a graph has an Eulerian circuit (in terms of degrees of vertices). There is no simple characterization of when a graph has a Hamiltonian cycle. On the contrary, the issue of whether or not a graph has a Hamiltonian cycle is notoriously difficult to resolve in general.

As we already mentioned, connectivity issues in digraphs are much more difficult than in graphs. A digraph is *strongly connected* if, for every two vertices v and w there is a directed path from v to w. From any digraph D, we can construct a simple graph $S(D)$ on the same set of vertices by letting $\{v, w\}$ be an edge of $S(D)$ if and only if at least one of (u, v) and (v, u) is an edge of D. You should be able to show that if D is strongly connected then $S(D)$ is connected. The converse is false. As an example, take $D = (V, E)$ to be the simple digraph where $V = \{1, 2\}$ and $E = \{(1, 2)\}$. There is no directed path from 2 to 1, but clearly $S(D) = (V, \{\{1, 2\}\})$ is connected.

Other issues for digraphs analogous to those for graphs work out pretty well, but are more technical. An example is the notion of degree for vertices. For any subset U of the vertices V of a directed graph $D = (V, E)$, define $d_{\text{in}}(U)$ to be the number of edges e of D with $\phi(e)$ of the form (w, u) where $u \in U$ and $w \notin U$. Define $d_{\text{out}}(U)$ similarly. If $U = \{v\}$ consists of just one vertex, $d_{\text{in}}(U)$ is usually written simply as $d_{\text{in}}(v)$ rather than

Basic Concepts in Graph Theory

the more technically correct $d_{in}(\{v\})$. Similarly, we write $d_{out}(v)$. You should compute $d_{in}(v)$ and $d_{out}(v)$ for the vertices v of the graph of Example 9. You should be able to show that $\sum d_{in}(v) = \sum d_{out}(v) = |E|$, where the sums range over all $v \in V$. See the Exercises for Section 1 for the idea.

Example 14 (Bicomponents of graphs) Let $G = (V, E, \phi)$ be a graph. For $e, f \in E$ write $e \sim f$ if either $e = f$ or there is a cycle of G that contains both e and f. We claim that this is an equivalence relation. The reflexive and symmetric parts are easy. Suppose that $e \sim f \sim g$. If $e = g$, then $e \sim g$, so suppose that $e \neq g$. Let $\phi(e) = \{v_1, v_2\}$. Let $C(e, f)$ be the cycle containing e and f and $C(f, g)$ the cycle containing f and g. In $C(e, f)$ there is a path P_1 from v_1 to v_2 that does not contain e. Let x and $y \neq x$ be the first and last vertices on P_1 that lie on the cycle containing f and g. We know that there must be such points because the edge f is on P_1. Let P_2 be the path in $C(e, f)$ from y to x containing e. In $C(f, g)$ there is a path P_3 from x to y containing g. We claim that P_2 followed by P_3 defines a cycle containing e and g.

Some examples may help. Consider a graph that consists of two disjoint cycles that are joined by an edge. There are three bicomponents — each cycle and the edge joining them. Now consider three cycles that are disjoint except for one vertex that belongs to all three of them. Again there are three bicomponents — each of the cycles.

Since $\sim$ is an equivalence relation on the edges of G, it partitions them. If the partition has only one block, then we say that G is a *biconnected graph*. If E' is a block in the partition, the subgraph of G induced by E' is called a *bicomponent* of G. Note that the bicomponents of G are not necessarily disjoint: Bicomponents may have vertices in common (but *never* edges). There are four bicomponents in the following graph. Two are the cycles, one is the edge $\{C, O\}$, and the fourth consists of all of the rest of the edges.

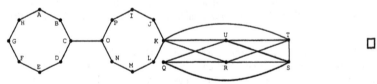

Exercises for Section 2

2.1. A graph $G = (V, E)$ is called *bipartite* if V can be partitioned into two sets C and S such that each edge has one vertex in C and one vertex in S. As a specific example, let C be the set of courses at the university and S the set of students. Let $V = C \cup S$ and let $\{s, c\} \in E$ if and only if student s is enrolled in course c.

(a) Prove that $G = (V, E)$ is a simple graph.

(b) Prove that every cycle of G has an even number of edges.

2.2. In each of the following graphs, find the longest trail (most edges) and longest circuit. If the graph has an Eulerian circuit or trail, say so.

Section 2: Digraphs, Paths, and Subgraphs

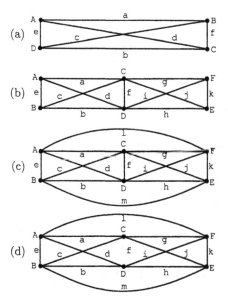

2.3. For each of the following graphs $G = (V, E, \phi)$, find a cycle in G of maximum length. State whether or not the graph is Hamiltonian.

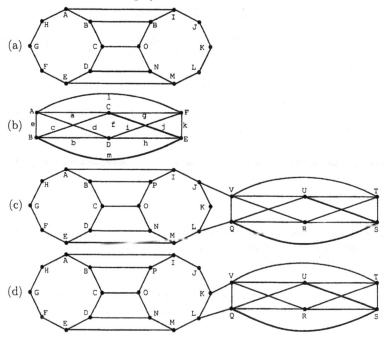

2.4. We are interested in the number of simple digraphs with $V = \underline{n}$

Basic Concepts in Graph Theory

  (a) Find the number of them.

  (b) Find the number of them with no loops.

  (c) In both cases, find the number of them with exactly q edges.

2.5. An *oriented simple graph* is a simple graph which has been converted to a digraph by assigning an orientation to each edge. The orientation of $\{u,v\}$ can be thought of as a mapping of it to either (u,v) or (v,u).

  (a) Give an example of a simple digraph that has no loops but is not an oriented simple graph

  (b) Find the number of oriented simple digraphs.

  (c) Find the number of them with exactly q edges.

2.6. A binary relation R on S is an *order relation* if it is reflexive, antisymmetric, and transitive. R is *antisymmetric* if for all $(x,y) \in R$ with $x \neq y$, $(y,x) \notin R$. Given an order relation R, the covering relation H of R consists of all $(x,z) \in R$, $x \neq z$, such that there is no y, distinct from both x and z, such that $(x,y) \in R$ and $(y,z) \in R$. A pictorial representation of the covering relation as a directed graph is called a "Hasse diagram" of H.

  (a) Show that the divides relation on
  $$S = \{2,3,4,5,6,7,8,9,10,11,12,13,14,15,16\}$$
  is an order relation. By definition, (x,y) is in the divides relation on S is x is a factor of y. Thus, $(4,12)$ is in the divides relation. $x|y$ is the standard notation for x is a factor of y.

  (b) Find and draw a picture of the directed graph of the covering relation of the divides relation.
  Hint: You must find all pairs $(x,z) \in S \times S$ such that $x|z$ but there does not exist any y, $x < y < z$, such that $x|y$ and $y|z$.

Section 3: Trees

Trees play an important role in a variety of algorithms. We have used decision trees to enhance our understanding of recursion. In this section, we define trees precisely and look at some of their properties.

Definition 13 (Tree) *If G is a connected graph without any cycles then G is called a tree. (If $|V| = 1$, then G is connected and hence is a tree.) A tree is also called a free tree.*

The graph of Example 2 is connected but is not a tree. It has many cycles, including $(\{A,B,C\},\{a,e,c\})$. The subgraph of this graph induced by the edges $\{a,e,g\}$ is a tree. If

Section 3: Trees

G is a tree, then ϕ is an injection since if $e_1 \neq e_2$ and $\phi(e_1) = \phi(e_2)$, then $\{e_1, e_2\}$ induces a cycle. In other words, any graph with parallel edges is not as tree. Likewise, a loop is a cycle, so a tree has no loops. Thus, we can think of a tree as a simple graph when we are not interested in names of the edges.

Since the notion of a tree is so important, it will be useful to have some equivalent definitions of a tree. We state them as a theorem

Theorem 4 (Alternative definitions of a tree) *If G is a connected graph, the following are equivalent.*

(a) *G is a tree.*

(b) *G has no cycles.*

(c) *For every pair of vertices $u \neq v$ in G, there is exactly one path from u to v.*

(d) *Removing any edge from G gives a graph which is not connected.*

(e) *The number of vertices of G is one more than the number of edges of G.*

Proof: We are given that G is connected, thus, by the definition of a tree, (a) and (b) are equivalent.

Theorem 3 can be used to prove that (b) implies (c). We leave that as an exercise (show **not** (c) implies **not** (b)).

If $\{u,v\}$ is an edge, it follows from (c) that the edge is the only path from u to v and so removing it disconnects the graph. Hence (c) implies (d).

We leave it as an exercise to prove that (d) implies (b) (show **not** (b) implies **not** (d)).

Thus far, we have shown (a) and (b) are equivalent, and we have shown that (b) implies (c) implies (d) implies (b), so (a), (b), (c), and (d) are all equivalent. All that remains is to include (e) in this equivalence class of statements. Do do this, all we have to do is show that (e) implies any of the equivalent statements (a), (b), (c), and (d) and, conversely, some one of (a), (b), (c), and (d) implies (e). We shall show that (b) implies (e) and that (e) implies (a).

We first show that (b) implies (e). We will use induction on the number of vertices of G. If G has one vertex, it has no edges and (e) is satisfied. Otherwise, we claim that G has a vertex u of degree 1; that is, it lies on only one edge $\{u, w\}$. We prove this claim shortly. Remove u and $\{u, v\}$ to obtain a graph H with one less edge and one less vertex. Since G is connected and has no cycles, the same is true of H. By the induction hypothesis, H has one less edge than vertex. Since we got from G to H by removing one vertex and one edge, G must also have one less edge than vertex. By induction, the proof is done. It remains to prove the existence of u. Suppose no such u exists; that is, suppose that each vertex lies on at least two edges. We will derive a contradiction. Start at any vertex v_1 of G leave v_1 by some edge e_1 to reach another vertex v_2. Leave v_2 by some edge e_2 different from the edge used to reach v_2. Continue with this process. Since each vertex lies on at least two edges, the process never stops. Hence we eventually repeat a vertex, say

$$v_1, e_1, v_2, \ldots, v_k, e_k, \ldots, v_n, e_n, v_{n+1} = v_k.$$

[169]

Basic Concepts in Graph Theory

The edges $e_k, \ldots, e_n$ form a cycle, which is a contradiction.

Having shown that (b) implies (e), we now show that (e) implies (a). We use the contrapositive and show that **not** (a) implies **not** (e). Thus we assume G is not a tree. Hence, by (d) we can remove an edge from G to get a new graph which is still connected. If this is not a tree, repeat the process and keep doing so until we reach a tree T. For a tree T, we trivially satisfy (a) which implies (b) and (b) implies (e). Thus, the number of vertices is now one more than the number of edges in the graph T. Since, in going from G to T, we removed edges from G but did not remove vertices, G must have at least as many edges as vertices. This shows **not** (a) implies **not** (e) and completes the proof. ☐

Definition 14 (Forest) *A forest is a graph all of whose connected components are trees. In particular, a forest with one component is a tree. (Connected components were defined following Definition 10.)*

Example 15 (A relation for forests) Suppose a forest has v vertices, e edges and c (connected) components. What values are possible for the triple of numbers (v, e, c)? It might seem at first that almost anything is possible, but this is not so. In fact $v - c = e$ because of Theorem 4(e). Why? Let the forest consist of trees $T_1, \ldots, T_c$ and let the triples for T_i be (v_i, e_i, c_i). Since a tree is connected, $c_i = 1$. By the theorem, $e_i = v_i - 1$. Since $v = v_1 + \cdots + v_c$ and $e = e_1 + \ldots + e_c$ we have

$$e = (v_1 - 1) + (v_2 - 1) + \cdots + (v_c - 1) = (v_1 + \cdots + v_c) - c = v - c.$$

Suppose a forest has $e = 12$ and $v = 15$. We know immediately that it must be made up of three trees because $c = v - e = 15 - 12$.

Suppose we know that a graph $G = (V, E, \phi)$ has $v = 15$ and $c = 3$, what is the fewest edges it could have? For each component of G, we can remove edges one by one until we cannot remove any more without breaking the component into two components. At this point, we are left with each component a tree. Thus we are left with a forest of $c = 3$ trees that still has $v = 15$ vertices. By our relation $v - c = e$, this forest has 12 edges. Since we may have removed edges from the original graph to get to this forest, the original graph has at least 12 edges.

What is the maximum number of edges that a graph $G = (V, E, \phi)$ with $v = 15$ and $c = 3$ could have? Since we allow multiple edges, a graph could have an arbitrarily large number of edges for a fixed v and c — if e is an edge with $\phi(e) = \{u, v\}$, add in as many edges e_i with $\phi(e_i) = \{u, v\}$ as you wish. Hence we will have to insist that G be a simple graph.

What is the maximum number of edges that a simple graph G with $v = 15$ and $c = 3$ could have? This is a bit trickier. Let's start with a graph where c is not specified. The edges in a simple graph are a subset of $\mathcal{P}_2(V)$ and since $\mathcal{P}_2(V)$ has $\binom{v}{2}$ elements, a simple graph with v vertices has at most $\binom{v}{2}$ edges.

Now let's return to the case when we know there must be three components in our simple graph. Suppose the number of vertices in the components are v_1, v_2 and v_3. Since there are no edges between components, we can look at each component by itself. Using

Section 3: Trees

the result in the previous paragraph for each component, the maximum number of possible edges is $\binom{v_1}{2} + \binom{v_2}{2} + \binom{v_3}{2}$. We don't know v_1, v_2, v_3. All we know is that they are strictly positive integers that sum to v. It turns out that the maximum occurs when one of v_i is as large as possible and the others equal 1, but the proof is beyond this course. Thus the answer is $\binom{v-2}{2}$, which in our case is $\binom{13}{2} = 78$. In general, if there were c components, $c-1$ components would have one vertex each and the remaining component would have $v - (c-1) = v + 1 - c$ vertices. Hence there can be no more than $\binom{v+1-c}{2}$ edges.

Reviewing what we've done, we see:

- There is no graph $G = (V, E, \phi)$ with $v - c > e$.

- If $v - c = e$, the graph is a forest of c trees and any such forest will do as an example.

- If $v - c < e$, there are many examples, none of which are forests.

- If $v - c < e$ and we have a *simple* graph, then we must have $e \leq \binom{v+1-c}{2}$. □

Recall that decision trees, as we have used them, have some special properties. First, they have a starting point. Second, the edges (decisions) out of each vertex are ordered. We now formalize these concepts.

Definition 15 (Rooted graph) *A pair (G, v), consisting of a graph $G = (V, E, \phi)$ and a specified vertex v, is called a rooted graph with root v.*

Definition 16 (Parent, child, sibling and leaf) *Let (T, r) be a rooted tree. If w is any vertex other than r, let $r = v_0, v_1, \ldots, v_k, v_{k+1} = w$, be the list of vertices on the unique path from r to w. We call v_k the parent of w and call w a child of v_k. Parents and children are also called fathers and sons. Vertices with the same parent are siblings. A vertex with no children is a leaf. All other vertices are internal vertices of the tree.*

Definition 17 (Rooted plane tree) *Let (T, r) be a rooted tree. For each vertex, order the children of the vertex. The result is a rooted plane tree, which we abbreviate to RP-tree. RP-trees are also called ordered trees. An RP-tree is also called, in certain contexts, a decision tree, and, when there is no chance of misunderstanding, simply a tree.*

Since almost all trees in computer science are rooted and plane, computer scientists usually call a rooted plane tree simply a tree. It's important to know what people mean!

Example 16 (A rooted plane tree) Below is a picture of a rooted plane tree $T = (V, E, \phi)$. In this case $V = \underline{11}$ and $E = \{a, \ldots, j\}$. There are no parallel edges or loops, as required by the definition of a RP-tree. The root is $r = 1$. For each vertex, there is a unique path from the root to that vertex. Since ϕ is an injection, once ϕ has been defined (as it is in the picture), that unique path can be specified by the vertex sequence alone. Thus, the path from the root to 6 is $(1, 3, 6)$. The path from the root to 9 is $(1, 3, 6, 9)$. Sometimes computer scientists refer to the path from the root to a vertex v as the "stack" of v.

Basic Concepts in Graph Theory

In the tree below, the vertex 6 is the parent of the vertex 9. The vertices 8, 9, 10, and 11 are the children of 6 and, they are siblings of each other. The leaves of the tree are 4, 5, 7, 8, 9, 10, and 11. All other vertices (including the root) are internal vertices of the tree.

Remember, an RP-tree is a tree with added properties. Therefore, it must satisfy (a) through (e) of Theorem 4. In particular, T has no cycles. Also, there is a unique path between any two vertices (e.g., the path from 5 to 8 is $(5, 2, 1, 3, 6, 8)$). Removing any edge gives a graph which is not connected (e.g., removing j disconnects T into a tree with 10 vertices and a tree with 1 vertex; removing e disconnects T into a tree with 6 vertices and one with 5 vertices). Finally, the number of edges (10) is one less than the number of vertices.

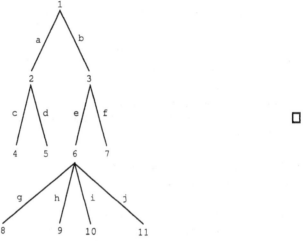

□

Example 17 (Traversing a rooted plane tree) Just as in the case of decision trees, one can define the notion of *depth first* traversals of a RP-tree.

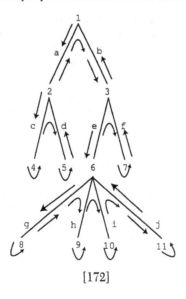

Imagine going around ("traversing") the above RP-tree following arrows. Start at the root, 1, go down edge a to vertex 2, etc. Here is the sequence of vertices as encountered in this process: 1, 2, 4, 2, 5, 2, 1, 3, 6, 8, 6, 9, 6, 10, 6, 11, 6, 3, 7, 3, 1. This sequence of vertices is called the *depth first vertex sequence*, DFV(T), of the RP-tree T. The number of times each vertex appears in DFV(T) is one plus the number of children of that vertex. For edges, the corresponding sequence is a, c, c, d, d, a, b, e, g, g, h, h, i, i, j, j, e, f, f, b. This sequence is the *depth first edge sequence*, DFE(T), of the tree. Every edge appears exactly twice in DFE(T). If the vertices of the RP-tree are read left to right, top to bottom, we obtain the sequence 1, 2, 3, 4, 5, 6, 7, 8, 9, 10, 11. This is called the *breadth first vertex sequence*, BFV(T). Similarly, the *breadth first edge sequence*, BFE(T), is a, b, c, d, e, f, g, h, i, j.

The sequences BFV(T) and BFE(T) are linear orderings of the vertices and edges of the RP-tree T (i.e., each vertex or edge appears exactly once in the sequence). We also associate linear orderings with DFV(T) called the *preorder sequence of vertices* of T, PREV(T), and the *postorder sequence of vertices* of T, POSV(T).

PREV(T) = 1, 2, 4, 5, 3, 6, 8, 9, 10, 11, 7 is the sequence of *first* occurrences of the vertices of T in DFV(T).

POSV(T) = 4, 5, 2, 8, 9, 10, 11, 6, 7, 3, 1 is the sequence of *last* occurrences of the vertices of T in DFV(T).

Notice that the order in which the leaves of T appear, 4, 5, 8, 9, 10, 11, is the same in both PREV(T) and POSV(T). Can you see why this is always true for any tree? □

*Example 18 (The number of labeled trees)** How many n-vertex labeled trees are there? In other words, count the number of trees with vertex set $V = \underline{n}$. The answer has been obtained in a variety of ways. We will do it by establishing a correspondence between trees and functions by using digraphs.

Suppose f is a function from V to V. We can represent this as a simple digraph (V, E) where the edges are $\{(v, f(v)) \mid v \in V\}$. The function

$$\begin{pmatrix} 1 & 2 & 3 & 4 & 5 & 6 & 7 & 8 & 9 & 10 & 11 \\ 1 & 10 & 9 & 2 & 8 & 2 & 2 & 5 & 1 & 6 & 11 \end{pmatrix}$$

corresponds to the directed graph

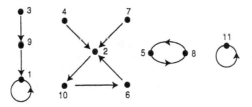

Such graphs are called *functional digraphs*. You should be able to convince yourself that a functional digraph consists of cycles (including loops) with each vertex on a cycle being the root of a tree of noncyclic edges. The edges of the trees are directed toward the roots. In the previous figure,

Basic Concepts in Graph Theory

- 1 is the root of the tree with vertex set $\{1,3,9\}$,
- 2 is the root of the tree with vertex set $\{2,4,7\}$,
- 5 is the root of the tree with vertex set $\{5\}$,
- 6 is the root of the tree with vertex set $\{6\}$,
- 8 is the root of the tree with vertex set $\{8\}$,
- 10 is the root of the tree with vertex set $\{10\}$ and
- 11 is the root of the tree with vertex set $\{11\}$.

In a tree, there is a unique path from the vertex 1 to the vertex n. Remove all the edges on the path and list the vertices on the path, excluding 1 and n, in the order they are encountered. Interpret this list as a permutation in 1 line form. Draw the functional digraph for the cycle form, adding the cycles (1) and (n). Add the trees that are attached to each of the cycle vertices, directing their edges toward the cycle vertices. Consider the following figure.

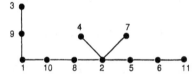

The one line form is $10, 8, 2, 5, 6$. In two line form it is $\begin{pmatrix} 2 & 5 & 6 & 8 & 10 \\ 10 & 8 & 2 & 5 & 6 \end{pmatrix}$. Thus the cycle form is $(2,10,6)(5,8)$. When we add the two cycles (1) and (11) to this, draw the directed graph, and attach the directed trees, we obtain the functional digraph pictured earlier.

We leave it to you to convince yourself that this gives us a one-to-one correspondence between trees with $V = \underline{n}$ and functions $f : \underline{n} \to \underline{n}$ with $f(1) = 1$ and $f(n) = n$. In creating such a function, there are n choices for each of $f(2), \ldots, f(n-1)$. Thus there are n^{n-2} such functions and hence n^{n-2} trees. $\square$

Spanning Trees

Trees are not only important objects of study per se, but are important as special subgraphs of general graphs. A *spanning tree* is one such subgraph. For notational simplicity, we shall restrict ourselves to simple graphs, $G = (V, E)$, in the following discussion. The ideas we discuss extend easily to graphs $G = (V, E, \phi)$, even allowing loops.

Definition 18 (Spanning tree) *A spanning tree of a (simple) graph $G = (V, E)$ is a subgraph $T = (V, E')$ which is a tree and has the same set of vertices as G.*

Section 3: Trees

Example 19 (Connected graphs and spanning trees) Since a tree is connected, a graph with a spanning tree must be connected. On the other hand, it is not hard to see that every connected graph has a spanning tree. Any simple graph $G = (V, E)$ has a subgraph that is a tree, $T' = (V', E')$. Take $V' = \{v\}$ to be one vertex and E' empty. Suppose that $T' = (V', E')$ is the largest such "subtree." If T' is not a spanning tree then there is a vertex w of G that is not a vertex of T'. If G is connected, choose a vertex u in T' and a path $w = x_1, x_2, \ldots, e_k = u$ from w to u. Let j, $1 < j \leq k$, be the first integer such that x_j is a vertex of T'. Then adding the edge $\{x_{j-1}, x_j\}$ and the vertex x_{j-1} to T' creates a subtree T of G that is larger than T', a contradiction of the maximality of T'. We have, in fact, shown that a graph is connected if and only if every maximal subtree is a spanning tree. Thus we have: A graph is connected if and only if it has a spanning tree. It follows that, if we had an algorithm that was guaranteed to find a spanning tree whenever such a tree exists, then this algorithm could be used to decide if a graph is connected. ☐

Example 20 (Minimum spanning trees) Suppose we wish to install "lines" to link various sites together. A site may be a computer installation, a town, or a factory. A line may be a digital communication channel, a rail line or, a shipping route for supplies. We'll assume that

(a) a line operates in both directions;

(b) it must be possible to get from any site to any other site using lines;

(c) each possible line has a cost (rental rate, construction cost, or shipping cost) independent of each other line's cost;

(d) we want to choose lines to minimize the total cost.

We can think of the sites as vertices V in a (simple) graph, the possible lines as edges E and the costs as a function λ from the edges to the positive real numbers. Because of (a) and (b), the lines $E' \subseteq E$ we actually choose will be such that $T = (V, E')$ is connected. Because of (d), T will be a spanning tree since, if it had more edges, we could delete some, but if we delete any from a tree it will not be connected by Theorem 4. ☐

We now formalize these ideas in a definition:

Definition 19 (Weights in a graph) Let $G = (V, E)$ be a simple graph and let λ be a function from E to the positive real numbers. We call $\lambda(e)$ the weight of the edge e. If $H = (V', E')$ is a subgraph of G, then $\lambda(H)$, the weight of H, is the sum of $\lambda(e')$ over all $e' \in E'$.

A *minimum weight spanning tree* for a connected graph G is a spanning tree such that $\lambda(T) \leq \lambda(T')$ whenever T' is another spanning tree.

How can we find a minimum weight spanning tree T? One approach is to construct T by adding an edge at a time in a greedy way. Since we want to minimize the weight, "greedy" means keeping the weight of each edge we add as low as possible. Here's such an algorithm.

Theorem 5 (Minimum weight spanning tree: Prim's algorithm) Let $G = (V, E)$ be a simple graph with edge weights given by λ. If the algorithm stops with $V' \neq V$, G has no spanning tree; otherwise, (V, E') is a minimum weight spanning tree for G.

Basic Concepts in Graph Theory

1. **Start:** Let $E' = \emptyset$ and let $V' = \{v_0\}$ where v_0 is any vertex in V.

2. **Possible Edges:** Let $F \subseteq E$ be those edges $f = \{x, y\}$ with one vertex in V' and one vertex not in V'. If $F = \emptyset$, stop.

3. **Choose Edge Greedily:** Let $f = \{x, y\}$ be such that $\lambda(f)$ is a minimum over all $f \in F$. Replace V' with $V' \cup \{x, y\}$ and E' with $E' \cup \{f\}$. Go to Step 2.

Proof: We begin with the first part; i.e, if the algorithm stops with $V' \neq V$, then G has no spanning tree. The argument is similar to that used in Example 19. Suppose that $V' \neq V$ and that there is a spanning tree. We will prove that the algorithm does not stop at V'. Choose $u \in V - V'$ and $v \in V'$. Since G is connected, there must be a path from u to v. Each vertex on the path is either in V' or not. Since $u \notin V'$ and $v \in V'$, there must be an edge f on the path with one end in V' and one end not in V'. But then $f \in F$ and so the algorithm does not stop at V'.

We now prove that, if G has a spanning tree, then (V, E') is a minimum weight spanning tree. One way to do this is by induction: We will prove that at each step there is a minimum weight spanning tree of G that contains E'.

The starting case for the induction is the first step in the algorithm; i.e., $E' = \emptyset$. Since G has a spanning tree, it must have a minimum weight spanning tree. The edges of this tree obviously contain the empty set, which is what E' equals at the start.

We now carry out the inductive step of the proof. Let V' and E' be the values going into Step 3 and let $f = \{x, y\}$ be the edge chosen there. By the induction hypothesis, there is a minimum weight spanning tree T of G that contains the edges E'. If it also contains the edge f, we are done. Suppose it does not contain f. We will prove that we can replace an edge in the minimum weight tree with f and still achieve minimum weight.

Since T contains all the vertices of G, it contains x and y and, also, some path P from x to y. Suppose $x \in V'$ and $y \notin V'$, this path must contain an edge $e = \{u, v\}$ with $u \in V'$ and $v \notin V'$. We now prove that removing e from T and then adding f to T will still give a minimum spanning tree.

By the definition of F in Step 2, $e \in F$ and so, by the definition of f, $\lambda(e) \geq \lambda(f)$. Thus the weight of the tree does not increase. If we show that the result is still a tree, this will complete the proof.

The path P together with the edge f forms a cycle in G. Removing e from P and adding f still allows us to reach every vertex in P and so the altered tree is still connected. It is also still a tree because it contains no cycles — adding f created only one cycle and removing e destroyed it. This completes the proof that the algorithm is correct. $\square$

The algorithm for finding a minimum weight spanning tree that we have just proved is sometimes referred to as *Prim's Algorithm*. A variation on this algorithm, proved in a similar manner, is called *Kruskal's algorithm*. In Kruskal's algorithm, step 2 of Prim's algorithm is changed to

2'. **Possible Edges:** Let $F \subseteq E$ be those edges $f = \{x, y\}$ where x and y do not belong to the same component of (V, E'). If $F = \emptyset$, stop.

Intuitively, $f \notin F$ if f forms a cycle with any collection of edges from E'. Otherwise, $f \in F$. This extra freedom is sometimes convenient. Our next example gives much less freedom in

Section 3: Trees

choosing new edges to add to the spanning tree, but produces a type of spanning tree that is useful in many algorithms applicable to computer science.

Example 21 (Algorithm for lineal or depth-first spanning trees) We start with a rooted simple graph $G = (V, E)$ with v_0 as root. The algorithmic process constructs a spanning tree rooted at v_0. It follows the same general form as Theorem 5. The weights, if there, are ignored.

1. **Start:** Let $E' = \emptyset$ and let $V' = \{v_0\}$ where v_0 is the root of G. Let $T' = (V', E')$ be the starting subtree, rooted at v_0.

2. **Possible New Edge:** Let v be the last vertex added to V' where $T' = (V', E')$ is the subtree thus far constructed, with root v_0. Let x be the first vertex on the unique path from v to v_0 for which there is an edge $f = \{x, y\}$ with $x \in V'$ and $y \notin V'$. If there is no such x, stop.

3. **Add Edge:** Replace V' with $V' \cup \{y\}$ and E' with $E' \cup \{f\}$ to obtain $T' = (V', E')$ as the new subtree thus far constructed, with root v_0. (Note: y is now the last vertex added to V'.) Go to Step 2.

Here is an example. We are going to find a lineal spanning tree for the graph below, root a. The result is shown on the right where the original vertices have been replaced by the order in which they have been added to the "tree thus far constructed" in the algorithm.

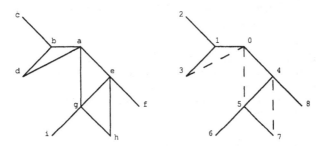

When there is a choice, we choose the left or upward vertex. For example, at the start, when b, d, e and g are all allowed, we choose b. When vertex 2 was added, the path to the root was $(2, 1, 0)$. We went along this path towards the root and found that at 1, a new edge to 3 could be added. Now the path to the root became $(3, 1, 0)$ and we had to go all of the way to 0 to add a new edge (the edge to 4). You should go through the rest of the algorithm. Although there are some choices, the basic rule of step 2 of the algorithm must always be followed.

There are two extremely important properties that this algorithm has

1. When the rooted spanning tree T for G has been constructed, there may be edges of G not in the spanning tree. In the above picture, there are three such edges, indicated by dashed lines. If $\{x, y\}$ is such an edge, then either x lies on the path from y to the root or the other way around. For example, the edge $\{4, 7\}$ in the example has 4 on the path from 7 to the root 0. This is the "lineal" property from which the spanning trees of this class get their name.

[177]

GT-33

Basic Concepts in Graph Theory

2. If, when the rooted spanning tree T has been constructed, the vertices of T are labeled in the order added by the algorithm **AND** the children of each vertex of T are ordered by the same numbering, then an RP-tree is the result. For this RP tree, the numbers on the vertices correspond to preorder, PREV(T), of vertices on this tree (starting with the root having value 0). Check this out for the above example.

We will not prove that the algorithm we have presented has properties 1 and 2. We leave it to you to study the example, construct other examples, and come to an intuitive understanding of these properties. ☐

Property 1 in the preceding example is the basis for the formal definition of a lineal spanning tree:

Definition 20 (Lineal or depth-first spanning tree) *Let x and y be two vertices in a rooted tree with root r. If x is on the path connecting r to y, we say that y is a descendant of x. (In particular, all vertices are descendants of r.) If one of u and v is a descendant of the other, we say that $\{u,v\}$ is a lineal pair. A lineal spanning tree or depth-first spanning tree of a connected graph $G = (V, E)$ is a rooted spanning tree of G such that each edge $\{u,v\}$ of G is a lineal pair.*

In our example, vertices $\{6, 7\}$ are not a lineal pair relative to the rooted tree constructed. But $\{4, 7\}$, which is an edge of G, is a lineal pair. Trivially, the vertices of any edge of the tree T form a lineal pair.

We close this section by proving a theorem using lineal spanning trees. We don't "overexplain" this theorem to encourage you to think about the properties of lineal spanning trees that make the proof much simpler than what we might have come up with without lineal spanning trees. Recall that a graph $G = (V, E)$ is called *bipartite* if V can be partitioned into two sets C and S such that each edge has one vertex in C and one vertex in S (Exercises for Section 2).

Theorem 6 (Bipartite and cycle lengths) *Let $G = (V, E)$ be a simple graph. G is bipartite if and only if every cycle has even length.*

Proof: If G has a cycle of odd length, label each vertex with the block of some proposed bipartite partition $\{C, S\}$. For example, if , (x_1, x_2, x_3) are the vertices, in some order, of a cycle of length three, then the block labels (start with C) would be (C, S, C) This would mean that the edge $\{x_1, x_3\}$ would have both vertices in block C. This violates the definition of a bipartite graph. Since this problem happens for any cycle of odd length, a bipartite graph can never contain a cycle of odd length.

To prove the converse, we must show that if every cycle of G has even length, then G is bipartite. Suppose every cycle of G has even length. Choose a vertex v_0 as root of G and construct a lineal spanning tree T for G with root v_0. Label the root v_0 of T with C, all vertices of T of distance 1 from v_0 with S, all of distance 2 from v_0 with C, etc. Put vertices labeled C into block C of a partition $\{C, S\}$ of V, put all other vertices into block S. If $f = \{x, y\}$ is an edge of T then x and y are in different blocks of the partition $\{C, S\}$

Section 3: Trees

by construction. If $f = \{x, y\}$ is an edge of G not in T then the two facts (1) T is lineal and (2) every cycle has even length, imply that x and y are in different blocks of the partition $\{C, S\}$. This completes the proof. □

Exercises for Section 3

3.1. In this exercise, we study how counting edges and vertices in a graph can establish that cycles exist. For parts (a) and (b), let $G = (V, E, \phi)$ be a graph with loops allowed.

(a) Using induction on n, prove:
If $n \geq 0$, G is connected and G has v vertices and $v + n$ edges, then G has at least $n + 1$ cycles.

(b) Prove that, if G has v vertices, e edges and c components, then G has at least $c + e - v$ cycles.
Hint: Use (a) for each component.

(c) Show that (a) is best possible, even for simple graphs. In other words, for each n construct a simple graph that has n more edges than vertices but has only $n + 1$ cycles.

3.2. Let $T = (V, E)$ be a tree and let $d(v)$ be the degree of a vertex

(a) Prove that $\sum_{v \in V}(2 - d(v)) = 2$.

(b) Prove that, if T has a vertex of degree $m \geq 2$, then it has at least m vertices of degree 1.

(c) Give an example for all $m \geq 2$ of a tree with a vertex of degree m and only m leaves.

3.3. Give an example of a graph that satisfies the specified condition or show that no such graph exists.

(a) A tree with six vertices and six edges

(b) A tree with three or more vertices, two vertices of degree one and all the other vertices with degree three or more.

(c) A disconnected graph with 10 vertices and 8 edges.

(d) A disconnected graph with 12 vertices and 11 edges and no cycle.

(e) A tree with 6 vertices and the sum of the degrees of all vertices 12.

(f) A connected graph with 6 edges, 4 vertices, and exactly 2 cycles.

(g) A graph with 6 vertices, 6 edges and no cycles.

3.4. The *height* of a rooted tree is the maximum height of any leaf. The length of the unique path from a leaf of the tree to the root is, by definition, the height of that

Basic Concepts in Graph Theory

leaf. A rooted tree in which each non-leaf vertex has at most two children is called a *binary tree*. If each non-leaf vertex has exactly two children, the tree is called a *full binary tree*.

(a) If a binary tree has l leaves and height h prove that $l \leq 2^h$. (Taking logarithms gives $\log_2(l) \leq h$.)

(b) A binary tree has l leaves. What can you say about the maximum value of h?

(c) Given a full binary tree with l leaves, what is the maximum height h?

(d) Given a full binary tree with l leaves, what is the minimum height h?

(e) Given a binary tree of l leaves, what is the minimum height h?

3.5. In each of the following cases, state whether or not such a tree is possible.

(a) A binary tree with 35 leaves and height 100.

(b) A full binary tree with 21 leaves and height 21.

(c) A binary tree with 33 leaves and height 5.

(d) A rooted tree of height 5 where every internal vertex has 3 children and there are 365 vertices.

3.6. For each of the following graphs:

(1)

(2)

(3)

(a) Find all spanning trees.

(b) Find all spanning trees up to isomorphism.

(c) Find all depth-first spanning trees rooted at A.

(d) Find all depth-first spanning trees rooted at B.

3.7. For each of the following graphs:

(1)

(2)

(3)

(a) Find all minimum spanning trees.

(b) Find all minimum spanning trees up to isomorphism.

(c) Among all depth-first spanning trees rooted at A, find those of minimum weight.

(d) Among all depth-first spanning trees rooted at B, find those of minimum weight.

3.8. In the following graph, the edges are weighted either 1, 2, 3, or 4.

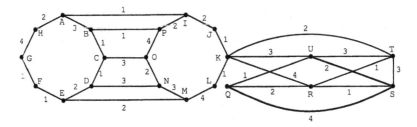

Referring to Theorem 5 and the discussion following of Kruskal's algorithm:

(a) Find a minimum spanning tree using Prim's algorithm

(b) Find a minimum spanning tree using Kruskal's algorithm.

(c) Find a depth-first spanning tree rooted at K.

Section 4: Rates of Growth and Analysis of Algorithms

Suppose we have an algorithm and someone asks us "How good is it?" To answer that question, we need to know what they mean. They might mean "Is it correct?" or "Is it understandable?" or "Is it easy to program?" We won't deal with any of these.

They also might mean "How fast is it?" or "How much space does it need?" These two questions can be studied by similar methods, so we'll just focus on speed. Even now, the question is not precise enough. Does the person mean "How fast is it on this particular problem and this particular machine using this particular code and this particular compiler?" We could answer this simply by running the program! Unfortunately, that doesn't tell us what would happen with other machines or with other problems that the algorithm is designed to handle.

We would like to answer a question such as "How fast is Algorithm 1 for finding a spanning tree?" in such a way that we can compare that answer to "How fast is Algorithm 2 for finding a spanning tree?" and obtain something that is not machine or problem dependent. At first, this may sound like an impossible goal. To some extent it is; however, quite a bit can be said.

How do we achieve machine independence? We think in terms of simple machine operations such as multiplication, fetching from memory and so on. If one algorithm uses fewer of these than another, it should be faster. Those of you familiar with computer

Basic Concepts in Graph Theory

instruction timing will object that different basic machine operations take different amounts of time. That's true, but the times are not wildly different. Thus, if one algorithm uses a *lot fewer* operations than another, it should be faster. It should be clear from this that we can be a bit sloppy about what we call an operation; for example, we might call something like $x = a + b$ one operation. On the other hand, we can't be so sloppy that we call $x = a_1 + \cdots + a_n$ one operation if n is something that can be arbitrarily large.

Example 22 (Finding the maximum) Let's look at how long it takes to find the maximum of a list of n integers where we know nothing about the order they are in or how big the integers are. Let $a_1, \ldots, a_n$ be the list of integers. Here's our algorithm for finding the maximum.

$max = a_1$
For $i = 2, \ldots, n$
  If $a_i > max$, then $max = a_i$.
End for
Return max

Being sloppy, we could say that the entire comparison and replacement in the "If" takes an operation and so does the stepping of the index i. Since this is done $n - 1$ times, we get $2n - 2$ operations. There are some setup and return operations, say s, giving a total of $2n - 2 + s$ operations. Since all this is rather sloppy all we can really say is that for large n and actual code on an actual machine, the procedure will take about Cn "ticks" of the machine's clock. Since we can't determine C by our methods, it will be helpful to have a notation that ignores it. We use $\Theta(f(n))$ to designate any function that behaves like a constant times $f(n)$ for arbitrarily large n. Thus we would say that the "If" takes time $\Theta(n)$ and the setup and return takes time $\Theta(1)$. Thus the total time is $\Theta(n) + \Theta(1)$. Since n is much bigger than 1 for large n, the total time is $\Theta(n)$. □

We need to define Θ more precisely and list its most important properties. We will also find it useful to define O, read "big oh."

Definition 21 (Notation for Θ and O) Let f, g and h be functions from the positive integers to the nonnegative real numbers. We say that $g(n)$ is $\Theta(f(n))$ if there exist positive constants A and B such that $Af(n) \leq g(n) \leq Bf(n)$ for all sufficiently large n. In this case we say that f and g grow at the same rate. We say that $h(n)$ is $O(f(n))$ if there exists a positive constant B such that $h(n) \leq Bf(n)$ for all sufficiently large n. In this case we say that h grows no faster than f or, equivalently, that f grows at least as fast as h.

The phrase "$S(n)$ is true for all sufficiently large n" means that there is some integer N such that $S(n)$ is true whenever $n \geq N$. Saying that something is $\Theta(f(n))$ gives an idea of *how big it is* for large values of n. Saying that something is $O(f(n))$ gives an idea of *an upper bound on how big it is* for all large values of n. (We said "idea of" because we don't know what the constants A and B are.)

Theorem 7 (Some properties of Θ and O) We have

(a) If $g(n)$ is $\Theta(f(n))$, then $g(n)$ is $O(f(n))$.

Section 4: Rates of Growth and Analysis of Algorithms

(b) $f(n)$ is $\Theta(f(n))$ and $f(n)$ is $O(f(n))$.

(c) If $g(n)$ is $\Theta(f(n))$ and C and D are positive constants, then $Cg(n)$ is $\Theta(Df(n))$.
If $g(n)$ is $O(f(n))$ and C and D are positive constants, then $Cg(n)$ is $O(Df(n))$.

(d) If $g(n)$ is $\Theta(f(n))$, then $f(n)$ is $\Theta(g(n))$.

(e) If $g(n)$ is $\Theta(f(n))$ and $f(n)$ is $\Theta(h(n))$, then $g(n)$ is $\Theta(h(n))$.
If $g(n)$ is $O(f(n))$ and $f(n)$ is $O(h(n))$, then $g(n)$ is $O(h(n))$.

(f) If $g_1(n)$ is $\Theta(f_1(n))$, $g_2(n)$ is $\Theta(f_2(n))$, then $g_1(n)+g_2(n)$ is $\Theta(\max(f_1(n), f_2(n)))$.
If $g_1(n)$ is $O(f_1(n))$, $g_2(n)$ is $O(f_2(n))$, then $g_1(n)+g_2(n)$ is $O(\max(f_1(n), f_2(n)))$.

Note that as a consequence of properties (b), (d) and (e) above, the statement "$g(n)$ is $\Theta(f(n))$" defines an equivalence relation on the set of functions from the positive integers to the nonnegative reals. As with any equivalence relation, we can think of it globally as partition into equivalence classes or locally as a relation between pairs of elements in the set on which the equivalence relation is defined. In the former sense "$g(n)$ is $\Theta(f(n))$" means that "$g(n)$ belongs to the equivalence class $\Theta(f(n))$ associated with f." In the latter sense, "$g(n)$ is $\Theta(f(n))$" means $g \sim_\Theta f$ where $\sim_\Theta$ is an equivalence relation called "is Θ."

Proof: Most of the proofs are left as an exercise. We'll do (e) for Θ. We are given that there are constants A_i and B_i such that

$$A_1 f(n) \leq g(n) \leq B_1 f(n)$$

and

$$A_2 h(n) \leq f(n) \leq B_2 h(n)$$

for all sufficiently large n. It follows that

$$A_1 A_2 h(n) \leq A_1 f(n) \leq g(n) \leq B_1 f(n) \leq B_1 B_2 h(n)$$

for all sufficiently large n. With $A = A_1 A_2$ and $B = B_1 B_2$, it follows that $g(n)$ is $\Theta(h(n))$. $\square$

Example 23 (Additional observations on Θ and O) In this example, we have collected some additional information about our notation.

Functions which are not always positive. Our definitions of Θ and O are only for functions whose values are nonnegative. The definitions can be extended to arbitrary functions by using absolute values; e.g., $A|f(n)| \leq |g(n)| \leq B|f(n)|$ means $g(n) = \Theta(f(n))$. All the results in the theorem still hold except (f) for Θ. This observation is most often applied to the case where the function f is "eventually" nonnegative ($\exists M$ such that $\forall n > M, f(n) \geq 0$). This is the case, for example with any polynomial in n with positive coefficient for the highest power of n.

Taking limits. When comparing two well-behaved functions $f(n)$ and $g(n)$, limits can be helpful:

$$\lim_{n \to \infty} \frac{g(n)}{f(n)} = C > 0 \quad \text{implies} \quad g(n) \text{ is } \Theta(f(n))$$

[183]

Basic Concepts in Graph Theory

and
$$\lim_{n\to\infty} \frac{g(n)}{f(n)} = C \geq 0 \quad \text{implies} \quad g(n) \text{ is } O(f(n)).$$

We assume here that the function f is never zero past some integer N so that the ratio is defined. The constants A and B of the definition can, in the first case, be taken to be $C - \epsilon$ and $C + \epsilon$, where ϵ is any positive number ($\epsilon = 1$ is a simple choice). In the second case, take B to be $C + \epsilon$. If, in the first case, $C = 1$, then f and g are said to be *asymptotic* or asymptotically equal. This is written $f \sim g$. If, in the second case, $C = 0$, then g is said to be *little oh* of f (written $g = o(f)$). We will not use the "asymptotic" and "little oh" concepts.

Polynomials. In particular, you can take any polynomial as $f(n)$, say $f(n) = a_k n^k + \cdots + a_0$, and any other polynomial as $g(n)$, say $g(n) = b_k n^k + \cdots + b_0$. For f and g to be eventually positive we must have both a_k and b_k positive. If that is so, then $g(n)$ is $\Theta(f(n))$. Note in particular that we must have $g(n)$ is $\Theta(n^k)$.

Logarithms. Two questions that arise concerning logarithms are (a) "What base should I use?" and (b) "How fast do they grow?"

The base does not matter because $\log_a x = (\log_a b)(\log_b x)$ and constant factors like $\log_a b$ are ignored in $\Theta(\ )$ and $O(\ )$.

It is known from calculus that $\log n \to \infty$ as $n \to \infty$ and that $\lim_{n\to\infty}(\log n)/n^\epsilon = 0$ for every $\epsilon > 0$. Thus logarithms grow, but they grow slower than powers of n. For example, $n \log n$ is $O(n^{3/2})$ but $n^{3/2}$ is not $O(n \log n)$.

A proof. How do we prove

$$\lim_{n\to\infty} \frac{g(n)}{f(n)} = C > 0 \quad \text{implies} \quad g(n) \text{ is } \Theta(f(n))?$$

By definition, the limit statement means that for any $\epsilon > 0$ there exists N such that for all $n > N$, $|\frac{g(n)}{f(n)} - C| < \epsilon$. If $\epsilon \geq C$, replace it with a smaller ϵ. From $|\frac{g(n)}{f(n)} - C| < \epsilon$, for all $n > N$,

$$C - \epsilon < \frac{g(n)}{f(n)} < C + \epsilon, \quad \text{or} \quad (C - \epsilon)f(n) < \frac{g(n)}{f(n)} < (C + \epsilon)f(n).$$

Take $A = (C - \epsilon)$ and $B = (C + \epsilon)$ in the definition of Θ. □

Example 24 (Using Θ) To illustrate these ideas, we'll consider three algorithms for evaluating a polynomial $p(x)$ of degree n at some point r; i.e., computing $p_0 + p_1 r + \cdots + p_n r^n$. We are interested in how fast they are when n is large. Here are the procedures. You should convince yourself that they work.

```
Poly1(n, p, r)
    S = p_0
    For i = 1, ..., n    S = S + p_i * Pow(r, i).
    Return S
End
```

GT-40

Section 4: Rates of Growth and Analysis of Algorithms

$\text{Pow}(r, i)$
    $P = 1$
    For $j = 1, \ldots, n$ $P = P * r$.
    Return P
End

$\text{Poly2}(n, p, r)$
    $S = p_0$
    $P = 1$
    For $i = 1, \ldots, n$
        $P = P * r$.
        $S = S + p_i * P$
    End for
    Return S
End

$\text{Poly3}(n, p, r)$
    $S = p_n$
    For $i = n, \ldots, 2, 1$ $S = S * r + p_{i-1}$
    Return S
End

Let $T_n(\text{Name})$ be the time required for the procedure Name. Let's analyze Poly1. The "**For**" loop in Pow is executed i times and so takes Ci operations for some constant C. The setup and return in Pow takes some constant number of operations D. Thus $T_n(\text{Pow}) = Ci + D$ operations. As a result, the ith iteration of the "**For**" loop in Poly1 takes $Ci + E$ operations for some constants C and $E > D$. Adding this over $i = 1, 2, \ldots, n$, we see that the total time spent in the "**For**" loop is $\Theta(n^2)$ since $\sum_{i=1}^{n} i = n(n+1)/2$. (This requires using some of the properties of Θ. You should write out the details.) Since the rest of Poly1 takes $\Theta(1)$ time, $T_n(\text{Poly1})$ is $\Theta(n^2)$.

The amount of time spent in the "**For**" loop of Poly2 is constant and the loop is executed n times. It follows that $T_n(\text{Poly2})$ is $\Theta(n)$. The same analysis applies to Poly3.

What can we conclude from this about the comparative speed of the algorithms? By the definition of Θ, there are positive reals A and B so that $An^2 \leq T_n(\text{Poly1})$ and $T_n(\text{Poly2}) \leq Bn$ for sufficiently large n. Thus $T_n(\text{Poly2})/T_n(\text{Poly1}) \leq B/An$. As n gets larger, Poly2 looks better and better compared to Poly1.

Unfortunately, the crudeness of Θ does not allow us to make any distinction between Poly2 and Poly3. What we can say is that $T_n(\text{Poly2})$ is $\Theta(T_n(\text{Poly3}))$; i.e., $T_n(\text{Poly2})$ and $T_n(\text{Poly3})$ grow at the same rate. A more refined estimate can be obtained by counting the actual number of operations involved. □

So far we have talked about how long an algorithm takes to run as if this were a simple, clear concept. In the next example we'll see that there's an important point that we've ignored.

*Example 25 (What is average running time?) Let's consider the problem of (a) deciding whether or not a simple graph can be properly colored with four colors and, (b) if a proper coloring exists, producing one. A *proper coloring* of a simple graph

[185]

Basic Concepts in Graph Theory

$G = (V, E)$ is a function $\lambda: V \to C$, the set of "colors," such that, if $\{u, v\}$ is an edge, then $\lambda(u) \neq \lambda(v)$. We may as well assume that $V = \underline{n}$ and that the colors are c_1, c_2, c_3 and c_4.

Here's a simple algorithm to determine a λ by using backtracking to go lexicographically through possible colorings $\lambda(1), \lambda(2), \ldots, \lambda(n)$.

1. **Initialize:** Set $v = 1$ and $\lambda(1) = c_1$.

2. **Advance in decision tree:** If $v = n$, stop with λ determined; otherwise, set $v = v + 1$ and $\lambda(v) = c_1$.

3. **Test:** If $\lambda(i) \neq \lambda(v)$ for all $i < v$ for which $\{i, v\} \in E$, go to Step 2.

4. **Select next decision:** Let j be such that $\lambda(v) = c_j$. If $j < 4$, set $\lambda(v) = c_{j+1}$ and go to Step 3.

5. **Backtrack:** If $v = 1$, stop with coloring impossible; otherwise, set $v = v - 1$ and go to Step 4.

How fast is this algorithm? Obviously it will depend on the graph. Here are two extreme cases:

- Suppose the subgraph induced by the first five vertices is the complete graph K_5 (i.e., all of the ten possible edges are present). The algorithm stops after trying to color the first five vertices and discovering that there is no proper coloring. Thus the running time does not depend on n and so is in $\Theta(1)$.

- Suppose that the first $n - 5$ vertices have no edges and that the last five vertices induce K_5. The algorithm tries all possible assignments of colors to the first $n - 5$ vertices and, for each of them, discovers that it cannot properly color the last five because they form K_5. Thus the algorithm makes between 4^{n-5} and 4^n assignments of colors and so its running time is $\Theta(4^n)$ — a much faster growing time than $\Theta(1)$.

What should we do about studying the running time of such an algorithm? It's reasonable to talk about the *average* time the algorithm takes if we expect to give it lots of graphs to look at. Most n vertex graphs will have many sets of five vertices that induce K_5. (We won't prove this.) As a result, the algorithm has running time in $\Theta(1)$ for most graphs. In fact, it can be proved that the average number of assignments of the form $\lambda(v) = c_k$ that are made is $\Theta(1)$ and so the average running time is $\Theta(1)$. This means that the average running time of the algorithm is bounded for all n, which is quite good!

Now suppose you give this algorithm to a friend, telling her that the average running time is bounded. She thanks you profusely for such a wonderful algorithm and puts it to work coloring randomly generated "planar" graphs. These are a special class of graphs whose pictures can be drawn in the plane without edges crossing each other. (All trees are planar, but K_5 is not planar.) By a famous theorem called the *Four Color Theorem*, every planar graph can be properly colored with four colors, so the algorithm will find the coloring. To do so it must make assignments of the form $\lambda(v) = c_k$ for each vertex v. Thus it must make at least n assignments. (Actually it will almost surely make *many, many* more.) Your friend soon comes back to you complaining that your algorithm takes a long time to run. What went wrong?

You were averaging over all simple graphs with n vertices. Your friend was averaging over all simple planar graphs with n vertices. The average running times are *very* different! There is a lesson here:

> You must be VERY clear what you are averaging over.

Section 4: Rates of Growth and Analysis of Algorithms

Because situations like this do occur in real life, computer scientists are careful to specify what kind of running time they are talking about; either the average of the running time over some reasonable, clearly specified set of problems or the worst (longest) running time over all possibilities. □

You should be able to see that saying something is $\Theta(\ )$ leaves a lot out because we have no idea of the constants that are omitted. How can we compare two algorithms? Here are two rules of thumb.

- If one algorithm is $\Theta(f(n))$ and the other is $\Theta(g(n))$, the algorithm with the slower growing function (f or g) is probably the better choice.

- If both algorithms are $\Theta(f(n))$, the algorithm with the simpler data structures is probably better.

These rules are far from foolproof, but they provide some guidance.

---

*Polynomial Time Algorithms

Computer scientists talk about "polynomial time algorithms." What does this mean? Suppose that the algorithm can handle arbitrarily large problems and that it takes $\Theta(n)$ seconds on a problem of "size" n. Then we call it a linear time algorithm. More generally, if there is a (possibly quite large) integer k such that the worst case running time on a problem of "size" n is $O(n^k)$, then we say the algorithm is polynomial time.

You may have noticed the quotes around size and wondered why. It is necessary to specify what we mean by the size of a problem. Size is often interpreted as the number of bits required to specify the problem in binary form. You may object that this is imprecise since a problem can be specified in many ways. This is true; however, the number of bits in one "reasonable" representation doesn't differ too much from the number of bits in another. We won't pursue this further.

If the worst case time for an algorithm is polynomial, theoretical computer scientists think of this as a good algorithm. (This is because polynomials grow relatively slowly; for example, exponential functions grow much faster.) The problem that the algorithm solves is called *tractable*.

Do there exist *intractable problems*; i.e., problems for which no polynomial time algorithm can ever be found? Yes, but we won't study them here. More interesting is the fact that there are a large number of practical problems for which

- no polynomial time algorithm is known and

- no one has been able prove that the problems are intractable.

We'll discuss this a bit.

Consider the following problems.

- **Coloring problem:** For any $c > 2$, devise an algorithm whose input can be any simple graph and whose output answers the question "Can the graph be properly colored in c colors?"

Basic Concepts in Graph Theory

- **Traveling salesman problem:** For any B, devise an algorithm whose input can be any $n > 0$ and any real valued edge labeling, $\lambda \colon \mathcal{P}_2(\underline{n}) \to \mathbb{R}$, for K_n, the complete graph on n vertices. The algorithm must answer the question "Is there a cycle through all n vertices with cost B or less?" (The cost of a cycle is the sum of $\lambda(e)$ over all e in the cycle.)

- **Clique problem:** Given a simple graph $G = (V, E)$ and an integer s, is there a subset $S \subseteq V$, $|S| = s$, whose induced subgraph is the complete graph on S (i.e., a subgraph of G with vertex set S and with $\binom{s}{2}$ edges)?

No one knows if these problems are tractable, but it is known that, if one is tractable, then they all are. There are hundreds more problems that people are interested in which belong to this particular list in which all or none are tractable. These problems are called *NP-complete problems* Many people regard deciding if the NP-complete problems are tractable to be the foremost open problem in theoretical computer science.

The NP-complete problems have an interesting property which we now discuss. If the algorithm says "yes," then there must be a specific example that shows why this is so (an assignment of colors, a cycle, an automaton). There is no requirement that the algorithm actually produce such an example. Suppose we somehow obtain a coloring, a cycle or an automaton which is claimed to be such an example. Part of the definition of NP-complete requires that we be able to check the claim in polynomial time. Thus we can check a purported example quickly but, so far as is known, it may take a long time to determine if such an example exists. In other words, I can check your guesses quickly but I don't know how to tell you quickly if any examples exist.

There are problems like the NP-complete problems where no one knows how to do any checking in polynomial time. For example, modify the traveling salesman problem to ask for the minimum cost cycle. No one knows how to verify in polynomial time that a given cycle is actually the minimum cost cycle. If the modified traveling salesman problem is tractable, so is the one we presented above: You need only find the minimum cost cycle and compare its cost to B. Such problems are called *NP-hard* because they are at least as hard as NP-complete problems. A problem which is tractable if the NP-complete problems are tractable is called *NP-easy*.

Some problems are both NP-easy and NP-hard but may not be NP-complete. Why is this? NP-complete problems must ask a "yes/no" type of question and it must be possible to check a specific example in polynomial time as noted in the previous paragraph. We discuss an example.

*Example 26 (Chromatic number) The *chromatic number* $\chi(G)$ of a graph G is the least number of colors needed to properly color G. The problem of deciding whether a graph can be properly colored with c colors is NP-complete. The problem of determining $\chi(G)$ is NP-hard. If we know $\chi(G)$, then we can determine if c colors are enough by checking if $c \geq \chi(G)$.

The problem of determining $\chi(G)$ is also NP-easy. You can color G with c colors if and only if $c \geq \chi(G)$. We know that $0 \leq \chi(G) \leq n$ for a graph with n vertices. Ask if c colors suffice for $c = 0, 1, 2, \ldots$. The least c for which the answer is "yes" is $\chi(G)$. Thus the worst case time for finding $\chi(G)$ is at most n times the worst case time for the NP-complete problem. Hence one time is O of a polynomial in n if and only if the other is. ∎

Section 4: Rates of Growth and Analysis of Algorithms

What can we do if we cannot find a good algorithm for a problem? There are three main types of partial algorithms:

1. **Almost good:** It is polynomial time for all but a very small subset of possible problems. (If we are interested in all graphs, our coloring algorithm in Example 25 is almost good for any fixed c.)

2. **Almost correct:** It is polynomial time but in some rare cases does not find the correct answer. (If we are interested in all graphs and a fixed c, automatically reporting that a large graph can't be colored with c colors is almost correct — but it is rather useless.) In some situations, a fast almost correct algorithm can be useful.

3. **Close:** It is a polynomial time algorithm for a minimization problem and comes close to the true minimum. (There are useful close algorithms for approximating the minimum cycle in the Traveling Salesman Problem.)

Some of the algorithms make use of random number generators in interesting ways. Unfortunately, further discussion of these problems is beyond the scope of this text.

---

*A Theorem for Recursive Algorithms

Some algorithms, such as merge sorting, call themselves. This is known as a *recursive algorithm* or a *divide and conquer algorithm*.

When we try estimate the running time of such algorithms, we obtain a recursion. In Section 2 of Unit DT, we examined the problem of solving recursions. We saw that finding exact solutions to recursions is difficult. The recursions that we obtain for algorithms are not covered by the methods in that section. Furthermore, the recursions are often not known exactly because we may only be able to obtain an estimate of the form $\Theta(\ )$ for some of the work. The next example illustrates this problem.

*Example 27 (Sorting by recursive merging)** Given a list L of n items, we wish to sort it. Here is the merge sorting algorithm from Section 3 of Unit DT.

```
Sort(L)
    If length is 1, return L
    Else
        Split L into two lists L1 and L2
        S1 = Sort(L1)
        S2 = Sort(L2)
        S = Merge(L1, L2)
        Return S
    End if
End
```

We need to be more specific about how the lists are split. Let m be $n/2$ rounded down, let L1 be the first m items in L and let L2 be the last $n - m$ items in L.

Basic Concepts in Graph Theory

One way to measure the running time of Sort(L) is to count the number of comparisons that are required. Let this number be $T(n)$. We would like to know how fast $T(n)$ grows as a function of n so we can tell how good the algorithm is. For example, is $T(n) = \Theta(n)$? is $T(n) = \Theta(n^2)$? or does it behave differently?

We now start work on this problem. Since the sorting algorithm is recursive (calls itself), we will end up with a recursion. This is a general principle for recursive algorithms. You should see why after the next two paragraphs.

All comparisons are done in Merge(L1,L2). It can be shown that the number of comparisons in Merge is between m and $n-1$. We take that fact as given.

Three lines of code are important:

S1 = Sort(L1)	a recursive call, so it gives us $T(m)$;
S2 = Sort(L2)	a recursive call, so it gives us $T(n-m)$;
S = Merge(L1, L2)	where the comparisons are, so it gives us a_n with $m \leq a_n \leq n = 1$.

We obtain $T(n) = T(m) + T(n-m) + a_n$ where all we know about a_n is that it is between m and $n-1$. What can we do?

Not only is this a type of recursion we haven't seen before, we don't even know the recursion fully since all we have is upper and lower bounds for a_n. The next theorem solves this problem for us. ∎

The following theorem provides an approximate solution to an important class of approximate recursions that arise in divide and conquer algorithms. We'll apply it to merge sorting. In the theorem

- $T(n)$ is the running time for a problem of size n.

- If the algorithm calls itself at w places in the code, then the problem is divided into w smaller problems of the same kind and $s_1(n), \ldots, s_w(n)$ are the sizes of the smaller problems.

- The constant c measures how much smaller each of these problems is.

- The time needed for the rest of the code is a_n.

Theorem 8 (Master Theorem for Recursions)   *Suppose that there are*

   (i) *numbers $N, b, w \geq 1$ and $0 < c < 1$ that do not depend on n*

   (ii) *a sequence $a_1, a_2, \ldots,$*

   (iii) *functions $s_1, s_2, \ldots, s_w,$ and T*

such that

   (a) *$T(n) > 0$ for all $n > N$ and $a_n \geq 0$ for all $n > N$;*

---

* This is not the most general version of the theorem; however, this version is easier to understand and is usually sufficient. For a more general statement and a proof, see any thorough text on the analysis of algorithms.

Section 4: Rates of Growth and Analysis of Algorithms

(b) $T(n) = a_n + T(s_1(n)) + T(s_2(n)) + \cdots + T(s_w(n))$ for all $n > N$;

(c) a_n is $\Theta(n^b)$ (If $a_n = 0$ for all large n, set $b = -\infty$.);

(d) $|s_i(n) - cn|$ is $O(1)$ for $i = 1, 2, \ldots, w$.

Let $d = -\log(w)/\log(c)$. Then

$$T(n) \text{ is } \begin{cases} \Theta(n^d) & \text{if } b < d, \\ \Theta(n^d \log n) & \text{if } b = d, \\ \Theta(n^b) & \text{if } b > d. \end{cases}$$

Note that $b = 0$ corresponds to a_n being in $\Theta(1)$ since $n^0 = 1$. In other words, a_n is bounded by nonzero constants for all large n: $0 < C_1 \leq a_n \leq C_2$.

Let's apply the theorem to our recursion for merge sorting:

$$T(n) = a_n + T(s_1(n)) + T(s_2(n))$$

where

$$s_1(n) = \lfloor n/2 \rfloor, \quad s_2(n) = \lfloor n - n/2 \rfloor \quad \text{and} \quad s_1(n) \leq a_n \leq n - 1.$$

Note that $s_1(n)$ and $s_2(n)$ differ from $n/2$ by at most $1/2$ and that $a_n = \Theta(n)$. Thus we can apply the theorem with $w = 2$, $b = 1$ and $c = 1/2$. We have

$$d = -\log(2)/\log(1/2) = \log(2)/\log(2) = 1.$$

Since $b = d = 1$, we conclude that $T(n)$ is $\Theta(n \log n)$.

How do we use the theorem on divide and conquer algorithms? First, we must find a parameter n that measures the size of the problem; for example, the length of a list to be sorted, the degree of polynomials that we want to multiply, the number of vertices in a graph that we want to study. Then use the interpretation of the various parameters that was given just before the theorem.

Our final example is more difficult because the algorithm that we study is more complicated. It was believed for some time that the quickest way to multiply polynomials was the "obvious" way that is taught when polynomials are first studied. That is not true. The next example contains an algorithm for faster multiplication of polynomials. There are also faster algorithms for multiplying matrices.

*Example 28 (Recursive multiplication of polynomials) Suppose we want to multiply two polynomials of degree at most n, say

$$P(x) = p_0 + p_1 x + \cdots + p_n x^n \quad \text{and} \quad Q(x) = q_0 + q_1 x + \cdots + q_n x^n.$$

The natural way to do this is to use the distributive law to generate $(n+1)^2$ products $p_0 q_0, p_0 q_1 x, p_0 q_2 x^2, \ldots, p_n q_n x^{2n}$ and then collect the terms that have the same powers of x. This involves $(n+1)^2$ multiplications of coefficients and, it can be shown, n^2 additions of coefficients. Thus, the amount of work is $\Theta(n^2)$. Unless we expect $P(x)$ or $Q(x)$ to have

[191]

GT-47

Basic Concepts in Graph Theory

some coefficients that are zero, this seems to be best we can do. Not so! We now present and analyze a faster recursive algorithm.

The algorithm depends on the following identity which you should verify by checking the algebra.

Identity: If $P_L(x)$, $P_H(x)$, $Q_L(x)$ and $Q_H(x)$ are polynomials, then

$$(P_L(x) + P_H(x)x^m)(Q_L(x) + Q_H(x)x^m) = A(x) + (C(x) - A(x) - B(x))x^m + B(x)x^{2m}$$

where

$$A(x) = P_L(x)Q_L(x), \quad B(x) = P_H(x)Q_H(x),$$

and

$$C(x) = (P_L(x) + P_H(x))(Q_L(x) + Q_H(x))$$

We can think of this identity as telling us how to multiply two polynomials $P(x)$ and $Q(x)$ by splitting them into lower degree terms ($P_L(x)$ and $Q_L(x)$) and higher degree terms ($P_H(x)x^m$ and $Q_H(x)x^m$):

$$P(x) = D(P_L(x) + P_H(x)x^m \quad \text{and} \quad Q(x) = Q_L(x) + Q_H(x)x^m.$$

The identity requires three polynomial multiplications to compute $A(x)$, $B(x)$ and $C(x)$. This leads naturally to two questions:

- Haven't things gotten worse — three polynomial multiplications instead of just one? No. The three multiplications involve polynomials of much lower degrees. We will see that this leads to a gain in speed.

- How should we do these three polynomial multiplications? Apply the identity to each of them. In other words, design a recursive algorithm. We do that now.

Here is the algorithm for multiplying two polynomials $P(x) = p_0 + p_1 x \cdots + p_n x^n$ and $Q(x) = q_0 + q_1 x + \cdots + q_n x^n$ of degree at most n.

```
MULT(P(x), Q(x), n)
    If (n=0) Return p₀q₀
    Else
        Let m = n/2 rounded up.
        P_L(x) = p₀ + p₁x + ··· p_{m-1} x^{m-1}
        P_H(x) = p_m + p_{m+1} x + ··· p_n x^{n-m}
        Q_L(x) = q₀ + p₁x + ··· q_{m-1} x^{m-1}
        Q_H(x) = q_m + q_{m+1} x + ··· q_n x^{n-m}
        A(x) = MULT(P_L(x), Q_L(x), m − 1)
        B(x) = MULT(P_H(x), Q_H(x), n − m)
        C(x) = MULT(P_L(x) + P_H(x), Q_L(x) + Q_H(x), n − m)
        D(x) = A(x) + (C(x) − A(x) − B(x))x^m + B(x)x^{2m}
        Return D(x)
    End if
End
```

As is commonly done, we imagine a polynomial stored as a vector of coefficients. The amount of work required is then the number of times we have two multiply or add two

Section 4: Rates of Growth and Analysis of Algorithms

coefficients. For simplicity, we just count multiplications. Let that number be $T(n)$. You should be able to see that $T(0) = 1$ and

$$T(n) = T(m-1) + T(n-m) + T(n-m) \quad \text{for} \quad n > 0.$$

We can write this as

$$T(n) = T(m-1) + T(n-m) + T(n-m) + a_n, \quad a_0 = 1 \text{ and } a_n = 0 \text{ for } n > 0.$$

Note that, since both $m-1$ and $n-m$ differ from $n/2$ by at most 1, $w = 3$ and $c = 1/2$. Also $b = -\infty$.

We have $d = \log 3/\log 2 > b$. Thus $T(n)$ is $\Theta(n^{\log 3/\log 2})$. Since $\log 3/log2$ is about 1.6 which is less than 2, this is less work than the straightforward method when n is large enough. (Recall that the work there was in $\Theta(n^2)$.) $\square$

Exercises for Section 4

4.1. We have three algorithms for solving a problem for graphs. Suppose algorithm A takes n^2 milliseconds to run on a graph with n vertices, algorithm B takes $100n$ milliseconds and algorithm C takes $100(2^{n/10} - 1)$ milliseconds.

   (a) Compute the running times for the three algorithms with $n = 5, 10, 30, 100$ and 300. Which algorithm is fastest in each case? slowest?

   (b) Which algorithm is fastest for all very large values of n? Which is slowest?

4.2. Let $p(x)$ be a polynomial of degree k with positive leading coefficient and suppose that $a > 1$. Prove the following.

   (a) $\Theta(p(n))$ is $\Theta(n^k)$.

   (b) $O(p(n))$ is $O(n^k)$.

   (c) $\lim_{n \to \infty} p(n)/a^n = 0$. (Also, what does this say about the speed of a polynomial time algorithm versus one which takes exponential time?)

   (d) Unless $p(x) = p_1 x^k + p_2$ for some p_1 and p_2, there is no C such that $a^{p(n)}$ is $\Theta(a^{Cn^k})$.

4.3. In each case, prove that $g(n)$ is $\Theta(f(n))$ using the definition of "g is $\Theta(f)$". (See Definition 21.)

   (a) $g(n) = n^3 + 5n^2 + 10$, $f(n) = 20n^3$.

   (b) $g(n) = n^2 + 5n^2 + 10$, $f(n) = 200n^2$

4.4. In each case, show that the given series has the indicated property.

Basic Concepts in Graph Theory

    (a) $\sum_{i=1}^{n} i^2$ is $\Theta(n^3)$.

    (b) $\sum_{i=1}^{n} i^3$ is $\Theta(n^4)$.

    (c) $\sum_{i=1}^{n} i^{1/2}$ is $\Theta(n^{3/2})$.

4.5. Show each of the following

    (a) $\sum_{i=1}^{n} i^{-1}$ is $\Theta(\log_b(n))$ for any base $b > 1$.

    (b) $\log_b(n!)$ is $O(n \log_b(n))$ for any base $b > 1$.

    (c) $n!$ is $\Theta((n/e)^{n+1/2})$.

*4.6. The following algorithm multiplies two $n \times n$ matrices A and B and puts the answer in C. Let $T(n)$ be the running time of the algorithm Find a simple function $f(n)$ so that is $\Theta(f(n))$.

```
MATRIXMULT(n,A,B,C)
    For i=1,...,n
        For j=1,...,n
            C(i,j)=0
            For k=1,...,n
                C(i,j) = C(i,j) + A(i,k)*B(k,j)
            End for
        End for
    End for
End
```

*4.7. The following algorithm computes x^n for n a positive integer, where x is a complicated object (e.g., a large matrix). MULT(x, y) is a procedure that multiplies two such objects. Let $T(n)$ be the number of times MULT is called. Find a simple function $f(n)$ so that $T(n)$ is $\Theta(f(n))$.

```
POW(x,n)
    If (n=1) Return x
    Else
        Let q be n/2 rounded down and r = n - 2q.
        y = MULT(x,x)
        z = POW(y,q)
        If (r=0) Return z
        Else
            w = MULT(x,z)
            Return w
        End if
    End if
End
```

Review Questions

Multiple Choice Questions for Review

Some of the following questions assume that you have done the exercises.

1. Indicate which, if any, of the following five graphs $G = (V, E, \phi)$, $|V| = 5$, is not isomorphic to any of the other four.

   (a) $\phi = \begin{pmatrix} A & B & C & D & E & F \\ \{1,3\} & \{2,4\} & \{1,2\} & \{2,3\} & \{3,5\} & \{4,5\} \end{pmatrix}$

   (b) $\phi = \begin{pmatrix} f & b & c & d & e & a \\ \{1,2\} & \{1,2\} & \{2,3\} & \{3,4\} & \{3,4\} & \{4,5\} \end{pmatrix}$

   (c) $\phi = \begin{pmatrix} b & f & e & d & c & a \\ \{4,5\} & \{1,3\} & \{1,3\} & \{2,3\} & \{2,4\} & \{4,5\} \end{pmatrix}$

   (d) $\phi = \begin{pmatrix} 1 & 2 & 3 & 4 & 5 & 6 \\ \{1,2\} & \{2,3\} & \{2,3\} & \{3,4\} & \{4,5\} & \{4,5\} \end{pmatrix}$

   (e) $\phi = \begin{pmatrix} b & a & e & d & c & f \\ \{4,5\} & \{1,3\} & \{1,3\} & \{2,3\} & \{2,5\} & \{4,5\} \end{pmatrix}$

2. Indicate which, if any, of the following five graphs $G = (V, E, \phi)$, $|V| = 5$, is not connected.

   (a) $\phi = \begin{pmatrix} 1 & 2 & 3 & 4 & 5 & 6 \\ \{1,2\} & \{1,2\} & \{2,3\} & \{3,4\} & \{1,5\} & \{1,5\} \end{pmatrix}$

   (b) $\phi = \begin{pmatrix} b & a & e & d & c & f \\ \{4,5\} & \{1,3\} & \{1,3\} & \{2,3\} & \{2,5\} & \{4,5\} \end{pmatrix}$

   (c) $\phi = \begin{pmatrix} b & f & e & d & c & a \\ \{4,5\} & \{1,3\} & \{1,3\} & \{2,3\} & \{2,4\} & \{4,5\} \end{pmatrix}$

   (d) $\phi = \begin{pmatrix} a & b & c & d & e & f \\ \{1,2\} & \{2,3\} & \{1,2\} & \{2,3\} & \{3,4\} & \{1,5\} \end{pmatrix}$

   (e) $\phi = \begin{pmatrix} a & b & c & d & e & f \\ \{1,2\} & \{2,3\} & \{1,2\} & \{1,3\} & \{2,3\} & \{4,5\} \end{pmatrix}$

3. Indicate which, if any, of the following five graphs $G = (V, E, \phi)$, $|V| = 5$, have an Eulerian circuit.

   (a) $\phi = \begin{pmatrix} F & B & C & D & E & A \\ \{1,2\} & \{1,2\} & \{2,3\} & \{3,4\} & \{4,5\} & \{4,5\} \end{pmatrix}$

   (b) $\phi = \begin{pmatrix} b & f & e & d & c & a \\ \{4,5\} & \{1,3\} & \{1,3\} & \{2,3\} & \{2,4\} & \{4,5\} \end{pmatrix}$

   (c) $\phi = \begin{pmatrix} 1 & 2 & 3 & 4 & 5 & 6 \\ \{1,2\} & \{1,2\} & \{2,3\} & \{3,4\} & \{4,5\} & \{4,5\} \end{pmatrix}$

   (d) $\phi = \begin{pmatrix} b & a & e & d & c & f \\ \{4,5\} & \{1,3\} & \{1,3\} & \{2,3\} & \{2,5\} & \{4,5\} \end{pmatrix}$

   (e) $\phi = \begin{pmatrix} a & b & c & d & e & f \\ \{1,3\} & \{3,4\} & \{1,2\} & \{2,3\} & \{3,5\} & \{4,5\} \end{pmatrix}$

4. A graph with $V = \{1, 2, 3, 4\}$ is described by $\phi = \begin{pmatrix} a & b & c & d & e & f \\ \{1,2\} & \{1,2\} & \{1,4\} & \{2,3\} & \{3,4\} & \{3,4\} \end{pmatrix}$. How many Hamiltonian cycles does it have?

   (a) 1   (b) 2   (c) 4   (d) 16   (e) 32

[195]

GT-51

Basic Concepts in Graph Theory

5. A graph with $V = \{1, 2, 3, 4\}$ is described by $\phi = \begin{pmatrix} a & b & c & d & e & f \\ \{1,2\} & \{1,2\} & \{1,4\} & \{2,3\} & \{3,4\} & \{3,4\} \end{pmatrix}$.
   It has weights on its edges given by $\lambda = \begin{pmatrix} a & b & c & d & e & f \\ 3 & 2 & 1 & 2 & 4 & 2 \end{pmatrix}$. How many minimum spanning trees does it have?

   (a) 2   (b) 3   (c) 4   (d) 5   (e) 6

6. Define an RP-tree by the parent-child adjacency lists as follows:

   (i) Root B: J, H, K;   (ii) H: P, Q, R;   (iii) Q: S, T;   (iv) K: L, M, N.

   The postorder vertex sequence of this tree is

   (a) J, P, S, T, Q, R, H, L, M, N, K, B.
   (b) P, S, T, J, Q, R, H, L, M, N, K, B.
   (c) P, S, T, Q, R, H, L, M, N, K, J, B.
   (d) P, S, T, Q, R, J, H, L, M, N, K, B.
   (e) S, T, Q, J, P, R, H, L, M, N, K, B.

7. Define an RP-tree by the parent-child adjacency lists as follows:

   (i) Root B: J, H, K;   (ii) J: P, Q, R;   (iii) Q: S, T;   (iv) K: L, M, N.

   The preorder vertex sequence of this tree is

   (a) B, J, H, K, P, Q, R, L, M, N, S, T.
   (b) B, J, P, Q, S, T, R, H, K, L, M, N.
   (c) B, J, P, Q, S, T, R, H, L, M, N, K.
   (d) B, J, Q, P, S, T, R, H, L, M, N, K.
   (e) B, J, Q, S, T, P, R, H, K, L, M, N.

8. For which of the following does there exist a graph $G = (V, E, \phi)$ satisfying the specified conditions?

   (a) A tree with 9 vertices and the sum of the degrees of all the vertices 18.
   (b) A graph with 5 components 12 vertices and 7 edges.
   (c) A graph with 5 components 30 vertices and 24 edges.
   (d) A graph with 9 vertices, 9 edges, and no cycles.
   (e) A connected graph with 12 edges 5 vertices and fewer than 8 cycles.

9. For which of the following does there exist a simple graph $G = (V, E)$ satisfying the specified conditions?

   (a) It has 3 components 20 vertices and 16 edges.
   (b) It has 6 vertices, 11 edges, and more than one component.

(c) It is connected and has 10 edges 5 vertices and fewer than 6 cycles.

(d) It has 7 vertices, 10 edges, and more than two components.

(e) It has 8 vertices, 8 edges, and no cycles.

10. For which of the following does there exist a tree satisfying the specified constraints?

    (a) A binary tree with 65 leaves and height 6.

    (b) A binary tree with 33 leaves and height 5.

    (c) A full binary tree with height 5 and 64 total vertices.

    (d) A full binary tree with 23 leaves and height 23.

    (e) A rooted tree of height 3, every vertex has at most 3 children. There are 40 total vertices.

11. For which of the following does there exist a tree satisfying the specified constraints?

    (a) A full binary tree with 31 leaves, each leaf of height 5.

    (b) A rooted tree of height 3 where every vertex has at most 3 children and there are 41 total vertices.

    (c) A full binary tree with 11 vertices and height 6.

    (d) A binary tree with 2 leaves and height 100.

    (e) A full binary tree with 20 vertices.

12. The number of simple digraphs with $|V| = 3$ is

    (a) 2^9 (b) 2^8 (c) 2^7 (d) 2^6 (e) 2^5

13. The number of simple digraphs with $|V| = 3$ and exactly 3 edges is

    (a) 92   (b) 88   (c) 80   (d) 84   (e) 76

14. The number of oriented simple graphs with $|V| = 3$ is

    (a) 27   (b) 24   (c) 21   (d) 18   (e) 15

15. The number of oriented simple graphs with $|V| = 4$ and 2 edges is

    (a) 40   (b) 50   (c) 60   (d) 70   (e) 80

16. In each case the depth-first sequence of an ordered rooted spanning tree for a graph G is given. Also given are the non-tree edges of G. Which of these spanning trees is a depth-first spanning tree?

    (a) 123242151 and $\{3,4\}, \{1,4\}$

    (b) 123242151 and $\{4,5\}, \{1,3\}$

    (c) 123245421 and $\{2,5\}, \{1,4\}$

    (d) 123245421 and $\{3,4\}, \{1,4\}$

    (e) 123245421 and $\{3,5\}, \{1,4\}$

Basic Concepts in Graph Theory

17. $\sum_{i=1}^{n} i^{-1/2}$ is
    (a) $\Theta((\ln(n))^{1/2})$ (b) $\Theta(\ln(n))$ (c) $\Theta(n^{1/2})$ (d) $\Theta(n^{3/2})$ (e) $\Theta(n^2)$

18. Compute the total number of bicomponents in all of the following three simple graphs, $G = (V, E)$ with $|V| = 5$. For each graph the edge sets are as follows:

    $$E = \{\{1,2\}, \{2,3\}, \{3,4\}, \{4,5\}, \{1,3\}, \{1,5\}, \{3,5\}\}$$
    $$E = \{\{1,2\}, \{2,3\}, \{3,4\}, \{4,5\}, \{1,3\}\}$$
    $$E = \{\{1,2\}, \{2,3\}, \{4,5\}, \{1,3\}\}$$

    (a) 4   (b) 5   (c) 6   (d) 7   (e) 8

19. Let $b > 1$. Then $\log_b((n^2)!)$ is
    (a) $\Theta(\log_b(n!))$
    (b) $\Theta(\log_b(2\ n!))$
    (c) $\Theta(n \log_b(n))$
    (d) $\Theta(n^2 \log_b(n))$
    (e) $\Theta(n \log_b(n^2))$

20. What is the total number of additions and multiplications in the following code?

    ```
    s := 0
    for i := 1 to n
        s:= s + i
        for j:= 1 to i
            s := s + j*i
        next j
    next i
    s := s+10
    ```

    (a) n (b) n^2 (c) $n^2 + 2n$ (d) $n(n+1)$ (e) $(n+1)^2$

Answers: 1 (a), 2 (e), 3 (e), 4 (c), 5 (b), 6 (a), 7 (b), 8 (b), 9 (d), 10 (e), 11 (d), 12 (a), 13 (d), 14 (a), 15 (c), 16 (c), 17 (c), 18 (c), 19 (d), 20 (e).

Solutions for Basic Counting and Listing

CL-1.1 This is a simple application of the Rules of Sum and Product.

(a) Choose a discrete math text OR a data structures text, etc. This gives $5 + 2 + 6 + 3 = 16$.

(b) Choose a discrete math text AND a data structures text, etc. This gives $5 \times 2 \times 6 \times 3 = 180$.

CL-1.2 We can form n digit numbers by choosing the leftmost digit AND choosing the next digit AND $\cdots$ AND choosing the rightmost digit. The first choice can be made in 9 ways since a leading zero is not allowed. The remaining $n-1$ choices can each be made in 10 ways. By the Rule of Product we have $9 \times 10^{n-1}$. To count numbers with at most n digits, we could sum up $9 \times 10^{k-1}$ for $1 \le k \le n$. The sum can be evaluated since it is a geometric series. This does not include the number 0. Whether we add 1 to include it depends on our interpretation of the problem's requirement that there be no leading zeroes. There is an easier way. We can pad out a number with less than n digits by adding leading zeroes. The original number can be recovered from any such n digit number by stripping off the leading zeroes. Thus we see by the Rule of Product that there are 10^n numbers with at most n digits. If we wish to rule out 0 (which pads out to a string of n zeroes), we must subtract 1.

CL-1.3 For each element of S you must make one of two choices: "x is/isn't in the subset." To visualize the process, list the elements of the set in any order: $a_1, a_2, \ldots, a_{|S|}$. We can construct a subset by

  including a_1 or not AND

  including a_2 or not AND

  $\cdot$ $\cdot$ $\cdot$

  including $a_{|S|}$ or not.

CL-1.4 (a) By the Rule of Product, we have $9 \times 10 \times \cdots \times 10 = 9 \times 10^{n-1}$.

(b) By the Rule of Product, we have 9^n.

(c) By the Rule of Sum, (answer)$+9^n = 9 \times 10^{n-1}$ and so the answer is $9(10^{n-1} - 9^{n-1})$

CL-1.5 (a) This is like the previous exercise. There are 26^4 4-letter strings and there are $(26-5)^4$ 4-letter strings that contain no vowels. Thus we have $26^4 - 21^4$.

(b) We can do this in two ways:
First way: Break the problem into 4 problems, depending on where the vowel is located. (This uses the Rule of Sum.) For each subproblem, choose each letter in the list and use the Rule of Product. We obtain one factor equal to 5 and three factors equal to 21. Thus we obtain 5×21^3 for each subproblem and $4 \times 5 \times 21^3$ for the final answer.
Second way: Choose one of the 4 positions for the vowel, choose the vowel and choose each of the 3 consonants. By the Rule of Product we have $4 \times 5 \times 21 \times 21 \times 21$.

CL-1.6 The only possible vowel and consonant pattern satisfying the two nonadjacent vowels and initial and terminal consonant conditions is CVCVC. By the Rule of Product, there are $3 \times 2 \times 3 \times 2 \times 3 = 108$ possibilities.

Solutions for Basic Counting and Listing

CL-1.7 To form a composition of n, we can write n ones in a row and insert either "$\oplus$" or "," in the spaces between them. This is a series of 2 choices at each of $n-1$ spaces, so we obtain 2^{n-1} compositions of n. The compositions of 4 are

$$4 = 3 \oplus 1 = 2 \oplus 2 = 2 \oplus 1 \oplus 1 = 1 \oplus 3 = 1 \oplus 2 \oplus 1 = 1 \oplus 1 \oplus 2 = 1 \oplus 1 \oplus 1 \oplus 1.$$

The compositions of 5 with 3 parts are

$$3 \oplus 1 \oplus 1 = 2 \oplus 2 \oplus 1 = 2 \oplus 1 \oplus 2 = 1 \oplus 3 \oplus 1 = 1 \oplus 2 \oplus 2 = 1 \oplus 1 \oplus 3.$$

CL-1.8 The allowable letters in alphabetic order are A, I, L, S, and T. There are 216 words that begin with L, and the same number that begin with S, and with T. The word we are asked to find is the last one that begins with L. Thus the word is of the form $LVCVCC$, $LVCCVC$, or $LCVCVC$. Since all of the consonants in our allowable-letters list come after the vowels, we want a word of the form $LCVCVC$. We need to start off $LTVCVC$. The next letter, a vowel, needs to be I (bigger than A in the alphabet). Thus we have $LTICVC$. Continuing in this way we get $LTITIT$. The next name in dictionary order starts off with S and is of the form $SVCVCC$. We now choose the vowels and consonants as small as possible: $SALALL$. But, this word doesn't satisfy the condition that adjacent consonants must be different. Thus the next legal word is $SALALS$.

CL-1.9 The ordering on the C_i is as follows:

$$C_1 = ((2,4), (2,5), (3,5)) \quad C_2 = (AA, AI, IA, II)$$

$$C_3 = (LL, LS, LT, SL, SS, ST, TL, TS, TT) \quad C_4 = (LS, LT, SL, ST, TL, TS).$$

The first seven are

$$(2,4)(AA)(LL)(LS),\ (2,4)(AA)(LL)(LT),\ (2,4)(AA)(LL)(SL),$$

$$(2,4)(AA)(LL)(ST),\ (2,4)(AA)(LL)(TL),$$

$$(2,4)(AA)(LL)(TS),\ (2,4)(AA)(LS)(LS).$$

The last 7 are

$$(3,5)(II)(TS)(TS),\ (3,5)(II)(TT)(LS),\ (3,5)(II)(TT)(LT),$$

$$(3,5)(II)(TT)(SL),\ (3,5)(II)(TT)(ST),$$

$$(3,5)(II)(TT)(TL),\ (3,5)(II)(TT)(TS).$$

The actual names can be constructed by following the rules of construction from these strings of symbols (e.g, $(3,5)(II)(TT)(LS)$ says place the vowels II in positions 3,5, the nonadjacent consonants are TT and the adjacent consonants are LS to get LSITIT).

CL-1.10 (a) One way to do this is to list all the possible multisets in some order. If you do this carefully, you will find that there are 15 of them. Unfortunately, it is easy to miss something if you do not choose the order carefully. One way to do this is to first write

Solutions for Basic Counting and Listing

all the a's in the multiset, then all the b's and then all the c's. For example, we would write the multiset $\{a,b,c,a\}$ as $aabc$. We can now list these in lex order:

$$aaaa, \ aaab, \ aaac, \ aabb, \ aabc, \ aacc, \ abbb, \ abbc,$$
$$abcc, \ accc, \ bbbb, \ bbbc, \ bbcc, \ bccc, \ cccc$$

For (b), the answer is that there are an infinite number because an element can be repeated any number of times. In fact, an infinite number of multisets can be formed by using just a.

CL-2.1 (a) We can arrange n people in $n!$ ways. Use $n = 7$.

(b) Arrange b boys ($b!$ ways) AND arrange g girls ($g!$ ways) AND choose which list comes first (2 ways). Thus we have $2(b!\ g!)$. Here $b = 3$ and $g = 4$ and the answer is 288.

(c) As in (b), we arrange the girls and the boys separately, AND then we interleave the two lists as GBGBGBG. Thus we get $4!\ 3! = 144$.

CL-2.2 This refers to the previous solution.

(a) Use $n = 6$.

(b) $b = g = 3$ and the answer is 72.

(c) We can interleave in two ways, as BGBGBG or as GBGBGB and so we get $2(3!\ 3!) = 72$.

CL-2.3 For (a) we have the circular list discussed in the text and the answer is therefore $n!/n = (n-1)!$.

For (b), note that each circular list gives two ordinary lists — one starting with the girls and the other with the boys. Hence the answer is $2(b!\ g!)/2 = b!\ g!$. For the two problems we have $4!\ 3! = 144$ and $3!\ 3! = 36$.

For (c), it is impossible if $b < g$ since this forces two girls to sit together. If we have $b = g$, circular lists are possible. As in the unrestricted case, each circular list gives $n = b + g = 2g$ linear lists by cutting it arbitrarily. Thus we get $2(g!)^2/2g = g!\ (g-1)!$, which in this case is $3!\ 2! = 12$.

CL-2.4 Each of the 7 letters ABMNRST appears once and each of the letters CIO appears twice. Thus we must form a list of length k from the 10 distinct letters. The solutions are

$$k = 2: \quad 10 \times 9 = 90$$
$$k = 3: \quad 10 \times 9 \times 8 = 720$$
$$k = 4: \quad 10 \times 9 \times 8 \times 7 = 5040$$

CL-2.5 Each of the 7 letters ABMNRST appears once and each of the letters CIO appears twice.

- For $k = 2$, the letters are distinct OR equal. There are $(10)_2 = 90$ distinct choices. Since the only repeated letters are CIO, there are 3 ways to get equal letters. This gives 93.

- For $k = 3$, we have either all distinct ($(10)_3 = 720$) OR two equal. The two equal can be worked out as follows

Solutions for Basic Counting and Listing

choose the repeated letter (3 ways) AND

choose the positions for the two copies of the letter (3 ways) AND

choose the remaining letter $(10 - 1 = 9$ ways).

By the Rules of Sum and Product, we have $720 + 3 \times 9 \times 3 = 801$.

CL-2.6 (a) The letters are EILST. The number or 3-words is $(5)_3 = 60$.

(b) The answer is $5^3 = 125$.

(c) The letters are EILST, with T occurring 3-times, L occurring 2-times. Either the letters are distinct OR one letter appears twice OR one letter appears three times. We have seen that the first can be done in 60 ways. To do the second, choose one of L and T to repeat, choose one of the remaining 4 different letters and choose where that letter is to go, giving $2 \times 4 \times 3 = 24$. To do the third, use T. Thus, the answer is $60 + 24 + 1 = 85$.

CL-2.7 (a) Stripping off the initial R and terminal F, we are left with a list of at most 4 letters, at least one of which is an L. There is just 1 such list of length 1. There are $3^2 - 2^2 = 5$ lists of length 2, namely all those made from E, I and L minus those made from just E and I. Similarly, there are $3^3 - 2^3 = 19$ of length 3 and $3^4 - 2^4 = 65$. This gives us a total of 90.

(b) The letters used are E, F, I, L and R in alphabetical order. To get the word before RELIEF, note that we cannot change just the F and/or the E to produce an earlier word. Thus we must change the I to get the preceding word. The first candidate in alphabetical order is F, giving us RELF. Working backwards in this manner, we come to RELELF, RELEIF, RELEF and, finally, RELEEF.

CL-2.8 (a) If there are 4 letters besides R and F, then there is only one R and one F, for a total of 65 spellings by the previous problem. If there are 3 letters besides R and F, we may have R$\cdots$F, R$\cdots$FF or RR$\cdots$F, which gives us $3 \times 19 = 57$ words by the previous problem. We'll say there are 3 RF patterns, namely RF, RFF and RRF. If there 2 letters besides R and F, there are 6 RF patterns, namely the three just listed, RFFF, RRFF and RRRF. This gives us $6 \times 5 = 30$ words. Finally, the last case has the 6 RF patterns just listed as well as RFFFF, RRFFF, RRRFF and RRRRF for a total of 10 patterns. This give us 10 words since the one remaining letter must be L. Adding up all these cases gives us $65 + 57 + 30 + 10 = 162$ possible spellings. Incidentally, there is a simple formula for the number of n long RF patterns, namely $n - 1$. Thus there are

$$1 + 2 + \ldots + (n-1) = n(n-1)/2$$

of length at most n. This gives our previous counts of 1, 3, 6 and 10.

(b) Reading toward the front of the dictionary from RELIEF we have RELIEF, RELFFF, RELFF, RELF, RELELF, RELEIF, RELEFF,..., and so the spelling five before RELIEF is RELEIF.

CL-2.9 There are $n!/(n-k)!$ lists of length k. The total number of lists (not counting the empty list) is

$$\frac{n!}{(n-1)!} + \frac{n!}{(n-2)!} + \cdots + \frac{n!}{1!} + \frac{n!}{0!}$$
$$= n!\left(\frac{1}{0!} + \frac{1}{1!} + \cdots + \frac{1}{(n-1)!}\right)$$
$$= n!\sum_{i=0}^{n-1} \frac{1^i}{i!}.$$

Since $e = e^1 = \sum_{i=0}^{\infty} 1^i/i!$, it follows that the above sum is close to e.

CL-3.1 Choose values for pairs
AND choose suits for the lowest value pair
AND choose suits for the middle value pair
AND choose suits for the highest value pair.
This gives $\binom{13}{3}\binom{4}{2}^3 = 61,776$.

CL-3.2 Choose the lowest value in the straight (A to 10) AND choose a suit for each of the 5 values in the straight. This gives $10 \times 4^5 = 10240$.

Although the previous answer is acceptable, a poker player may object since a "straight flush" is better than a straight — and we included straight flushes in our count. Since a straight flush is a straight all in the same suit, we only have 4 choices of suits for the cards instead of 4^5. Thus, there are $10 \times 4 = 40$ straight flushes. Hence, the number of straights which are not straight flushes is $10240 - 40 = 10200$.

CL-3.3 If there are n 1's in the sequence, there are $n-1$ spaces between the 1's. Thus, there are 2^{n-1} compositions of n. A composition of n with k parts has $k-1$ commas The number of ways to insert $k-1$ commas into $n-1$ positions is $\binom{n-1}{k-1}$.

CL-3.4 Note that EXERCISES contains 3 E's, 2 S's and 1 each of C, I, R and X. We can use the multinomial coefficient

$$\binom{n}{m_1, m_2, \ldots, m_k} = \frac{n!}{m_1!\, m_2! \cdots m_k!}$$

where $n = m_1 + m_2 + \ldots + m_k$. Take $n = 9$, $m_1 = 3$, $m_2 = 2$ and $m_3 = m_4 = m_5 = m_6 = 1$. This gives $9!/3!\,2! = 30240$. This calculation can also be done without the use of a multinomial coefficient as follows. Choose 3 of the 9 possible positions to use for the three E's AND choose 2 of the 6 remaining positions to use for the two S's AND put a permutation of the remaining 4 letters in the remaining 4 places. This gives us $\binom{9}{3} \times \binom{6}{2} \times 4!$.

CL-3.5 An arrangement is a list formed from 13 things each used 4 times. Thus we have $n = 52$ and $m_i = 4$ for $1 \le i \le 13$ in the multinomial coefficient

$$\binom{n}{m_1, m_2, \ldots, m_k} = \frac{n!}{m_1!\, m_2! \cdots m_k!}.$$

CL-3.6 (a) The first 4 names in dictionary order are LALALAL, LALALAS, LALALAT, LALALIL.

(b) The last 4 names in dictionary order are TSITSAT, TSITSIL, TSITSIS, TSITSIT.

Solutions for Basic Counting and Listing

(c) To compute the names, we first find the possible consonant vowel patterns. They are CCVCCVC, CCVCVCC, CVCCVCC and CVCVCVC. The first three each contain two pairs of adjacent consonants, one isolated consonant and two vowels. Thus each corresponds to $(3 \times 2)^2 \times 3 \times 2^2$ names. The last has four isolated consonants and three vowels and so corresponds to $3^4 \times 2^3$ names. In total, there are 1,944 names.

CL-3.7 The first identity can be proved by writing the binomial coefficients in terms of factorials. It can also be proved from the definition of the binomial coefficient: Choosing a set of size k from a set of size n is equivalent to choosing a set of size $n-k$ to throw away, namely the things not chosen.

The total number of subsets of an n element set is 2^n. On the other hand, we can divide the subsets into collections T_j, where T_i contains all the i element subsets. The number of subsets in T_i is $\binom{n}{i}$. Apply the Rule of Sum.

CL-3.8 $S(n,n) = 1$: The only way to partition an n element set into n blocks is to put each element in a block by itself, so $S(n,n) = 1$.

$S(n,n-1) = \binom{n}{2}$: The only way to partition an n element set into $n-1$ blocks is to choose two elements to be in a block together and put the remaining $n-2$ elements in $n-2$ blocks by themselves. Thus it suffices to choose the 2 elements that appear in a block together and so $S(n,n-1) = \binom{n}{2}$.

$S(n,1) = 1$: The only way to partition a set into one block is to put the entire set into the block.

$S(n,2) = (2^n - 2)/2$: We give two solutions. Note that $S(n,k)$ is the number of k-sets S where the entries in S are nonempty subsets of a given n-set T and each element of T appears in exactly one entry of S. We will count k-lists, which is $k!$ times the number of k-sets. We choose a subset for the first block (first list entry) and use the remaining set elements for the second block. Since an n-set has 2^n, this would seem to give $2^n/2$; however, we must avoid empty blocks. In the ordered case, there are two ways this could happen since either the first or second list entry could be the empty set. Thus, we must have $2^n - 2$ instead of 2^n. The answer is $(2^n - 2)/2$.

Here is another way to compute $S(n,2)$. Look at the block containing n. Once it is determined, the entire two block partition is determined. The block containing n can be gotten by starting with n and adjoining one of the $2^{n-1} - 1$ proper subsets of $\{1, 2, \ldots, n-1\}$.

CL-3.9 We use the hint. Choose i elements of $\{1, 2, \cdots, n\}$ to be in the block with $n+1$ AND either do nothing else if $i = n$ OR partition the remaining elements. This gives $\binom{n}{n}$ if $i = n$ and $\binom{n}{i} B_{n-i}$ otherwise. If we set $B_0 = 1$, the second formula applies for $i = n$, too. Since $i = 0$ OR $i = 1$ OR $\cdots$ OR $i = n$, the result follows.

(b) To calculate B_n for $n \leq 5$: We have $B_0 = 1$ from (a). Using the formula in (a) for $n = 0, 1, 2, 3, 4$ in order, we obtain $B_1 = 1$, $B_2 = 2$, $B_3 = 5$, $B_4 = 15$ and $B_5 = 52$.

CL-3.10 (a) There is exactly one arrangement — 1,2,3,4,5,6,7,8,9.

(b) We do this by counting those arrangements that have $a_i \leq a_{i+1}$ except, perhaps, for $i = 5$. Then we subtract off those that also have $a_5 < a_6$. In set terms:

- S is the set of rearrangements for which $a_1 < a_2 < a_3 < a_4 < a_5$ and $a_6 < a_7 < a_8 < a_9$,

Solutions for Basic Counting and Listing

- T is the set of rearrangements for which $a_1 < a_2 < a_3 < a_4 < a_5 < a_6 < a_7 < a_8 < a_9$, and
- we want $|S \setminus T| = |S| - |T|$.

An arrangement in S is completely determined by specifying the set $\{a_1, \ldots, a_5\}$, of which there are $\binom{9}{5} = 126$. In (a), we saw that $|T| = 1$. Thus the answer is $126 - 1 = 125$.

CL-4.1 Let the probability space consist of all $\binom{6}{2} = 15$ pairs of horses and use the uniform probability. Thus each pair has probability $1/15$. Since each horse is in exactly 5 pairs, the probability of your choosing the winner is $5/15 = 1/3$, regardless of which horse wins.

Here is another way. You could choose your first horse and your second horse, so the space consists of 6×5 choices. The probability that your first choice was the winner is $1/6$. The probability that your second choice was the winner is also $1/6$. Since these events are disjoint, the probability of picking the winner is $1/6 + 1/6 = 1/3$.

Usually the probability of winning a bet on a horse race depends on picking the fastest horse after much study. The answer to this problem, $1/3$, doesn't seem to have anything to do with studying the horses? Why?

CL-4.2 The sample space is $\{0, 1, \ldots, 36, 00\}$. We have $P(0) = P(1) = \cdots = P(36)$ and $P(00) = 1.05 P(0)$. Thus

$$1 = P(0) + \cdots + P(36) + P(00) = 38.05 P(0).$$

Hence $P(0) = 1/38.05$ and so $P(00) = 1.05/38.05 = 0.0276$.

CL-4.3 Let the event space be $\{A, B\}$, depending on who finds the key. Since Alice searches 20% faster than Bob, it is reasonable to assume that $P(A) = 1.2\, P(B)$. The odds that Alice finds the key are $P(A)/P(B) = 1.2$, that is, 1.2:1, which can also be written as 6:5. Combining $P(A) = 1.2\, P(B)$ with $P(A) + P(B) = 1$, we find that $P(A) = 1.2/2.2 = 0.545$.

CL-4.4 Let A be the event that you pick the winner and B the probability that you pick the horse that places. From a previous exercise, $P(A) = 1/3$ Similarly, $P(B) = 1/3$. We want $P(A \cup B)$. By the principle of inclusion and exclusion, this is $P(A) + P(B) - P(A \cap B)$. Of all $\binom{6}{2} = 15$ choices, only one is in $A \cap B$. Thus $P(A \cap B) = 1/15$ and the answer is $1/3 + 1/3 - 1/15 = 3/5$.

CL-4.5 Since probabilities are uniform, we simply count the number of events that satisfy the conditions and divide by the total number of events, which is m^n for n balls and m boxes. First we will do the problems in an ad hoc manner, then we'll discuss a systematic solution. We use (a')–(c') to denote the answers for (d).

(a) We place one ball in the first box AND one in the second AND so on. Since this can be done in $4!$ ways, the answer is $4!/4^4 = 3/32$.

(a') We must have one box with two balls and one ball in each of the other three boxes. We choose one box to contain two balls AND two balls for the box AND distribute the three remaining balls into three boxes as in (a). This gives us $4 \times \binom{5}{2} \times 3! = 240$. Thus the answer is $240/4^5 = 15/64$.

Solutions for Basic Counting and Listing

(b) This is somewhat like (a'). Choose a box to be empty AND choose a box to contain two balls AND choose two balls for the box AND distribute the other two balls into the other two boxes. This gives $4 \times 3 \times \binom{4}{2} \times 2! = 144$. Thus the answer is $144/4^4 = 9/16$.

(b') This is more complicated since the ball counts can be either 3,1,1,0 or 2,2,1,0. As in (b), there are $4 \times 3 \times \binom{5}{3} \times 2! = 240$ to do the first. In the second, there are $\binom{4}{2} \times 2 = 12$ ways to designate the boxes and $\binom{5}{2} \times \binom{3}{2} = 30$ ways to choose the balls for the boxes that contain two each. Thus there are 360 ways and the answer is $(240 + 360)/4^5 = 75/128$.

(c) Simply subtract the answer for (a) from 1 since we are asking for the complementary event. This gives 29/32. For (c') we have 39/64.

We now consider a systematic approach. Suppose we want to assign n balls to m boxes so that exactly $k \leq m$ of the boxes contain balls. Call the balls $1, 2, \ldots, n$ First partition the set of n balls into k blocks. This can be done in $S(n, k)$ ways, where $S(n,k)$ is the Stirling number discussed in Section 3. List the blocks in some order (pick your favorite; e.g., numerical order based on the smallest element in the block). Assign the first block to a box AND assign the second block to a box AND, etc. This can be done in $m(m-1)\cdots(m-k+1) = m!/(m-k)!$ ways. Hence the number of ways to distribute the balls is $S(n,k)m!/(m-k)!$ and so the probability is $S(n,k)m!/(m-k)!m^n$. For our particular problems, the answers are

(a) $S(4,4)4!/0! \, 4^4 = 3/32$ (a') $S(5,4)4!/0! \, 4^5 = 15/64$
(b) $S(4,3)4!/1! \, 4^4 = 9/16$ (b') $S(5,3)4!/1! \, 4^5 = 75/128$.

The moral here is that if you can think of a systematic approach to a class of problems, it is likely to be easier than solving each problem separately.

CL-4.6 (a) Since the die is thrown k times, the sample space is S^k, where $S = \{1,2,3,4,5,6\}$. Since the die is fair, all 6^k sequences in S^k are equally likely. We claim that exactly half have an even sum and so $P(E) = 1/2$. Why do half have an even sum? Here are two proofs.

- Let $N_o(n)$ be the number of odd sums in the first n throws and let $N_e(n)$ be the number of even sums. We have

$$N_e(k) = 3N_e(k-1) + 3N_o(k-1) \quad \text{and} \quad N_o(k) = 3N_o(k-1) + 3N_e(k-1)$$

because an even sum is obtained from an even by throwing 2, 4, or 6 and from an odd by throwing 1, 3, or 5; and similarly for an odd sum. Thus $N_e(k) = N_o(k)$. Since the probability on S^k is uniform, the probability of an even sum is 1/2.

- Let S_o be all the k-lists in S^k with odd sum and let S_e be those with even sum. Define the function $f : S^k \to S^k$ as follows

$$f(x_1, x_2 \ldots, x_k) = \begin{cases} (x_1 + 1, x_2, \ldots, x_k), & \text{if } x_1 \text{ is odd;} \\ (x_1 - 1, x_2, \ldots, x_k), & \text{if } x_1 \text{ is even.} \end{cases}$$

We leave it to you to convince yourself that this function is a bijection between S_o and S_e. (A bijection is a one-to-one correspondence between elements of S_o and S_e.)

Solutions for Basic Counting and Listing

(b) The sample space for drawing cards n times is S^n where S is the Cartesian product

$$\{A, 2, 3, \ldots, 10, J, Q, K\} \times \{\clubsuit, \diamond, \heartsuit, \spadesuit\}.$$

The probability of any point in S^n is $(1/52)^n$. The number of draws with no king is $(52-4)^n$ and so the probability of none is $(48/52)^n = (12/13)^n$. The probability of at least one king is $1 - (12/13)^n$.

(c) The equiprobable sample space is gotten by distinguishing the marbles $M = \{w_1, w_2, w_3, r_1, \ldots\}$ and defining the sample space by

$$S = \{(m, m') : m \text{ and } m' \text{ are distinct elements of } M\}.$$

If E_r is the event that both m and m' are red, then $P(E_r) = 4*3/|S|$ where $|S| = 12*11$.

RELATED PROBLEMS TO THINK ABOUT: What is the probability of two white and two blue marbles being drawn if four marbles are drawn without replacement? Of two white and two blue marbles being drawn if four marbles are drawn with replacement?

CL-4.7 This is nearly identical to the example on hypergeometric probabilities. The answer is $C(5,3)C(10,3)/C(15,6)$.

CL-4.8 Let $B = \{1, 2, \ldots, 10\}$.

(a) The sample space S is the set of all subsets of B of size 2. Thus $|S| = \binom{10}{2} = 45$. Since each draw is equally likely, we just need to know how many pairs have an odd sum. One of the balls must have an odd label and the other an even label. The number of pairs with this property is 5×5 since there are 5 odd labels and 5 even labels. Thus the probability is $25/45 = 5/9$.

(b) The sample space S is the set of ordered pairs (b_1, b_2) with $b_1 \neq b_2$ both from B. Thus $|S| = 10 \times 9 = 90$. To get an odd sum, one of b_1 and b_2 must be even and the other odd. Thus there are 10 choices for b_1 AND then 5 choices for b_2. The probability is $50/90 = 5/9$.

(c) The sample space is $S = B \times B$ and $|S| = 100$. The number of pairs (b_1, b_2) is 50 as in (b). Thus the probability is $50/100 = 1/2$.

CL-4.9 This is an inclusion and exclusion type of problem. There are three ways to approach such problems:

- Have a variety of formulas handy that you can plug into. This, by itself, is not a good idea because you may encounter a problem that doesn't fit any of the formulas you know.

- Draw a Venn diagram and use the information you have to compute the probability of as many regions as you can. If there are more than 3 sets, the Venn diagram is too confusing to be very useful. With 2 or 3 sets, it is a good approach.

- Carry out the preceding idea without the picture. We do this here.

Suppose we are dealing with k sets, $A_1, \ldots, A_k$. We need to know what the regions in the Venn diagram are. Each region corresponds to $T_1 \cap \cdots \cap T_k$ where T_i is either A_i or A_i^c. In our case, $k = 2$ and so the probabilities of the regions are

$$P(A \cap B) \qquad P(A \cap B^c) \qquad P(A^c \cap B) \qquad P(A^c \cap B^c).$$

Solutions for Basic Counting and Listing

We get A by combining $A \cap B$ and $A \cap B^c$. We get B by combining $A \cap B$ and $A^c \cap B$. By properties of sets, $(A \cup B)^c = A^c \cap B^c$. Thus our data corresponds to the three equations

$$P(A \cap B) + P(A \cap B^c) = 3/8 \qquad P(A \cap B) + P(A^c \cap B) = 1/2 \qquad P(A^c \cap B^c) = 3/8.$$

We have one other equation: The probabilities of all four regions sum to 1. This gives us four equations in four unknowns whose solution is

$$P(A \cap B) = 1/4 \qquad P(A \cap B^c) = 1/8 \qquad P(A^c \cap B) = 1/4 \qquad P(A^c \cap B^c) = 3/8.$$

Thus the answer to the problem is 1/4.

When we are not asked for the probability of all regions, it is often possible to take shortcuts. That is the case here. From $P((A \cup B)^c) = 3/8$ we have $P(A \cup B) = 1 - 3/8 = 5/8$. Since $P(A \cup B) = P(A) + P(B) - P(A \cap B)$ and three of the four terms in this equation are known, we can easily solve for $P(A \cap B)$.

CL-4.10 This is another Venn diagram problem. This time we'll work with number of people instead of probabilities. Let C correspond to the set of computer science majors, W the set of women and S to the entire student body. We are given

$$|C| = 20\% \times 5{,}000 = 1{,}000$$
$$|W| = 58\% \times 5{,}000 = 2{,}900$$
$$|C \cap W| = 430.$$

(a) We want $|W \cap C^c|$, which equals $|W| - |W \cap C| = 2{,}470$. You should be able to see why this is so by the Venn diagram or by the method used in the previous problem.

(b) The number of men who are computer science majors is the number of computer science majors who are not women. This is $|C| - |C \cap W| = 1{,}000 - 430 = 570$. The number of men in the student body is $42\% \times 5{,}000 = 2{,}100$. Thus $2{,}100 - 570 = 1{,}530$ men are not computer science majors.

(c) The probability is $\frac{430}{5{,}000} = 0.086$.

(c) Since there are $58\% \times 5{,}000 = 2{,}900$ women, the probability is $\frac{430}{2{,}900}$.

CL-4.11 Since the coin is fair $P(H) = 1/2$, what about $P(W)$, the probability that Beatlebomb wins? Recall the meaning of the English phrase "the odds that it will occur." This is trivial but important, as the phrase is used often in everyday applications of probability. If you don't recall the meaning, see the discussion of odds in the text. From the definition of odds, you should be able to show that $P(W) = 1/101$. If we had studied "independent" events, you could immediately see that the answer to the questions is $(1/2) \times (1/101) = 1/202$, but we need a different approach which lets independent events sneak in through the back door.

Let the sample space be $\{H, T\} \times \{W, L\}$, corresponding to the outcome of the coin toss and the outcome of the race. From the previous paragraph $P(\{(H, W), (T, W)\}) = 1/101$. Since the coin is fair and the coin toss doesn't influence the race, we should have $P((H, W)) = P((T, W))$. Since

$$P(\{(H, W), (T, W)\}) = P((H, W)) + P((T, W)),$$

Solutions for Basic Counting and Listing

It follows after a little algebra that $P(H,W)) = 1/202$.

CL-4.12 This is another example of the hypergeometric probability. Do you see why? The answer is $C(37,11)C(2,2)/C(39,13)$.

CL-4.13 It may seem at first that you need to break up the problem according to what the other players have been dealt. Not so! You should be able to see that the results would have been the same if you had been dealt your fifth card *before* the other players had been dealt their cards. Now it's not hard to work things out. After you've been dealt 4 cards, there are 48 cards left. Of those, the fourth card in the 3 of a kind (4◇ in the example) and any of the 3 cards with the same value as your odd card (10♡ 10◇ 10♣ in the example) improve your hand. That's 4 cards out of 48, so the probability is $4/48 = 1/12$.

CL-4.14 (a) Let words of length 6 formed from three G's and three B's stand for the arrangements in terms of Boys and Girls; for example, BBGGBG or BBBGGG. There are $\binom{6}{3} = 6!/(3!\,3!) = 20$ such words. Four such words correspond to the three girls together: GGGBBB, BGGGBB, BBGGGB, BBBGGG. The probability of three girls being together is $4/20 = 1/5$.

(b) If they are then seated around a circular table, there are two additional arrangements that will result in all three girls sitting together: GGBBBG and GBBBGG. The probability is $6/20 = 3/10$.

CL-4.15 You can draw the Venn diagram for three sets and, for each of the eight regions, count how much a point in the region contributes to the addition and subtraction. This does not extend to the general case. We give another proof that does.

Let S be the sample space and let T be a subset of S Define the function χ_T with domain S by

$$\chi_T(s) = \begin{cases} 1 & \text{if } s \in T, \\ 0 & \text{if } s \notin T. \end{cases}$$

This is called the *characteristic function* of T.[1] We leave it to you to check that

$$\chi_{T^c}(s) = 1 - \chi_T(s), \quad \chi_{T \cap U}(s) = \chi_T(s)\chi_U(s), \quad \text{and} \quad P(S) = \sum_{s \in S} P(s)\chi_T(s).$$

---

[1] χ is a lower case Greek letter and is pronounced like the "ki" in "kind."

Solutions for Basic Counting and Listing

Using these equations and a little algebra, we have

$$P(A^c \cap B^c \cap C^c) = \sum_{s \in S} P(s) \chi_{A^c \cap B^c \cap C^c}(s)$$

$$= \sum_{s \in S} P(s)(1 - \chi_A(s))(1 - \chi_B(s))(1 - \chi_C(s))$$

$$= \sum_{s \in S} P(s) - \sum_{s \in S} P(s)\chi_A(s) - \sum_{s \in S} P(s)\chi_B(s) - \sum_{s \in S} P(s)\chi_C(s)$$

$$+ \sum_{s \in S} P(s)\chi_A(s)\chi_B(s) + \sum_{s \in S} P(s)\chi_A(s)\chi_C(s)$$

$$+ \sum_{s \in S} P(s)\chi_B(s)\chi_C(s) - \sum_{s \in S} P(s)\chi_A(s)\chi_B(s)\chi_C(s)$$

$$= 1 - P(A) - P(B) - P(C)$$
$$+ P(A \cap B) + P(A \cap C)$$
$$+ P(B \cap C) - P(A \cap B \cap C).$$

CL-4.16 Let the stick have unit length and let x be the distance from the end of the stick where the break is made. Thus $0 \leq x \leq 1$. The longer piece will be at least twice the length of the shorter if $x \leq 1/3$ or if $x \geq 2/3$. The probability of this is $1/3 + 1/3 = 2/3$. You should be able to fill in the details.

CL-4.17 Let x and y be the places where the stick is broken. Thus, (x,y) is chosen uniformly at random in the square $S = (0,1) \times (0,1)$. Three pieces form a triangle if the sum of the lengths of any two is always greater than the length of the third. We must determine which regions in S satisfy this condition.

Suppose $x < y$. The lengths are then x, $y - x$, and $1 - y$. The conditions are

$$x + (y - x) > 1 - y, \quad x + (1 - y) > y - x, \quad \text{and} \quad (y - x) + (1 - y) > x.$$

With a little algebra, these become

$$y > 1/2, \quad y < x + 1/2, \quad \text{and} \quad x < 1/2,$$

respectively. If you draw a picture, you will see that this is a triangle of area $1/8$.

If $x > y$, we obtain the same results with the roles of x and y reversed. Thus the total area is $1/8 + 1/8 = 1/4$. Since S has area 1, the probability is $1/4$.

CL-4.18 Look where the center of the coin lands. If it is within $d/2$ of a lattice point, it covers the lattice point. Thus, there is a circle of diameter d about each lattice point and the coin covers a lattice point if and only if it lands in one of the circles. We need to compute the fraction of the plane covered by these circles. Since the pattern repeats in a regular fashion, all we need to do is calculate the fraction of the square $\{(x,y) | 0 \leq x \leq 1, 0 \leq y \leq 1\}$ that contains parts of circles. There is a quarter circle about each of the points (0,0), (0,1), (1,0) and (1,1) inside the square. Since the circle has diameter at most 1, the quarter circles have no area in common and so their total area equals the area of the coin, $\pi d^2 / 4$. Since the area of the square is 1, the probability that the coin covers a lattice point is $\pi d^2 / 4$.

Solutions for Basic Counting and Listing

CL-4.19 Select the three points uniformly at random from the circumference of the circle and label them 1, 2, 3 going clockwise around the circle from the top of the circle. Let E_1 denote the event consisting of all such configurations where points 2 and 3 lie in the half circle starting at 1 and going clockwise (180 degrees). Let E_2 denote the event that points 2 and 1 lie in the half circle starting at 2 and going clockwise 180 degrees. Let E_3 be defined similarly. Note that the events E_1, E_2, and E_3 are mutually exclusive. (Draw a picture and think about this.) By our basic probability axioms, the probability of the union is the sum of the probabilities $P(E_1) + P(E_2) + P(E_3)$. To compute $P(E_1)$, imagine point 1 on the circle, consider its associated half circle and, before looking at the other two points, ask "What is the probability that they lie in that half circle?" Let x be the number of degrees clockwise from point 1 to point 2 and y the number from 1 to 3. Thus (x, y) is a point chosen uniformly at random in the square $[0, 360) \times [0, 360)$. For event E_1 to occur, (x, y) must lie in $[0, 180) \times [0, 180)$, which is 1/4 of the original square. Thus $P(E_1) = 1/4$. (This can also be done using independent events: the locations of points 2 and 3 are chosen independently so one gets $(1/2) \times (1/2)$.) The probabilities of E_2 and E_3 are the same for the same reason. Thus $P(E_1) + P(E_2) + P(E_3) = 3/4$.

What is the probability that k points selected uniformly at random on the circumference of a circle lie the same semicircle? Use the same method. The answer is $k/(2^{k-1})$.

Solutions for Functions

Fn-1.1 (a) We know the domain and range of f. f is not an injection. Since no order is given for the domain, the attempt to specify f in one-line notation is meaningless (the ASCII order $+, <, >, ?$, is a possibility, but is unusual enough in this context that explicitly specifying it would be essential). If the attempt at specification makes any sense, it tells us that f is a surjection. We cannot give it in two-line form since we don't know the function.

(b) We know the domain and range of f and the domain has an implicit order. Thus the one-line notation specifies f. It is an injection but not a surjection. In two-line form it is $\begin{pmatrix} 1 & 2 & 3 \\ ? & < & + \end{pmatrix}$.

(c) This function is specified and is an injection. In one-line notation it would be (4,3,2), and, in two-line notation, $\begin{pmatrix} 1 & 2 & 3 \\ 4 & 3 & 2 \end{pmatrix}$.

Fn-1.2 (a) If f is an injection, then $|A| \leq |B|$. **Solution:** Since f is an injection, every element of A maps to a different element of B. Thus B must have at least as many elements as A.

(b) If f is a surjection, then $|A| \geq |B|$. **Solution:** Since f is a surjection, every element of B is the image of at least one element of A. Thus A must have at least as many elements as B.

(c) If f is a bijection, then $|A| = |B|$. **Solution:** Combine the two previous results.

(d) If $|A| = |B|$, then f is an injection if and only if it is a surjection. **Solution:** Suppose that f is an injection and not a surjection. Then there is some $b \in B$ which is not the image of any element of A under f. Hence f is an injection from A to $B - \{b\}$. By (a), $|A| \leq |B - \{b\}| < |B|$, contradicting $|A| = |B|$.
Now suppose that f is a surjection and not an injection. Then there are $a, a' \in A$ such that $f(a) = f(a')$. Consider the function f with domain restricted to $A - \{a'\}$. It is still a surjection to B and so by (b) $|B| \leq |A - \{a'\}| < |A|$, contradicting $|A| = |B|$.

(e) If $|A| = |B|$, then f is a bijection if and only if it is an injection or it is a surjection. **Solution:** By the previous part, if f is either an injection or a surjection, then it is both, which is the definition of a bijection.

Fn-1.3 (a) Since ID numbers are unique and every student has one, this is a bijection.

(b) This is a function since each student is born exactly once. It is not a surjection since D includes dates that could not possibly be the birthday of any student; e.g., it includes yesterday's date. It is not an injection. Why? You may very well know of two people with the same birthday. If you don't, consider this. Most entering freshman are between 18 and 19 years of age. Consider the set F of those freshman and their possible birth dates. The maximum number of possible birth dates is $366 + 365$, which is smaller than the size of the set F. Thus, when we look a the function on F it is not injective.

(c) This is not a function. It is not defined for some dates because no student was born on that date. For example, D includes yesterday's date

Solutions for Functions

(d) This is not a function because there are students whose GPAs are outside the range 2.0 to 3.5. (We cannot *prove* this without student record information, but we can be sure it is true.)

(e) We cannot *prove* that it is a function without gaining access to student records; however, we can be sure that it is a function since we can be sure that each of the 16 GPAs between 2.0 and 3.5 will have been obtained by many students. It is not a surjection since the codomain is larger than the domain. It is an injection since a student has only one GPA.

Fn-1.4 $\{(1,a), (2,b), (3,c)\}$ is not a relation because $c \notin B$. The others are relations.
Among the relations, $\{(1,a), (2,b), (1,d)\}$ is not a functional relation because the value of the function at 3 is not defined and $\{(1,a), (2,b) (3,d), (1,b)\}$ is not a function because the value of the function at 1 is not uniquely defined. Thus only $\{(3,a), (2,b), (1,a)\}$ is a functional relation.
Only the inverse of $\{(1,a), (2,b), (1,d)\}$ is a functional relation. We omit the explanation.

Fn-2.1 (a) For $(1,5,7,8)$ $(2,3)$ (4) (6): $\begin{pmatrix} 1 & 2 & 3 & 4 & 5 & 6 & 7 & 8 \\ 5 & 3 & 2 & 4 & 7 & 6 & 8 & 1 \end{pmatrix}$ is the two-line form and $(5,3,2,4,7,6,8,1)$ is the one-line form. (We'll omit the two-line form in the future since it is simply the one-line form with $1,2,\ldots$ placed above it.) The inverse is $(1,8,7,5)$ $(2,3)$ (4) (6) in cycle form and $(8,3,2,4,1,6,5,7)$ in one-line form.

(b) For $\begin{pmatrix} 1 & 2 & 3 & 4 & 5 & 6 & 7 & 8 \\ 8 & 3 & 7 & 2 & 6 & 4 & 5 & 1 \end{pmatrix}$: The cycle form is $(1,8)$ $(2,3,7,5,6,4)$. Inverse: cycle form is $(1,8)$ $(2,4,6,5,7,3)$; one-line form is $(8,4,2,6,7,5,3,1)$.

(c) For $(5,4,3,2,1)$, which is in one-line form: The cycle form is $(1,5)$ $(2,4)$ (3). The permutation is its own inverse.

(d) $(5,4,3,2,1)$, which is in cycle form: This is not the standard form for cycle form. Standard form is $(1,5,4,3,2)$. The one-line form is $(5,1,2,3,4)$. The inverse is $(1,2,3,4,5)$ in cycle form and $(2,3,4,5,1)$ in one-line form.

Fn-2.2 Write one entire set of interchanges as a permutation in cycle form. The interchanges can be written as $(1,3)$, $(1,4)$ and $(2,3)$. Thus the entire set gives $1 \to 3 \to 2$, $2 \to 3$, $3 \to 1 \to 4$ and $4 \to 1$. In cycle form this is $(1,2,3,4)$. Thus five applications takes 1 to 2.

Fn-2.3 (a) Imagine writing the permutation in cycle form. Look at the cycle containing 1, starting with 1. There are $n-1$ choices for the second element of the cycle AND then $n-2$ choices for the third element AND $\cdots$ AND $(n-k+1)$ choices for the kth element. Prove that the number of permutations in which the cycle generated by 1 has length n is $(n-1)!$: The answer is given by the Rule of Product and the above result with $k = n$.

(b) For how many permutations does the cycle generated by 1 have length k? We write the cycle containing 1 in cycle form as above AND then permute the remaining $n-k$ elements of $\underline{n}$ in any fashion. For the k long cycle containing 1, the above result gives $\frac{(n-1)!}{(n-k)!}$ choices. There are $(n-k)!$ permutations on a set of size $n-k$. Putting this all together using the Rule of Product, we get $(n-1)!$, a result which does not depend on k.

Solutions for Functions

(c) Since 1 must belong to some cycle and the possible cycle lengths are $1, 2, \ldots, n$, summing the answer to (b) over $1 \le k \le n$ will count all permutations of $\underline{n}$ exactly once. In our case, the sum is $(n-1)! + \cdots + (n-1)! = n \times (n-1)! = n!$.

This problem has shown that if you pick a random element in a permutation of an n-set, then the length of the cycle it belongs to is equally likely to be any of the values from 1 to n.

Fn-2.4 Let e be the identity permutation of A. Since $e \circ f = f$ for any permutation of A, we have $e \circ e = e$. Applying this many times $e^k = e \circ e \circ \cdots \circ e = e$ for any $k > 0$. We will use this in discussing the solution.

(a) We can step around the cycle as in Example 8 and see that after 3 steps we are back where we started from. Three hundred steps simply does this one hundred times. Instead of phrasing it this way, we could say $(1,2,3)^3 = e$ and so $(1,2,3)^{300} = ((1,2,3)^3)^{100} = e^{100} e$.

(b) Since we step around each cycle separately,
$$\bigl((1,3)(2,5,4)\bigr)^{300} = (1,3)^{300}(2,5,4)^{300} = e^{300/2} e^{300/3} = e.$$

(c) A permutation of a k-set cannot have a cycle longer than k. Thus the possible cycle lengths for permutations of $\underline{5}$ are 1, 2, 3, 4 and 5. A cycle of any of these lengths raised to the 60th power is the identity. For example $(a,b,c,d)^{60} = ((a,b,c,d)^4)^{15} = e^{15} = e$. Thus $f^{60} = e$. Finally $f^{61} = f^{60} f = ef = f$.

Fn-3.1 (a) The domain and range of f are specified and f takes on exactly two distinct values. f is not an injection. Since we don't know the values f takes, f is not completely specified; however, it cannot be a surjection because it would have to take on all four values in its range.

(b) Since each block in the coimage has just one element, f is an injection. Since $|\text{Coimage}(f)| = 5 = |\text{range of } f|$, f is a surjection. Thus f is a bijection and, since the range and domain are the same, f is a permutation. In spite of all this, we don't know the function; for example, we don't know $f(1)$, but only that it differs from all other values of f.

(c) We know the domain and range of f. From $f^{-1}(2)$ and $f^{-1}(4)$, we can determine the values f takes on the union $f^{-1}(2) \cup f^{-1}(4) = \underline{5}$. Thus we know f completely. It is neither a surjection nor an injection.

(d) This function is a surjection, cannot be an injection and has no values specified.

(e) This specification is nonsense. Since the image is a subset of the range, it cannot have more than four elements.

(f) This specification is nonsense. The number of blocks in the coimage of f equals the number of elements in the image of f, which cannot exceed four.

Fn-3.2 (a) The coimage of a function is a partition of the domain with one block for each element of $\text{Image}(f)$.

(b) You can argue this directly or apply the previous result. In the latter case, note that since $\text{Coimage}(f)$ is a partition of A, $|\text{Coimage}(f)| = |A|$ if and only if each block

Solutions for Functions

of Coimage(f) contains just one element. On the other hand, f is an injection if and only if no two elements of A belong to the same block of Coimage(f).

(c) By the first part, this says that $|\text{Image}| = |B|$. Since Image(f) is a subset of B, it must equal B.

Fn-3.3 (a) The list is $321, 421, 431, 432, 521, 531, 532, 541, 542, 543$.

(b) The first number is $\binom{x_1-1}{3} + \binom{x_2-1}{2} + \binom{x_3-1}{1} + 1 = \binom{2}{3} + \binom{1}{2} + \binom{0}{1} + 1 = 1$. The last number is $\binom{4}{3} + \binom{3}{2} + \binom{2}{1} + 1 = 10$. The numbers $\binom{x_1-1}{3} + \binom{x_2-1}{2} + \binom{x_3-1}{1} + 1$ are, consecutively, $1, 2, \ldots 10$ and represent the positions of the corresponding strings $x_1 x_2 x_3$ in the list.

(c) The list is $123, 124, 125, 134, 135, 145, 234, 245, 345$.

(d) If, starting with the list of (c), you form the list $(6-x_1)(6-x_2)(6-x_3)$, you get $543, 542, 541, 532, 531, 521, 432, 431, 421, 321$ which is the list of (a) in reverse order. Thus the formula of (b) gives the positions $\rho(x_x, x_2, x_3)$ in reverse order of the list (c). Subtract $11 - \rho(x_x, x_2, x_3)$ to get the position in forward order.

(e) Successor: 98421. Predecessor: 97654.

(f) Let $x_1 = 9$, $x_2 = 8$, $x_3 = 3$, $x_2 = 2$ and $x_1 = 1$. Using the idea in part (b) of this exercise, the answer is

$$\binom{x_1-1}{5} + \binom{x_2-1}{4} + \binom{x_3-1}{3} + \binom{x_4-1}{2} + \binom{x_5-1}{1}$$
$$= \binom{8}{5} + \binom{7}{4} + \binom{2}{3} + \binom{1}{2} + \binom{0}{1}$$
$$= 56 + 35 + 0 + 0 + 0 = 91.$$

Fn-3.4 (a) The first distribution of balls to boxes corresponds to the strictly decreasing string 863. The next such string in lex order on all strictly decreasing strings of length 3 from $\underline{8}$ is 864. To get the corresponding distribution, place the three moveable box boundaries under positions 8, 6, and 4 and put balls under all other positions in $\underline{8}$. The predecessor to 863 is 862. The second distribution corresponds to 542. Its successor is 543, its predecessor is 541.

(b) The formula $p(x_1, x_2, x_3) = \binom{x_1-1}{3} + \binom{x_2-1}{2} + \binom{x_3-1}{1} + 1$ gives the position of the string $x_1 x_2 x_3$ in the list of decreasing strings of length three from $\underline{8}$. We solve the equation $p(x_1, x_2, x_3) = \binom{8}{3}/2 = 28$ for the variables x_1, x_2, x_3. Equivalently, find x_1, x_2, x_3 such that $\binom{x_1-1}{3} + \binom{x_2-1}{2} + \binom{x_3-1}{1} = 27$. First try to choose $x_1 - 1$ as large as possible so that $\binom{x_1-1}{3} \leq 27$. A little checking gives $x_1 - 1 = 6$, with $\binom{x_1-1}{3} = \binom{6}{3} = 20$. Subtracting, $27 - 20 = 7$. Now choose $x_2 - 1$ as large as possible so that $\binom{x_2-1}{2} \leq 7$. This gives $x_2 - 1 = 4$ with $\binom{x_2-1}{2} = \binom{4}{2} = 6$. Now subtract $7 - 6 = 1$ and choose $x_3 - 1 = 1$. Thus, $(x_1, x_2, x_3) = (7, 5, 2)$. The first element in the second half of the list is the next one in lex order after 752 which is 753. The corresponding distributions of ball into boxes can be obtained in the usual way.

Fn-3.5 (a) $2, 2, 3, 3$ is not a restricted growth (RG) function because it doesn't start with 1. $1, 2, 3, 3, 2, 1$ is a restricted growth function. It starts with 1 and the first occurrence of each integer is exactly one greater than the maximum of all previous integers.

Solutions for Functions

$1,1,1,3,3$ is not an RG function. The first occurrence of 3 is *two* greater than the max of all previous integers.
$1,2,3,1$ is an RG function.

(b) We list the blocks $f^{-1}(i)$ in order of i. Observe that all partitions of 4 occur exactly once as coimages of the RG functions.

$1111 \to \{1,2,3,4\}$	$1112 \to \{1,2,3\},\{4\}$	$1121 \to \{1,2,4\},\{3\}$
$1122 \to \{1,2\},\{3,4\}$	$1123 \to \{1,2\},\{3\},\{4\}$	$1211 \to \{1,3,4\},\{2\}$
$1212 \to \{1,3\},\{2,4\}$	$1213 \to \{1,3\},\{2\},\{4\}$	$1221 \to \{1,4\},\{2,3\}$
$1222 \to \{1\},\{2,3,4\}$	$1223 \to \{1\},\{2,3\},\{4\}$	$1231 \to \{1,4\},\{2\},\{3\}$
$1232 \to \{1\},\{2,4\},\{3\}$	$1233 \to \{1\},\{2\},\{3,4\}$	$1234 \to \{1\},\{2\},\{3\},\{4\}$

(c) $11111, 11112, 11121, 11122, 11123 \to \{\{1,2,3\},\{4\},\{5\}\}$
$11211, 11212, 11213, 11221, 11222 \to \{\{1,2\},\{3,4,5\}\}$
$11223, 11231, 11232, 11233, 11234 \to \{\{1,2\},\{3\},\{4\},\{5\}\}$

Fn-3.6 $S(6,3)(5)_3 = 90 \times 5 \times 4 \times 3 = 5400$.

Fn-3.7 The set B of balls is the domain and the set C of cartons is the range. Every function in C^B describes a different one of the ways to put balls into cartons. Since 2 cartons are to remain empty, we are interested in functions f with $|\text{Image}(f)| = 3$. Thus the answer to this exercise is exactly the same as for the previous exercise.

Fn-3.8 By the theorem in the text and Example 14, these are all the same. By the method in Example 14, the answer is $\binom{4+6-1}{6} = \binom{9}{6} = \binom{9}{3} = 84$.

Fn-4.1

$h_{X,Y}$	0	1	2	3	4	f_X
0	1/16	0	0	0	0	1/16
1	0	4/16	0	0	0	4/16
2	0	3/16	3/16	0	0	6/16
3	0	0	2/16	2/16	0	4/16
4	0	0	0	0	1/16	1/16
f_Y	1/16	7/16	5/16	2/16	1/16	

The row index is X and the column index is Y.

$E(X) = 2$, $\text{Var}(X) = \sigma_X = 1$ $E(Y) = 1.69$, $\text{Var}(Y) = 0.96$, $\sigma_Y = 0.98$

(c) $\text{Cov}(X,Y) = 0.87$

(d) $\rho(X,Y) = 0.87/(1)(0.98) = +0.89$ Since the correlation is close to 1, X and Y move up and down together. In fact, you can see from the table for the joint distribution that X and Y are often equal.

Fn-4.2 (a) You should be able to supply reasons for each of the following steps

$$\text{Cov}(aX + bY, aX - bY) = E[(aX+bY)(aX-bY)] - E[(aX+bY)]E[(aX-bY)]$$
$$= E[a^2X^2 - b^2Y^2] - [aE(X) - bE(Y)][aE(X) + bE(Y)]$$
$$= E[a^2X^2 - b^2Y^2] - [a^2(E(X))^2 - b^2(E(Y))^2]$$
$$= a^2[E(X^2) - (E(X))^2] - b^2[E(Y^2) - (E(Y))^2]$$
$$= a^2\text{Var}(X) - b^2\text{Var}(Y)$$

Alternatively, using the bilinear and symmetric properties of Cov:

$$\text{Cov}(aX + bY, aX - bY) = a^2\text{Cov}(X,X) - ab\text{Cov}(X,Y) + ba\text{Cov}(Y,X) + b^2\text{Cov}(Y,Y)$$
$$= a^2\text{Var}(X) - b^2\text{Var}(Y)$$

(b) Here is the calculation:

$$\text{Var}[(aX+bY)(aX-bY)] = \text{Var}[a^2X^2 - b^2Y^2)]$$
$$= a^4\text{Var}(X^2) - 2a^2b^2\text{Cov}(X^2,Y^2) + b^4\text{Var}(Y^2)$$

Fn-4.3 We begin our calculations with no assumptions about the distribution for X. Expand the argument of the expectation and then use linearity of expectation to obtain.

$$E((aX+b)^2) = E(a^2X^2 + 2abX + b^2)) = a^2E(X^2) + 2abE(X) + b^2 .$$

(The last term comes from the fact that $E(b^2) = b^2$ since b^2 is a constant.) By definition, $\text{Var}(X) + (E(X))^2 = E(X^2)$. Thus

$$E((aX+b)^2) = a^2\left(\text{Var}(X) + (E(X))^2\right) + 2abE(X) + b^2 .$$

With a little algebra this becomes,

$$E((aX+b)^2) = a^2\text{Var}(X) + (aE(X)+b)^2 .$$

Specializing to the particular distributions for parts (a) and (b), we have the following.

(a) $E((aX+b)^2) = a^2np(1-p) + (anp+b)^2$.

(b) $E((aX+b)^2) = a^2\lambda + (a\lambda+b)^2$.

Fn-4.4 We make the dubious assumption that the misprints are independent of one another. (This would not be the case if the person preparing the book was more careless at some times than at others.)

Focus your attention on page 8. Go one by one through the misprints m_1, m_2, ..., m_{200} asking the question, "Is misprint m_i on page 8?"

By the assumptions of the problem, the probability that the answer is "yes" for each m_i is $1/100$. Thus, we are dealing with the binomial distribution $b(k; 200, 1/100)$. The probability of there being less than four misprints on page 8 is

$$\sum_{k=0}^{3} b(k; 200, 1/100) = \sum_{k=0}^{3} \binom{200}{k}(1/100)^k(99/100)^{200-k}.$$

Using a calculator, we find the sum to be 0.858034.

Using the Poisson approximation, we set $\lambda = np = 2$ and compute the easier sum

$$e^{-2}2^0/0! + e^{-2}2^1/1! + e^{-2}2^2/2! + e^{-2}2^3/3!,$$

which is 0.857123 according to our calculator.

Fn-4.5 From the definition of Z and the independence of X and Y, Tchebycheff's inequality states that

$$P(|Z - aE(X) - bE(y)| \geq \epsilon) \leq \frac{\text{Var}(X) + \text{Var}(Y)}{\epsilon^2}.$$

Applying this to the two parts (a) and (b), we get

Solutions for Functions

(a) $P(|Z - a\gamma - b\delta| \geq \epsilon) \leq \dfrac{\gamma + \delta}{\epsilon^2}$.

(b) $P(|Z - anr - bns| \geq \epsilon) \leq \dfrac{nr(1-r) + ns(1-s)}{\epsilon^2}$.

Fn-4.6 We are dealing with $b(k; 1000, 1/10)$. The mean is $np = 100$ and the variance is $npq = 90$. The standard deviation is thus, 9.49. The exact solution is

$$\sum_{k=85}^{115} b(k; 1000, 1/10) = \sum_{k=85}^{115} \binom{1000}{k} (1/10)^k (9/10)^{1000-k}.$$

Using a computer with multi-precision arithmetic, the exact answer is 0.898. To apply the normal distribution, we would compute the probability of the event $[100, 115]$ using the normal distribution with mean 100 and standard deviation 9.49. In terms of the standard normal distribution, we compute the probability of the event $[0, (115 - 100)/9.49] = [0, 1.6]$ (rounded off). If you have access to values for areas under the standard normal distribution, you can find that the probability is 0.445. We double this to get the approximate answer: 0.89.

Fn-4.7 We have

$$E(\overline{X}) = E((1/n)(X_1 + \cdots + X_n)) = (1/n)E(X_1 + \cdots + X_n)$$
$$= (1/n)(E(X_1) + \cdots + E(X_n)) = (1/n)(\mu + \cdots + \mu) = \mu$$

$$\mathrm{Var}(\overline{X}) = \mathrm{Var}((1/n)(X+1+\cdots+X_n)) = (1/n)^2 \mathrm{Var}(X+1+\cdots+X_n)$$
$$= (1/n)^2 (\mathrm{Var}(X+1) + \cdots + \mathrm{Var}(X_n)) = (1/n)^2 (n\sigma^2) = \sigma^2/n.$$

Since $\overline{X}$ has mean μ, it is a reasonable approximation to μ. Of course, it's important to know something about the accuracy.

(c) Since $\mathrm{Var}(\overline{X}) = \sigma^2/n$, we have $\sigma_{\overline{X}} = \sigma/\sqrt{n}$. If we change from n to N, $\sigma_{\overline{X}}$ changes to $\sigma/\sqrt{N}$. Since we want to improve accuracy by a factor of 10, we want to have $\sigma/\sqrt{N} = (1/10)(\sigma/\sqrt{n})$. After some algebra, this gives us $N = 100n$. In other words we need to do 100 times as many measurements!

Solutions for Decision Trees and Recursion

DT-1.1 PREV: C, CC, CCV, CCVC, CCVCC, CCCVCV, CV, CVC, CVCC, CVCCV, CVCV, CVCVC, V, VC, VCC, VCCV, VCCVC, VCV, VCVC, VCVCC, VCVCV.
POSV: CCVCC, CCVCV, CCVC, CCV, CC, CVCCV, CVCC, CVCVC, CVCV, CVC, CV, C, VCCVC, VCCV, VCC, VCVCC, VCVCV, VCVC, VCV, VC, V.
BFV: C, V, CC, CV, VC, CCV, CVC, VCC, VCV, CCVC, CVCC, CVCV, VCCV, VCVC, CCVCC, CCVCV, CVCCV, CVCVC, VCCVC, VCVCC, VCVCV.

DT-1.2 You will need the decision trees for lex and insertion order for permutations of $\underline{3}$ and $\underline{4}$. The text gives the tree for insertion order for $\underline{4}$, from which the tree for $\underline{3}$ can be found — just stop one level above the leaves of $\underline{4}$. You should construct the tree for lex order.

(a) To answer this, compare the leaves. For $n = 3$, permutations $\sigma = 123, 132$, and 321 have $\text{RANK}_L(\sigma) = \text{RANK}_I(\sigma)$. For $n = 4$ the permutations $\sigma = 1234, 1243$, and 4321 have $\text{RANK}_L(\sigma) = \text{RANK}_I(\sigma)$.

(b) From the tree for (a), $\text{RANK}_L(2314) = 8$.
Rather than draw the large tree for $\underline{5}$, we use a smarter approach to compute $\text{RANK}_L(45321) = 95$. To see the latter, Note that all permutations on $\underline{5}$ that start with 1, 2, or 3 come before 45321. There are $3 \times 24 = 72$ of those. This leads us to the subtree for permutations of $\{1,2,3,5\}$ in lex order. It looks just like the decision tree for $\underline{4}$ with 4 replaced by 5. (Why is this?) Since $\text{RANK}_L(4321) = 23$, this makes a total of $72 + 23 = 95$ permutations that come before 45321 and so $\text{RANK}_L(45321) = 95$. If you find this unclear, you should try to draw a picture to help you understand it.

(c) $\text{RANK}_I(2314) = 16$. What about $\text{RANK}_I(45321)$? First does 1, then 2, and so on. After have done all but 5, we are at the rightmost leaf of the tree for $\underline{4}$. It has 23 leaves to the left of it. When we insert 5, each of these leaves is replaced by 5 new leaves because there are 5 places to insert 5. This gives us $5 \times 23 = 115$ leaves. Finally, of the 5 places we could insert 5 into 4321, we chose the 4th so there are 3 additional leaves to the left of it. Thus the rank is $115 = 3 = 118$.

(d) $\text{RANK}_L(3241) = 15$.

(e) $\text{RANK}_I(4213) = 15$.

(f) The first 24 permutations on $\underline{5}$ consist of 1 followed by a permutation on $\{2,3,4,5\}$. Since our goal is the permutation of rank 15, it is in this set. By (d), RANK_L of 3241 is 15 for $n = 4$. Thus $\text{RANK}_L(4352) = 15$ in the lex list of permutations on $\{2,3,4,5\}$.

Solutions for Decision Trees and Recursion

DT-1.3 Here is the tree

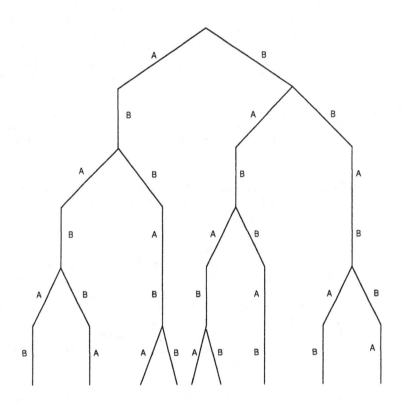

The list in lex order:

ABABAB ABABBA, ABBABA, ABBABB BABABA BABABB BABBAB BBABAB BBABBA

DT-1.4 Here is a decision tree for $D(\underline{6}^4)$. The leaves correspond to the elements of $D(\underline{6}^4)$ in

Solutions for Decision Trees and Recursion

lex order, obtained by reading the sequence of vertex labels from the root to the leaf.

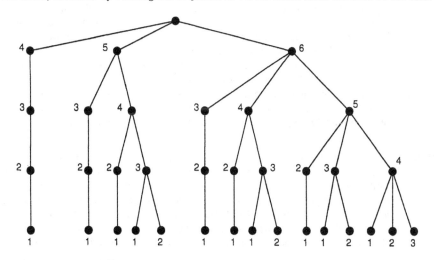

(a) The rank of 5431 is 3. The rank of 6531 is 10.

(b) 4321 has rank 0 and 6431 has rank 7.

(c) The first 5 leaves correspond to $D(\underline{5}^{\underline{4}})$.

(d) $D(\underline{6}^{\underline{4}})$ is bijectively equivalent to the set, $\mathbf{P}(\underline{6},4)$, of all subsets of $\underline{6}$ of size 4. Under this bijection, an element such as $5431 \in D(\underline{6}^{\underline{4}})$ corresponds to the set $\{1,3,4,5\}$.

DT-1.5 For PREV and POSV, omit Step 2. For PREV, begin Step 3 with the sentence

"If you have not used any edges leading out from the vertex, list the vertex."

For POSV, change Step 3 to

"If there are no unused edges leading out from the vertex, list the vertex and go to Step 4; otherwise, go to Step 5."

DT-1.6 The problem is that the eight hibachi grills, though different as domino coverings, are all equivalent or "isomorphic" once they are made into grills. All eight in the first row below can be gotten by rotating and/or turning over the first grill.

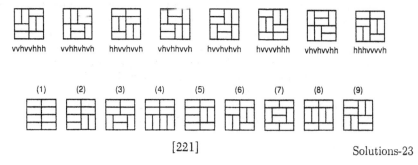

Solutions for Decision Trees and Recursion

There are nine different grills as shown in the picture. These nine might be called a "representative system" for the domino coverings up to "grill equivalence." Note that these nine representatives are listed in lex order according to their codes (starting with hhhhhhhh and ending with hvvhvvhh). They each have another interesting property: each one is lexicographically minimal among all patterns equivalent to it. The one we selected from the list of "screwup" grills (number (6)) has code hhhvvvvh and that is minimal among all codes on the first row of coverings.

This problem is representative of an important class of problems called "isomorph rejection problems." The technique we have illustrated, selecting a lex minimal system of representatives up to some sort of equivalence relation, is an important technique in this subject.

DT-2.1 We refer to the decision tree in Example 10. The permutation 87612345 specifies, by edge labels, a path from the root L($\underline{8}$) to a leaf in the decision tree. To compute the rank, we must compute the number of leaves "abandoned" by each edge just as was done in Example 14. There are eight edges in the path with the number of abandoned leaves equal to $7\times 7! + 6\times 6! + 5\times 5! + 0 + 0 + 0 + 0 + 0 = 35,280 + 4,320 + 600 = 40,200$. This is the RANK of 87612345 in the lex list of permutations on $\underline{8}$. Note that $8! = 40,320$, so the RANK 20,160 permutation is the first one of the second half of the list: 51234678.

DT-2.2 (a) The corresponding path in the decision tree is H(8, S, E, G), H(7, E, S, G), H(6, S, E, G), H(5, S, G, E), H(4, S, E, G), H(3, E, S, G), E $\xrightarrow{3}$ G.

(b) The move that produced the configuration of (a) was E $\xrightarrow{3}$ G. The configuration prior to that was Pole S: 6, 5, 2, 1; Pole E: 3; Pole G: 8, 7, 4.

(c) The move just prior to E $\xrightarrow{3}$ G was G $\xrightarrow{1}$ S. This is seen from the decision tree structure or from the fact that the smallest washer, number 1, moves every other time in the pattern S, E, G, S, E, G, etc. The configuration just prior to the move G $\xrightarrow{1}$ S was Pole S: 6, 5, 2; Pole E: 3; Pole G: 8, 7, 4, 1.

(d) The next move after E $\xrightarrow{3}$ G will be another move by washer 1 in its tiresome cycle S, E, G, S, E, G, etc. That will be S $\xrightarrow{1}$ E.

(e) The RANK of the move that produced (a) can be computed by summing the abandoned leaves associated with each edge of the path (a) in the decision tree. (See Example 14.) There are six edges in the path of part (a) with associated abandoned leaves being $2^7 = 128$, $2^6 = 64$, 0, 0, $2^3 = 8$, $2^2 - 1 = 3$. The total is 203.

DT-2.3 (a) 110010000 is preceded by 110010001 and is followed by 110110000. You can find this by first drawing the path from the root to 110110000. You will find that the last edge of the path goes to the right. Therefore, we can get the preceding element by going to the left instead. This changes the last element from 0 to 1 and all other elements remain fixed. To get the element that follows it, we want to branch to the right instead of the left. The last five edges to 110110000 all go to the right and the edge just before them, say e goes to the left. Instead of taking e, we take the edge that goes to the right. Now what? We must take edges to the left after this so that we end up as close to the original leaf 110010000 as possible. A trick: Since we are dealing with a Gray code, we know that there is only one change so that when we've found it we can just copy everything else. In this case we changed the underlined symbol in 110$\underline{0}$10000 (from 0 to 1) and so the others are the same.

Solutions for Decision Trees and Recursion

(b) The first element of the second half of the list corresponds to a path in the decision tree that starts with a right-sloping edge and has all of the remaining eight edges left-sloping. That element is 110000000.

(c) Each right-sloping edge abandons 2^{n-k} leaves, if the edge is the kth one in the path. For the path 111111111 the right-sloping edges are numbers 1, 3, 5, 7, and 9 (remember, after the first edge, a label 1 causes the direction of the path to change). Thus, the rank of 111111111 is $2^8 + 2^6 + 2^4 + 2^2 + 2^0 = 341$.

(d) To compute the element of RANK 372, we first compute the path in the decision tree that corresponds to the element. The first edge must be **(1) right sloping** (abandoning 256 leaves), since the largest rank of any leaf at the end of a path that starts left sloping is $2^8 - 1 = 255$. We apply this same reasoning recursively. The right sloping edge leads to 256 leaves. We wish to find the leaf of RANK $372 - 256 = 116$ in that list of 256 leaves. That means the second edge must be **(1) left sloping** (abandoning 0 leaves), so our path starts off **(1) right sloping, (1) left sloping**. This path can access 128 leaves. We want the leaf of RANK $116 - 0$ in this list. Thus we must access a leaf in the second half of the list of 128, so the third edge must be **(1) right sloping** (abandoning 64 leaves). In that second half we must find the leaf of RANK $116 - 64 = 52$.

Our path is now **(1) right sloping, (1) left sloping, (1) right sloping**. Following that path leads to 64 leaves of which we want the leaf of RANK 52. Thus, the fourth edge must be **(0) right sloping** (abandoning 32 leaves). This path of four edges leads to 32 leaves of which we must find the one of RANK $52 - 32 = 20$. Thus the fifth edge must also be **(0) right sloping** (abandoning 16 leaves). Thus we must find the leaf of RANK $20 - 16 = 4$. This means that the sixth edge must be **(1) left sloping** (abandoning 0 leaves), the seventh edge must be **(1) right sloping** (abandoning 4 leaves), and the last two edges must be left sloping: **(1) left sloping** (abandoning 0 leaves), **(0) left sloping** (abandoning 0 leaves). Thus the final path is 111001110.

DT-2.4 (a) Let $\mathcal{A}(n)$ be the assertion "H(n,S,E,G) takes the least number of moves." Clearly $\mathcal{A}(1)$ is true since only one move is required. We now prove $\mathcal{A}(n)$. Note that to do $S \xrightarrow{n} G$ we must first move all the other washers to pole E. They can be stacked only one way on pole E, so moving the washers from S to E requires using a solution to the Towers of Hanoi problem for $n-1$ washers. By $\mathcal{A}(n-1)$, this is done in the least number of moves by H($n-1$,S,G,E). Similarly, H($n-1$,E,S,G) moves these washers to G in the least number of moves.

(b) For $n = 1$, $f_1 = 1$: $S \xrightarrow{1} G$
For $n = 2$, $f_2 = 3$: $S \xrightarrow{1} E$, $S \xrightarrow{2} G$, $E \xrightarrow{1} G$
For $n = 3$, $f_3 = 5$: $S \xrightarrow{1} E$, $S \xrightarrow{2} F$, $S \xrightarrow{1} G$, $F \xrightarrow{2} G$, $E \xrightarrow{1} G$

(c) Let $s(p,q)$ be the number of moves for G(p, q, S, E, F, G). The recursive step in the problem is described for $p > 0$, so the simplest case is $p = 0$ and $s(0,q) = h(q) = 2^q - 1$. In that case, (i) tells us what to do.

Otherwise, the number of moves in (ii) is $s(p,q) = 2s(i,j) + h_q$. To find the minimum, we look at all allowed values of i and j, choose those for which $s(i,j)$ is a minimum. This choice of i and j, when used in (ii) tells us which moves to make. In the following table, numbers on the rows refer to p and those on the columns refer to q.

[223]

Solutions for Decision Trees and Recursion

Except for the s_p column, then entries are $s(p,q)$. The $p = 0$ row is h_q by (i). To find $s(p,q)$ for $p > 0$, we use (ii). To do this, we look along the diagonal whose indices sum to p, choose the minimum (It's location is (i,j).), double it and add h_q. For example, $s(5,2)$ is found by taking the minimum of the diagonal entries at (0,5), (1,4), (2,3), (3,2), and (4,1). Since these entries are 31, 17, 13, 13, and 19, the minimum is 13. Since this occurs at (2,3) and (3,2), we have a choice for (i,j). Either one gives us $2 \times 13 + h_2 = 29$ moves. To compute s_n we simply look along the $p+q = n$ diagonal and choose the minimum.

	s_p	1	2	3	4	5	6	(values of q)
0		1	3	7	15	31	63	($s(0,q) = h_q$)
1	1	3	5	9	17	33	65	
2	3	7	9	13	21	27		
3	5	11	13	17	25			
4	9	19	21	25				
5	13	27	29					
6	17	35						

Column labels are p.

(d) From the description of the algorithm,

- $s(p,q) = 2 \min s(i,j) + h_q$, where the minimum is over $i + j = p$ and
- $s_n = \min s(p,q)$, where the minimum is over $p + q = n$.

Putting these together gives us $s(p,q) = 2s_p + h_q$ and so $s_n = \min(2s_p + h_q)$. The initial condition is $s_0 = 0$. In summary

$$s_n = \begin{cases} 0 & \text{if } n = 0, \\ \min_{\substack{p+q=n \\ q>0}} (2s_p + h_q) & \text{if } n > 0. \end{cases}$$

(e) Change the recursive procedure in the algorithm to use the moves for f_p instead of using those for $s(p,q)$. It follows that we can solve the puzzle in $2f_{n-j} + h_j$ moves.

DT-3.1 When there is replacement, the result of the first choice does not matter since the ball is placed back in the box. Hence the answer to both parts of (a) is 3/7.

(b) If the first ball is green, we are drawing a ball from three white and three green and so the probability is $3/6 = 1/2$. If the first ball is white, we are drawing a ball from two white and four green and so the probability is $2/6 = 1/3$.

DT-3.2 There are five ways to get a total of six: $1+5$, $2+4$, $3+3$, $4+2$, and $5+1$. All five are equally likely and so each outcome has probability 1/5. We get the answers by counting the number that satisfy the given conditions and multiplying by 1/5:
(a) 1/5, (b) 2/5, (c) 3/5.

Solutions for Decision Trees and Recursion

DT-3.3 Here is the decision tree for this problem

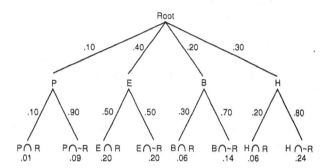

(a) We want to compute the conditional probability that a student is a humanities major, given that that student has read Hamlet. In the decision tree, if we follow the path from the Root to H to $H \cap R$, we get a probability of .06 at the leaf. We must divide this by the sum over all probabilities of such paths that end at $X \cap R$ (as opposed to $X \cap \sim R$). That sum is $0.01 + 0.20 + 0.06 + 0.06 = 0.33$. The answer is $0.06/0.33 = 0.182$.

(b) We compute the probabilities that a student has not read Hamlet and is a P (Physical Science) or E (Engineering) major: $0.09 + 0.20 = 0.29$. We must divide this by the sum over all probabilities of such paths that end at $X \cap \sim R$ (as opposed to $X \cap R$). The answer is $0.29/0.67 = 0.433$.

DT-3.4 Here is a decision tree where the vertices are urn compositions. The edges incident on the root are labeled with the outcome sets of the die and the probabilities that these sets occur. The edges incident on the leaves are labeled with the color of the ball drawn and the probability that such a ball is drawn. The leaves are labeled with the product of the probabilities on the edges leading from the root to that leaf.

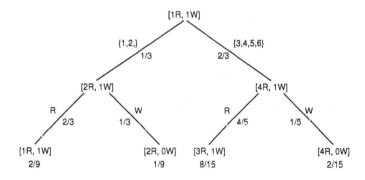

(a) To compute the conditional probability that a 1 or 2 appeared, given that a red ball was drawn, we take the probability 2/9 that a 1 or 2 appeared and a red ball was drawn and divide by the total probability that a red ball was drawn: $2/9 + 8/15 = 34/45$. The answer is $5/17 = 0.294$.

[225]

Solutions for Decision Trees and Recursion

(b) We divide the probability that a 1 or 2 appeared and the final composition had more than one red ball (1/9) by the sum of the probabilities where the final composition had more than one red ball : $1/9 + 8/15 + 2/15 = 7/9 = 0.78$.

DT-3.5 A decision tree is shown below. The values of the random variable X are shown just below the amount remaining in the pot associated with each leaf. To compute $E(X)$ we sum the values of X times the product of the probabilities along the path from the root to that value of X. Thus, we get

$$E(X) = 1 \times (1/2) + 2 \times (1/8) + (2 + 3 + 3 + 3 + 4 + 5) \times (1/16) = 2.$$

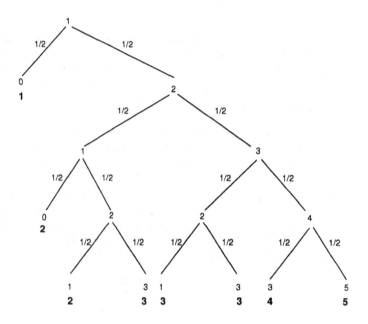

DT-3.6 A decision tree is shown below. Under the leaves is the length of the game (the height of the leaf). The expected length of the game is the sum of the products of the probabilities on the edges of each path to a leaf times the height of that leaf:

$$2((1/3)^2 + (2/3)^2) +$$

$$4((1/3)^3(2/3) + (1/3)^2(2/3)^2 + (1/3)^2(2/3)^2 + (1/3)(2/3)^3) +$$

$$3((1/3)(2/3)^2 + (1/3)^2(2/3)).$$

[226]

Solutions for Decision Trees and Recursion

The expected number of games is about 2.69.

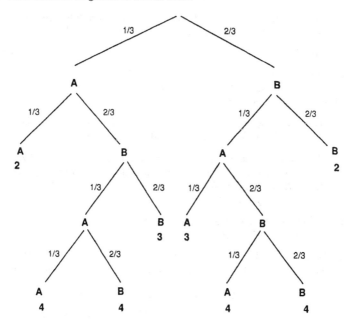

DT-3.7 We are given

$$P(F' \mid A) = 0.6, \quad P(F \mid A') = 0.8 \quad \text{and} \quad P(A) = 0.7.$$

You can draw a decision tree. The first level branches according as the air strike is successful (A) or not (A'). The probabilities, left to right, are 0.7 and $1 - 0.7 = 0.3$. The second level branches according as there is enemy fire (F) or not (F'). To compute the conditional probabilities on the edges, note that

$$P(F \mid A) = P(F' \mid A) = 1 - 0.6 = 0.4 \quad \text{and} \quad P(F' \mid A') = 1 - 0.8 = 0.2.$$

The leaves and their probabilities are

$$P(A \cap F) = 0.7 \times 0.4 = 0.28, \qquad P(A \cap F') = 0.7 \times 0.6 = 0.42,$$

$$P(A' \cap F) = 0.3 \times 0.8 = 0.24, \qquad P(A' \cap F') = 0.3 \times 0.2 = 0.06.$$

For (a), $P(F') = 0.42 + 0.06 = 0.48$ and for (b)

$$P(A \mid F') = \frac{P(A \cap F')}{P(F')} = \frac{0.42}{0.48} \approx 82\%.$$

DT-4.1 (a) $a_n = 1$ for all n.

(b) $a_0 = 0$, $a_1 = 0 + a_0 = 0$, $a_2 = 1 + a_1 = 1$, $a_3 = 1 + a_2 = 2$, $a_4 = 2 + a_3 = 4$.

Solutions for Decision Trees and Recursion

(c) $a_0 = 1$, $a_1 = 1 + a_0 = 2$, $a_2 = 2 + a_1 = 4$, $a_3 = 3 + a_1 = 5$, $a_4 = 4 + a_2 = 8$.

(d) $a_0 = 0$, $a_1 = 1$, $a_2 = 1 + a_1 a_1 = 2$, $a_3 = 1 + \min(a_1 a_2, a_2 a_1) = 1 + a_1 a_2 = 3$, $a_4 = 1 + \min(a_1 a_3, a_2 a_2, a_3, a_1) = 1 + \min(3, 4) = 4$.

DT-4.2 $a_n = \lfloor n/2 \rfloor$, $b_n = (-1)^n \lfloor 1 + (n/2) \rfloor = (-1)^n (1 + \lfloor n/2 \rfloor)$, $c_n = n^2 + 1$, $d_n = n!$.

DT-4.3 $x^2 - 6x + 5 = 0$ has roots $r_1 = 1$ and $r_2 = 5$
$x^2 - x - 2 = 0$ has roots $r_1 = -1$ and $r_2 = 2$
$x^2 - 5x - 5 = 0$ has roots $\frac{5 \pm \sqrt{45}}{2}$.

DT-4.4 The characteristic equation is $x^2 - 6x + 9 = 0$, which factors as $(x-3)^2 = 0$. Thus $r_1 = r_2 = 3$. We have $K_1 = a_0 = 0$ and $3K_2 = a_1 = 3$. Thus $a_n = n3^n$.

DT-4.5 Let $A_n = a_{n+2}$ so that $A_0 = 1$, $A_1 = 3$ and $A_n = 3A_{n-1} + 2A_{n-2}$ for $n > 2$. The characteristic equation is $x^2 - 3x - 2 = 0$ and has roots $r_1 = 1$, $r_2 = 2$. Thus $K_1 + K_2 = 1$ and $K_1 + 2K_2 = 3$ and so $K_1 = -1$ and $K_2 = 2$. We have $A_n = -1 + 2 \times 2^n = 2^{n+1} - 1$ and so $a_n = A_{n-2} = 2^{n-1} - 1$.

DT-4.6 The characteristic equation is $x^2 - 2x + 1 = (x-1)^2 = 0$. Thus $r_1 = r_2 = 1$ and so $K_1 = a_0 = 2$ and $K_1 + K_2 = a_1 = 1$. We have $K_2 = 1 - K_1 = -1$ and so $a_n = 2 - n$.

DT-4.7 (a) Let $\mathcal{A}(n)$ be the assertion that $G(n) = (1 - A^n)/(1 - A)$. When $n = 1$, $G(1) = 1$ and $(1 - A^n)/(1 - A) = 1$, so the base case is proved. For $n > 1$, we have

$$\begin{aligned}
G(n) &= 1 + A + A^2 + \ldots + A^{n-1} & \text{by definition,} \\
&= (1 + A + A^2 + \ldots + A^{n-2}) + A^{n-1} \\
&= \frac{1 - A^{n-1}}{1 - A} + A^{n-1} & \text{by } \mathcal{A}(n-1), \\
&= \frac{1 - A^n}{1 - A} & \text{by algebra.}
\end{aligned}$$

(b) The recursion can be found by looking at the definition or by examining the proof in (a). It is $G(1) = 1$ and, for $n > 1$, $G(n) = G(n-1) + A^{n-1}$.

(c) Applying the theorem is straightforward. The formula equals 1 when $n = 1$, which agrees with $G(1)$. By some simple algebra

$$\frac{1 - A^{n-1}}{1 - A} + A^{n-1} = \frac{(1 - A^{n-1}) + (A^{n-1} - A^n)}{1 - A} = \frac{1 - A^n}{1 - A},$$

and so the formula satisfies the recursion.

(d) Letting $A = y/x$ and cleaning up some fractions

$$\frac{1 - (y/x)^n}{1 - y/x} = \frac{y^n - x^n}{x - y} x^{n-1}.$$

Let $n = k + 1$, multiply by x^k and use the geometric series to obtain

$$\begin{aligned}
\frac{x^{k+1} D - y^{k+1}}{x - y} &= x^k \left(1 + (y/x) + (y/x)^2 + \cdots + (y/x)^k \right) \\
&= x^k y^0 + x^{k-1} y^1 + \cdots + x^0 y^k.
\end{aligned}$$

Solutions for Decision Trees and Recursion

DT-4.8 We will Theorem 7 to prove our conjectures are correct.

(a) Writing out the first few terms gives A, $A/(1+A)$, $A/(1+2A)$, $A/(1+3A)$, etc. It appears that $a_k = A/(1+kA)$. Since $A > 0$, the denominators are never zero. When $k = 0$, $A/(1+kA) = A$, which satisfies the initial condition. We check the recursion:

$$\frac{A/(1+(k-1)A)}{1+A/(1+(k-1)A)} = \frac{A}{(1+(k-1)A)+A} = A/(1+kA),$$

which is the conjectured value for a_k.

(b) Writing out the first few terms gives C, $AC+B$, $A^2C+AB+B$, $A^3C+A^2B+AB+B$, $A^4C+A^3B+A^2B+AB+B$, etc. Here is one possible formula:

$$a_k = A^kC + B(1 + A + A^2 + \ldots + A^{k-1}).$$

Here is a second possibility:

$$a_k = A^kC + B\left(\frac{1-A^k}{1-A}\right).$$

Using the previous exercise, you can see that they are equal. We leave it to you to give a proof of correctness for both formulas, without using the previous exercise.

DT-4.9 We use Theorem 7. The formula gives the correct value for $k = 0$. The recursion checks because

$$A + B(k-1)\bigl(((k-1)^2 - 1)/3\bigr) + Bk(k-1) = A + B(k-1)\bigl((k^2 - 2k + 1 - 1) - 3k\bigr)$$
$$= A + B(k-1)k(k+1)/3$$
$$= A + Bk(k^2 - 1)/3.$$

This completes the proof.

DT-4.10 (a) We apply Theorem 7, but there is a little complication: The formula starts at $k = 1$, so we cannot check the recursion for $k = 1$. Thus we need a_1 to be the initial condition. From the recursion, $a_1 = 2A - C$, which we take as our initial condition and use the recursion for $k > 1$. You should verify that the formula gives a_1 correctly and that the formula satisfies the recursion when $k > 1$.

(b) From the last part of Exercise 4.7 with $x = 2$ and $y = -1$, we obtain

$$a_k = A\left(\frac{2^{k+1} - (-1)^{k+1}}{3}\right) + (-1)^k(C - A).$$

Make sure you can do the calculations to derive this.

DT-4.11 Let p_k denote the probability that the gambler is ruined if he starts with $0 \leq k \leq Q$ dollars. Note that $p_0 = 1$ and $p_Q = 0$. Assume $1 < k \leq Q$. Then the recurrence relation $p_{k-1} = (1/2)p_k + (1/2)p_{k-2}$ holds. Solving for p_k gives $p_k = 2p_{k-1} - p_{k-2}$. This looks familiar. It is a two term linear recurrence relation. But the setup was a little strange! We would expect to know p_0 and p_1 and would expect the values of p_k

Solutions for Decision Trees and Recursion

to make sense for all $k \geq 0$. But here we have an interpretation of the p_k only for $0 \leq k \leq Q$ and we know p_0 and p_Q instead of p_0 and p_1. Such a situation is not for faint-hearted students.

We are going to keep going as if we knew what we were doing. The characteristic equation is $r^2 - 2r + 1 = 0$. There is one root, $r = 1$. That means that the sequence $a_k = 1$, for all $k = 0, 1, 2, \ldots$, is a solution and so is $b_k = k$, for $k = 0, 1, 2, \ldots$. We need to find A and B such that $Aa_0 + Bb_0 = 1$ and $Aa_Q + Bb_Q = 0$. We find that $A = 1$ and $B = -1/Q$. Thus we have the general solution

$$p_k = 1 - \frac{k}{Q} = \frac{Q-k}{Q} \qquad q_k = \frac{k}{Q}.$$

Note that p_k is defined for all $k \geq 0$ like it would be for any such linear two term recurrence. The fact that we are only interested in it for $0 \leq k \leq Q$ is no problem to the theory.

Suppose a rich student, Brently Q. Snodgrass the III, has 8,000 dollars and he wants to play the coin toss game to make 10,000 dollars so he has 2,000 his parents don't know about. His probability of being ruined is $(10,000 - 8000)/10000 = 1/5$. His probability of getting his extra 2000 dollars is 4/5. A poor student who only had 100 dollars and wanted to make 2000 dollars would have a probability of $(2,100 - 100)/2,100 = 0.95$ of being ruined. Life isn't fair.

There is one consolation. The expected number of times Brently will have to toss the coin to earn his 2,000 dollars is 16,000,000. It will take him 69.4 weeks tossing 40 hours per week, one toss every 10 seconds. If he does get his 2000 dollars, he will have been working as a "coin tosser" for over a year at a salary of 72 cents per hour. He should get a minimum wage job instead!

Solutions for Basic Concepts in Graph Theory

GT-1.1 To specify a graph we must choose $E \in \mathcal{P}_2(V)$. Let $N = |\mathcal{P}_2(V)|$. (Note that $N = \binom{n}{2}$.) There are 2^N subsets E of $\mathcal{P}_2(V)$ and $\binom{N}{q}$ of them have cardinality q. This proves (a) and answers (b).

GT-1.2 The sum is the number of ends of edges since, if x and y are the ends of an edge, the edge contributes 1 to the value of $d(x)$ and 1 to the value of $d(y)$. Since each edge has two ends, the sum is twice the number of edges.

Since $\sum_v d(v)$ is even if and only if the number of odd summands is even, it follows that there are an even number of v for which $d(v)$ is odd.

GT-1.3 (a) The graph is isomorphic to Q. The correspondence between vertices is given by

$$\phi = \begin{pmatrix} A & B & C & D & E & F & G & H \\ H & A & C & E & F & D & G & B \end{pmatrix}$$

where the top row corresponds to the vertices of Q.

(b) The graph Q' is not isomorphic to Q. It can be made isomorphic by deleting one edge and adding another. You should try to figure out which edges these are.

GT-1.4 (a) $(0,2,2,3,4,4,4,5)$ is the degree sequence of Q. (b) If a pictorial representation of R can be created by labeling $P'(Q)$ with the edges and vertices of R, then R has degree sequence $(0,2,2,3,4,4,4,5)$ because the degree sequence is determined by ϕ.

(c) This is the converse of (b). It is false. The following graph has degree sequence $(0,2,2,3,4,4,4,5)$ but cannot be morphed into the form $P'(Q)$.

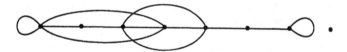

GT-1.5 (a) There is no graph Q with degree sequence $(1,1,2,3,3,5)$ since the sum of the degrees is odd. The sum of the degrees of a graph is $2|E|$ and must, therefore, be even.

(d) (answers (b) and (c) as well) There is a graph with degree sequence $(1,2,2,3,3,5)$, no loops or parallel edges allowed. Take

$$\phi = \begin{pmatrix} a & b & c & d & e & f & g & h \\ A & B & C & A & B & C & E & F \\ B & C & E & D & D & D & D & D \end{pmatrix}.$$

(e) (answers (f) as well) A graph with degree sequence $(3,3,3,3)$ has $(3+3+3+3)/2 = 6$ edges and, of course 4 vertices. That is the maximum $\binom{4}{2}$ of edges that a simple graph with 4 vertices can have. It is easy to construct such a graph. Draw the four vertices and make all possible connections. This graph is called the *complete* graph on 4 vertices.

(g) There is no simple graph (or graph without loops or parallel edges) with degree sequence $(3,3,3,5)$. See (f).

Solutions for Basic Concepts in Graph Theory

(h) Similar arguments to (f) apply to the complete graph with degree sequence $(4,4,4,4,4)$. Such a graph would have $20/2 = 10$ edges. But $\binom{5}{2} = 10$. To construct such a graph, use 5 vertices and make all possible connections.

(i) There is no such graph. See (h).

GT-1.6 Each of (a) and (c) has just one pair of parallel edges (edges with the same endpoints), while (b) and (d) each have two pairs of parallel edges. Thus neither (b) nor (d) is equivalent to (a) or (c). Vertex 1 of (b) has degree 4, but (d) has no vertices of degree 4. Thus (b) and (d) are not equivalent. It turns out that (a) and (c) are equivalent. Can you see how to make the forms correspond?

GT-1.7 (a) We know that the expected number of triangles behaves like $(np)^3/6$. This equals 1 when $p = 6^{1/3}/n$.

(b) By Example 6, the expected number of edges is $\binom{n}{2}p$, which behaves like $(n^2/2)p$ for large n. Thus we expect about $(6^{1/3}/2)n$

GT-1.8 Introduce random variables X_S, one for each $S \in \mathcal{P}_k(V)$. Reasoning as in the example, $E(X_S) = p^K$ where $K = \binom{k}{2}$, the number of edges that must be present. Thus the expected number of sets of k vertices with all edges present is $\binom{n}{k}p^K$.

For large n, this behaves like $n^k p^K/k!$, which will be 1 when $p = (k!/n^k)^{1/K}$. For large n, the expected number of edges behaves like $(n^2/2)(k!/n^k)^{1/K}$. This last number has the form Cn^α where $C = (k!)^{1/K}/2$ and $\alpha = 2 - k/K = 2 - 2/(k-1) = \frac{2(k-2)}{k-1}$.

GT-1.9 The first part comes from factoring out $\binom{n}{3}p^3$ from the last equation in Example 7. To obtain the inequality, replace $(1-p^3)$ with $(1-p^2)$, factor it out, and use $1+3(n-3) < 3n$.

GT-2.1 Since $E \subseteq \mathcal{P}_2(V)$, we have a simple graph. Regardless of whether you are in set C or S, following an edge takes you into the other set. Thus, following a path with an odd number of edges takes you to the opposite set from where you started while a path with an even number of edges takes you back to your starting set. Since a cycle returns to its starting vertex, it obviously returns to its starting set.

GT-2.2 (a) The graph is not Eulerian. The longest trail has 5 edges, the longest circuit has 4 edges.

(b) The longest trail has 9 edges, the longest circuit has 8 edges.

(c) The longest trail has 13 edges (an Eulerian trail starting at C and ending at D). The longest circuit has 12 edges (remove edge f).

(d) This graph has an Eulerian circuit (12 edges).

GT-2.3 (a) The graph is Hamiltonian.

(b) The graph is Hamiltonian.

(c) The graph is not Hamiltonian. There is a cycle that includes all vertices except K.

(d) The graph is Hamiltonian.

GT-2.4 (a) There are $|V \times V|$ potential edges to choose from. Since there are two choices for each edge (either in the digraph or not), we get 2^{n^2} simple digraphs.

Solutions for Basic Concepts in Graph Theory

(b) With loops forbidden, our possible edges include all elements of $V \times V$ except those of the form (v,v) with $v \in V$. Thus there are $2^{n(n-1)}$ loopless simple digraphs. An alternative derivation is to note that a simple graph has $\binom{n}{2}$ edges and we have 4 possible choices in constructing a digraph: (i) omit the edge, (ii) include the edge directed one way, (iii) include the edge directed the other way, and (iv) include two edges, one directed each way. This gives $4^{\binom{n}{2}} = 2^{n(n-1)}$. The latter approach is not useful in doing part (c).

(c) Given the set S of possible edges, we want to choose q of them. This can be done in $\binom{|S|}{q}$ ways. In the general case, the number is $\binom{n^2}{q}$ and in the loopless case it is $\binom{n(n-1)}{q}$.

GT-2.5 (a) Let $V = \{u,v\}$ and $E = \{(u,v),(v,u)\}$.

(b) For each $\{u,v\} \in \mathcal{P}_2(V)$ we have three choices: (i) select the edge (u,v), (ii) select the edge (v,u) or (iii) have no edge between u and v. Let $N = |\mathcal{P}_2(V)| = \binom{n}{2}$. There are 3^N oriented simple graphs.

(c) We can choose q elements of $\mathcal{P}_2(V)$ and then orient each of them in one of two ways. This gives us $\binom{N}{q}2^q$.

GT-2.6 (a) For all $x \in S$, $x|x$. For all $x,y \in S$, if $x|y$ and $x \neq y$, then y does not divide x. For all $x,y,z \in S$, $x|y$, $y|z$ implies that $x|z$.

(b) The covering relation is

$$H = \{(2,4),(2,6),(2,10),(2,14),(3,6),(3,9),(3,15),$$
$$(4,8),(4,12),(5,10),(5,15),(6,12),(7,14)\}.$$

We leave it to you to draw the picture!

GT-3.1 (a) Suppose G is a connected graph with v vertices and v edges. A connected graph is a tree if and only if the number of vertices is one more than the number of edges. Thus G is not a tree and must have at least one cycle. This proves the base case, $n = 0$. Suppose $n > 0$ and G is a graph with v vertices and $v+n$ edges. We know that the graph is not a tree and thus has a cycle. We know that removing an edge from a cycle does not disconnect the graph. However, removing the edge destroys any cycles that contain it. Hence the new graph G' contains one less edge and at least one less cycle than G. By the induction hypothesis, G' has at least n cycles. Thus G has at least $n+1$ cycles.

(b) Let G be a graph with components $G_1,\ldots,G_k$. With subscripts denoting components, G_i has v_i vertices, $e_i = v_i + n_i$ edges and at least $n_i + 1$ cycles. From the last two formulas, G_i has at least $1 + e_i - v_i$ cycles. Now sum over i.

(c) For each n we wish to construct a simple graph that has n more edges than vertices but has only $n+1$ cycles. There are many possibilities. Here's one solution. The vertices are v and, for $0 \le i \le n$, x_i and y_i. The edges are $\{v,x_i\}$, $\{v,y_i\}$, and $\{x_i,y_i\}$. (This gives $n+1$ triangles joined at v.) There are $1+2(n+1)$ vertices, $3(n+1)$ edges, and $n+1$ cycles.

GT-3.2 (a) $\sum_{v \in V} d(v) = 2|E|$. For a tree, $|E| = |V| - 1$. Since $2|V| = \sum_{v \in V} 2$,

$$2 = 2|V| - 2|E| = \sum_{v \in V}(2 - d(v)).$$

[233]

Solutions for Basic Concepts in Graph Theory

(b) Suppose that T is more than just a single vertex. Since T is connected, $d(v) \neq 0$ for all v. Let n_k be the number of vertices of T of degree k. By the previous result, $\sum_{k \geq 1}(2-k)n_k = 2$. Rearranging gives $n_1 = 2 + \sum_{k \geq 2}(k-2)n_k$. If $n_m \geq 1$, the sum is at least $m-2$.

(c) Let the vertices be u and v_i for $1 \leq i \leq m$. Let the edges be $\{u, v_i\}$ for $1 \leq i \leq m$.

GT-3.3 (a) No such tree exists. A tree with six vertices must have five edges.

(b) No such tree exists. Such a tree must have at least one vertex of degree three or more and hence at least three vertices of degree one.

(c) A graph with two connected components, each a tree, each with five vertices will have this property.

(d) No such graph exists.

(e) No such tree exists.

(f) Such a graph must have at least $c + e - v = 1 + 6 - 4 = 3$ cycles.

(g) No such graph exists. If the graph has no cycles, then each component is a tree. In such a graph, the number of vertices is strictly greater than the number of edges for each component and hence for the whole graph.

GT-3.4 (a) The idea is that for a rooted planar tree of height h, having at most 2 children for each non-leaf, the tree with the most leaves occurs when each non-leaf vertex has exactly 2 children. You should sketch some cases and make sure you understand this point. For this case $l = 2^h$ and so $\log_2(l) = h$. Any other rooted planar tree of height h, having most 2 children for each non-leaf, is a subtree (with the same root) of this maximal-leaf binary tree and thus has fewer leaves.

(b) Knowing the number of leaves does not bound the height of a tree — it can be arbitrarily large.

(c) The maximum height is $h = l - 1$. One leaf has height 1, one height 2, etc., one of height $l-2$ and, finally, two of height $l-1$.

(d) (answers (e) as well) $\lceil \log_2(l) \rceil$ is a lower bound for the height of *any* binary tree with l leaves. It is easy to see that you can construct a full binary tree with l leaves and height $\lceil \log_2(l) \rceil$.

GT-3.5 (a) A binary tree with 35 leaves and height 100 is possible.

(b) A full binary tree with 21 leaves can have height at most 20. So such a tree of height 21 is impossible.

(c) A binary tree of height 5 can have at most 32 leaves. So one with 33 leaves is impossible.

(d) No way! The total number of vertices is

$$\sum_{i=0}^{5} 3^5 = \frac{3^6 - 1}{2} = 364.$$

GT-3.6 (a) For *(1)* there are four spanning trees. For *(2)* there are 8 spanning trees. Note that there are $\binom{5}{3} = 10$ ways to choose three edges. Eight of these 10 choices result in spanning trees, the other two choices result in cycles (with vertex sequences (A, B, D) and (B, C, D)). For *(3)* there are 16 spanning trees.

(b) For *(1)* there is one. For *(2)* there are two. For *(3)* there are two.

(c) For *(1)* there are two. For *(2)* there are four. For *(3)* there are six.

(d) For *(1)* there are two. For *(2)* there are three. For *(3)* there are six.

GT-3.7 (a) For *(1)* there are three minimum spanning trees. For *(2)* there are two minimum spanning trees. For *(3)* there is one minimum spanning tree.

(b) For *(1)* there is one minimum spanning tree up to isomorphism. For *(2)* there are two. For *(3)* there is one.

(c) For *(1)* there is one. For *(2)* there is one. For *(3)* there are four.

(d) For *(1)* there are two. For *(2)* there is one. For *(3)* there are four.

GT-3.8 (a) (and (b)) There are 21 vertices, so the minimum spanning tree has 20 edges. Its weight is 30. We omit details.

(c) Note that K is a the only vertex in common to the two bicomponents of this graph. Whenever this happens (two bicomponents, common vertex), the depth-first spanning tree rooted at that common vertex has exactly two "principal subtrees" at the root. In other words, the root of the depth-first spanning tree has down-degree two (two children). The two children of K can be taken to be P and L. P is the root of a subtree consisting of 5 vertices, 4 with one child, one leaf. L is the root of a subtree consisting of 15 vertices, 14 with one child, one leaf.

GT-4.1 (a) The algorithm that has running time $100n$ is better than the one with running time n^2 for $n > 100$. $100n$ is better than $(2^{n/10} - 1)100$ for $n \geq 60$. For $1 \leq n < 10$, $(2^{n/10} - 1)100$ is worse than n^2. At $n = 10$ they are the same. For $10 < n < 43$, n^2 is worse than $(2^{n/10} - 1)100$. For $n \geq 43$, $(2^{n/10} - 1)100$ is worse than n^2. Here are the graphs:

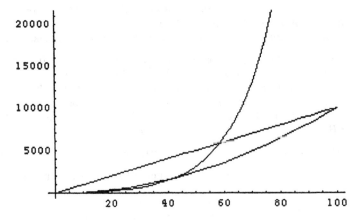

(b) When n is very large, B is fastest and C is slowest. This is because, of two polynomials the one with the lower degree is eventually faster and an exponential function grows faster than any polynomial.

Solutions for Basic Concepts in Graph Theory

GT-4.2 (a) The most direct way to prove this is to use Example 23. additional observations on Θ and O.
$$\lim_{n\to\infty} \frac{g(n)}{f(n)} = C > 0 \quad \text{implies} \quad g(n) \text{ is } \Theta(f(n))$$

Let $p(n) = \sum_{i=0}^{k} b_i n^i$ with $b_k > 0$. Take $f(n) = p(n)$, $g(n) = n^k$ and $C = b_k > 0$. Thus, $p(n)$ is $\Theta(n^k)$, hence the equivalence class of each is the same set: $\Theta(p(n))$ is $\Theta(n^k)$.

(b) $O(p(n))$ is $O(n^k)$ follows from (a).

(c) $\lim_{n\to\infty} p(n)/a^n = 0$. This requires some calculus. By applying l'Hospital's Rule k times, we see that the limit is $\lim_{n\to\infty} (k!/(\log(a))^k)/a^n$, which is 0. Any algorithm with exponential running time is eventually much slower than a polynomial time algorithm.

(d) For $p(n)$ to be $\Theta(a^{Cn^k})$, we must have positive constants A and B such that $A \leq a^{p(n)}/a^{Cn^k} \leq B$. Taking logarithms gives us $\log_a A \leq p(n) - Cn^k \leq \log_a B$. The center of this expression is a polynomial which is not constant unless $p(n) = Cn^k + D$ for some constant D, the case which is ruled out. Thus $p(n) - Cn^k$ is a nonconstant polynomial and so is unbounded.

GT-4.3 Here is a general method of working this type of problem:

Let $p(n) = \sum_{i=0}^{k} b_i n^i$ with $b_k > 0$. Show using definition that $\Theta(p(n))$ is $\Theta(n^k)$.
Let $s = \sum_{i=0}^{k-1} |b_i|$ and assume that $n \geq 2s/b_k$. We have

$$|p(n) - b_k n^k| \leq \left|\sum_{i=0}^{k-1} b_i n^i\right| \leq \sum_{i=0}^{k-1} |b_i| n^i \leq \sum_{i=0}^{k-1} |b_i| n^{k-1} = sn^{k-1} \leq b_k n^k/2.$$

Thus $|p(n)| \geq b_k n^k - b_k n^k/2 \geq (b_k/2) n^k$ and also $|p(n)| \leq b_k n^k + b_k n^k/2 \leq (3b_k/2) n^k$.

The definition is satisfied with $N = 2s/b_k$, $A = (b_k/2)$ and $B = (3b_k/2)$. If you want to show, using the definition, that $\Theta(p(n))$ is $\Theta(Kn^k)$ for some $K > 0$, replace A with $A' = A/K$ and B with $B' = B/K$.

In our particular cases we can be sloppy and it gets easier. Take (a) as an example.

(a) For $g(n) = n^3 + 5n^2 + 10$, choose N such that $n^3 > 5n^2 + 10$ for $n > N$. You can be ridiculous in the choice of N. $N^3 > 5N^2 + 10$ is valid if $1 > 5/N + 10/N^3$. $N = 10$ is plenty big enough. If $n^3 > 5n^2 + 10$ then $n^3 < g(n) < 2n^3$. So taking $A = 1$ and $B = 2$ works for the definition: $An^3 < g(n) < Bn^3$ showing g is $\Theta(n^3)$. If you want to use $f(n) = 20n^3$ as the problem calls for, replace these constants by $A' = A/20$ and $B' = B/20$. Thus, $A'(20n^3) < g(n) < B'(20n^3)$ for $n > N$.

This problem should make you appreciate the much easier approach of Example 23.

GT-4.4 (a) There is an explicit formula for the sum of the squares of integers.

$$\sum_{i=1}^{n} i^2 = \frac{n(n+1)(2n+1)}{6}.$$

This is a polynomial of degree 3, hence the sum is $\Theta(n^3)$.

Solutions for Basic Concepts in Graph Theory

(b) There is an explicit formula for the sum of the cubes of integers.

$$\sum_{i=1}^{n} i^3 = \left(\frac{n(n+1)}{2}\right)^2.$$

This is a polynomial of degree 4, hence the sum is $\Theta(n^4)$.

(c) To show the $\sum_{i=1}^{n} i^{1/2}$ is $\Theta(n^{3/2})$ it helps to know a little calculus. You can interpret the integral as upper and lower Riemann sum approximations to the integral of $f(x) = x^{1/2}$ with $\Delta x = 1$:

$$\int_0^n f(x)\, dx < \sum_{i=1}^{n} i^{1/2} = \sum_{i=1}^{n-1} i^{1/2} + n^{1/2} < \int_1^n f(x)\, dx + n^{1/2}.$$

Since $\int x^{1/2}\, dx = 2x^{3/2}/3 + C$. You can fill in the details to get $\Theta(n^{3/2})$.

The method used in (c) will also work for (a) and (b). The idea works in general: Suppose $f(x) \geq 0$ and $f'(x) > 0$. Let $F(x)$ be the antiderivative of $f(x)$. If $f(n)$ is $O(F(n))$, then $\sum_{i=0}^{n} f(n)$ is $\Theta(F(n))$. There is a similar result if $f'(x) < 0$: replace "$f(n)$ is $O(F(n))$" with "$f(1)$ is $O(F(n))$."

GT-4.5 (a) To show $\sum_{i=1}^{n} i^{-1}$ is $\Theta(\log_b(n))$ for any base $b > 1$ use the Riemann sum trick from the previous exercise. $\int_1^n x^{-1}\, dx = \ln(x)$. This shows that $\sum_{i=1}^{n} i^{-1}$ is $\Theta(\log_e(n))$. But, $\log_e(x) = \log_e(b) \log_b(x)$ (as we learned in high school). Thus, $\log_e(x)$ and $\log_b(x)$ belong to the same Θ equivalence class as they differ by a positive constant multiple $\log_e(b)$ (recall $b > 1$).

(b) First you need to note that $\log_b(n!) = \sum_{i=1}^{n} \log_b(i)$. Use the Riemann sum trick again.

$$\int_1^n \log_b(x)\, dx = \log_b(e) \int_1^n \log_e(x)\, dx = \log_b(e)\bigl(n\ln(n) - n + 1\bigr).$$

Thus, the sum is $\Theta(n\ln(n) - n + 1)$ which is $\Theta(n\ln(n))$ which is $\Theta(n\log_b(n))$.

(c) Use Stirling's approximation for $n!$, $n!$ is asymptotic to $(n/e)^n (2\pi n)^{1/2}$. Thus, $n!$ is $\Theta((n/e)^n (2\pi n)^{1/2})$, by Example 23. Do a little algebra to rearrange the latter expression to get $\Theta((n/e)^{n+1/2})$.

GT-4.6 A single execution of "C(i,j) = C(i,j) + A(i,k)*B(k,j)" takes a constant amount of time and so its time is $\Theta(1)$.
The loop on k is done n times and so its time is $n\Theta(1)$, which is $\Theta(n)$.
The loop on j is done n times and each time requires work that is $\Theta(n)$. Thus its time is $n\Theta(n)$, which is $\Theta(n^2)$.
The loop on I is done n times and so its time is $n\Theta(n^2)$, which is $\Theta(n^3)$.
Alternatively, you could notice that innermost loops take the most time and "C(i,j) = C(i,j) + A(i,k)*B(k,j)" is executed once for each value of i, j, and k. Thus it is done n^3 times and so the time for the algorithm is $\Theta(n^3)$.

GT-4.7 We use the Master Theorem. Since there is just one recursive call, $w = 1$ and $s_1(n) = q$. Since $0 \leq n/2 - q \leq 1/2$, $c = 1/2$. We have $T(n) = a_n + T(s_1(n))$ where a_n is 1 or 2. Thus a_n is $\Theta(n^0)$. In summary, $w = 1$, $c = 1/2$ and $b = 0$. Thus $d = -\log(1)/\log(1/2) = 0$ and so $T(n)$ is $\Theta(\log n)$.

Notation Index

(Page references herein refer to book's chapter numbering system.)

$\exists$ (there exists) Fn-4
$\forall$ (for all) Fn-4
$\ni$ (such that) Fn-4
B_n (Bell numbers) CL-27
$s \sim t$ (equivalence relation) GT-5
$\binom{n}{k}$ (binomial coefficient) CL-15
$\binom{n}{m_1, m_2, \ldots}$ (multinomial coefficient) CL-20
$BFE(T)$ (breadth first vertex sequence) DT-8, GT-29
$BFV(T)$ (breadth first vertex sequence) DT-8, GT-29
$C(n, k)$ (binomial coefficient) CL-15
$Cov(X, Y)$ (covariance) Fn-25
$DFV(T)$ (depth first vertex sequence) DT-8, GT-29
$x|y$ (x divides y) GT-24
$DFE(T)$ (depth first edge sequence) DT-8, GT-29
μ_X (expectation or mean) Fn-24
$E(X)$ (expectation) Fn-24
$f \circ g$ (composition) Fn-7
$(n)_k$ (falling factorial) CL-9
F_n (Fibonacci numbers) DT-48
$\lfloor x \rfloor$ (floor) DT-50
(V, E) (simple graph) GT-2
(V, E, ϕ) (graph) GT-2
$\mathbb{N}$ (natural numbers) CL-13
$\underline{n}$ (first n integers) Fn-1
$O(\ )$ (Big oh notation) GT-38
$o(\ )$ (little oh notation) GT-40
$\mathcal{P}_k(A)$ (k-subsets of A) CL-15, Fn-1
$S(A)$ (permutations of A) Fn-7
$PER(A)$ (permutations of A) Fn-7

Probability notation
  μ_X (expectation, or mean) Fn-24
  $\rho(X, Y)$ (correlation) Fn-25
  σ_X (standard deviation) Fn-25
  $E(X)$ (expectation) Fn-24
  $Cov(X, Y)$ (covariance) Fn-25
  $Var(X)$ (variance) Fn-25
  $P(A|B)$ (conditional probability) DT-27
$POSV(T)$ (postorder sequence of vertices) DT-8
$PREV(T)$ (preorder sequence of vertices) DT-8
$\mathbb{Q}$ (rational numbers) Fn-1
$\mathbb{R}$ (real numbers) CL-28, Fn-1
$\rho(X, Y)$ (correlation) Fn-25
Set notation
  $\sim A$ (complement) CL-14, Fn-1
  $\in$ and $\notin$ (in and not in) CL-14
  A' (complement) CL-14, Fn-1
  $A - B$ (difference) CL-14, Fn-1
  $A \cap B$ (intersection) CL-14, Fn-1
  $A \cup B$ (union) CL-14, Fn-1
  $A \oplus B$ (symmetric difference) Fn-1
  $A \setminus B$ (difference) CL-14, Fn-1
  $A \subseteq B$ (subset) CL-14
  $A \times B$ (Cartesian product) CL-4, Fn-1
  A^c (complement) CL-14, Fn-1
  $\mathcal{P}_k(A)$ (k-subsets of A) CL-15, Fn-1
  $|A|$ (cardinality) CL-3, CL-14
σ_X (standard deviation) Fn-25
$S(n, k)$ (Stirling numbers) CL-25
$\Theta(\ )$ (rate of growth) GT-38
$Var(X)$ (variance) Fn-25
$\mathbb{Z}$ (integers) CL-13, Fn-1

Subject Index

(Page references herein refer to book's chapter numbering system.)

Absorption rule CL-15
Adjacent vertices GT-3
Algebraic rules for
  sets CL-15
Algorithm
  backtracking DT-7
  divide and conquer DT-16,
    GT-45
  Kruskal's (minimum weight
    spanning tree) GT-32
  lineal (= depth-first) spanning
    tree GT-33
  partial GT-45
  polynomial time (tractable) GT-43
  Prim's (minimum weight
    spanning tree) GT-31
  which is faster? GT-43
Antisymmetric binary
  relation GT-24
Associative rule CL-15
Asymptotic GT-40
Average running time GT-41

Backtracking DT-7
Base (simplest) cases for
  induction DT-42
Bayes' Theorem DT-28, DT-32
Bell numbers CL-27
Bicomponents GT-22
Biconnected components GT-22
Bijection Fn-3
Binary relation GT-5
  antisymmetric GT-24
  covering GT-24
  equivalence relation GT-5
  order relation GT-24
  reflexive GT-5
  symmetric GT-5
  transitive GT-5

Binary tree GT-36
  full GT-36
Binomial coefficients CL-15
  recursion CL-23
Binomial distribution Fn-34
Binomial theorem CL-18
Bipartite graph GT-22
  cycle lengths of GT-34
Blocks of a partition CL-20, CL-25,
  Fn-15
Boolean variables DT-36
Breadth first vertex (edge)
  sequence DT-8, GT-29

Card hands
  and multinomial coefficients CL-23
  full house CL-19
  straight CL-26
  two pairs CL-19
Cardinality CL-3
Cardinality of a set CL-14
Cartesian product CL-4, Fn-1
Central Limit Theorem Fn-38
Characteristic equation DT-47
Characteristic function DT-23
Chebyshev's inequality Fn-27
Child vertex DT-2, GT-27
Chromatic number GT-41, GT-44
Circuit in a graph GT-18
  Eulerian GT-20
Clique GT-44
Clique problem GT-44
Codomain (range) of a
  function Fn-2
Coimage of a function Fn-14
Coloring a graph GT-41, GT-44
Coloring problem GT-43

[241]

Index

Commutative rule   CL-15
Comparing algorithms   GT-43
Complement of a set   Fn-1
Complete simple graph   GT-15
Component connected   GT-19
Composition of an integer   CL-8
Composition of functions   Fn-7
Conditional probability   DT-27
Conjunctive normal form   DT-36
Connected components   GT-19
Correlation   Fn-25
Covariance   Fn-25
Covering relation   GT-24
Cycle in a graph   GT-18
   Hamiltonian   GT-21
Cycle in a permutation   Fn-9

Decision tree   DT-1
   see also Rooted tree
   ordered tree is equivalent   GT-27
   probabilistic   DT-30
   RP-tree is equivalent   GT-27
   Towers of Hanoi   DT-18
   traversals   DT-8, GT-28
Decreasing (strictly) function or
   list   Fn-17
Decreasing (weakly) function or
   list   Fn-17
Degree of a vertex   DT-2, GT-4
Degree sequence of a graph   GT-4
DeMorgan's rule   CL-15
Density function   Fn-22
Depth first vertex (edge)
   sequence   DT-8, GT-29
Derangement   Fn-12
Deviation
   standard   Fn-25
Dictionary order   CL-4
Digraph   GT-14
   functional   GT-29

Direct (Cartesian) product   CL-4, Fn-1
Direct insertion order for
   permutations   DT-6
Directed graph   GT-14
Directed loop   GT-15
Disjunctive normal form   DT-36
Distribution   Fn-22
   binomial   Fn-34
   hypergeometric   CL-32
   joint   Fn-28
   marginal   Fn-28
   normal   Fn-36
   Poisson   Fn-35
   uniform   CL-28
Distribution function
   see Distribution
Distributive rule   CL-15
Divide and conquer   DT-16, GT-45
Domain of a function   Fn-2
Domino covering   DT-11
Double negation rule   CL-15
Down degree of a vertex   DT-2

Edge   DT-2, GT-2
   directed   GT-14
   incident on vertex   GT-3
   loop   GT-4, GT-11
   parallel   GT-11
Edge sequence
   breadth first   DT-8, GT-29
   depth first   DT-8, GT-29
Elementary event   CL-29
Envelope game   Fn-2
Equation
   characteristic   DT-47
Equivalence class   GT-5
Equivalence relation   GT-5
Error
   percentage   CL-10
   relative   CL-10
Eulerian circuit or trail   GT-20

Index

Event   CL-28, Fn-21
   elementary=simple   CL-29
   independent pair   Fn-29, DT-28
Expectation of a random
      variable   Fn-24

Factorial
   falling   CL-9
Factorial estimate (Stirling's
   formula)   CL-10
Falling factorial $(n)_k$ CL-9
Fibonacci recursion   DT-48
First Moment Method   DT-37
Full binary tree   GT-36
Function
   bijection   Fn-3
   characteristic   DT-23
   codomain (range) of   Fn-2
   coimage of   Fn-14
   composition of   Fn-7
   decreasing: decision tree   DT-14
   density   Fn-22
   distribution, see Distribution
   domain of   Fn-2
   generating   CL-16
   image of   Fn-14
   image of and Stirling numbers
     (set partitions)   Fn-15
   injective (one-to-one)   Fn-3
   inverse   Fn-3
   inverse image of   Fn-14
   monotone   Fn-17
   one-line notation   Fn-2
   partial   DT-3
   probability   Fn-21
   range of   Fn-2
   restricted growth and set
     partitions   Fn-20
   strictly decreasing   Fn-17
   strictly increasing   Fn-17
   surjective (onto)   Fn-3
   two-line notation   Fn-5
   weakly decreasing   Fn-17
   weakly increasing   Fn-17

Functional relation   Fn-4

Gambler's ruin problem   DT-51
Generating function   CL-16
Geometric probability   CL-34
Geometric series   DT-50
Graph   GT-2
   see also specific topic
   biconnected   GT-22
   bipartite   GT-22
   bipartite and cycle
     lengths   GT-34
   complete simple   GT-15
   connected   GT-19, GT-19
   directed   GT-14
   incidence function   GT-3
   induced subgraph (by edges or
     vertices)   GT-17
   isomorphism   GT-7
   oriented simple   GT-24
   random   GT-7
   rooted   GT-27
   simple   GT-2
   subgraph of   GT-17
Gray code for subsets   DT-23
Growth
   rate of, see Rate of growth

Hamiltonian cycle   GT-21
Hasse diagram   GT-24
Height of a tree   GT-35
Height of a vertex   DT-3
Hypergeometric probability   CL-32

Idempotent rule   CL-15
Identity permutation   Fn-7
Image of a function   Fn-14
   Stirling numbers (set partitions)
     and   Fn-15
Incidence function of a graph   GT-3

Index

Inclusion and exclusion   CL-31, CL-39
Increasing (strictly) function or list   Fn-17
Increasing (weakly) function or list   Fn-17
Independent events   Fn-29, DT-28
Independent random variables   Fn-29
Induced subgraph (by edges or vertices)   GT-17
Induction   Fn-8, DT-41
  base (simplest) cases   DT-42
  induction hypothesis   DT-42
  inductive step   DT-42
Inequality
  Tchebycheff   Fn-27
Injection   Fn-3
Internal vertex   DT-2, GT-27
Intersection of sets   Fn-1
Inverse image of a function   Fn-14
Involution   Fn-10
Isolated vertex   GT-11
Isomorph rejection   DT-14
Isomorphic graphs   GT-7

Joint distribution function   Fn-28

Kruskal's algorithm for minimum weight spanning tree   GT-32

Leaf vertex   DT-2, GT-27
  rank of   DT-4
Lexicographic order (lex order)   CL-4

List   CL-2
  circular   CL-10
  strictly decreasing   Fn-17
  strictly increasing   Fn-17
  weakly decreasing   Fn-17
  weakly increasing   Fn-17
  with repetition   CL-3
  without repetition   CL-3, CL-9
  without repetition are injections   Fn-3
Little oh notation   GT-40
Local description   DT-16
  Gray code for subsets   DT-25
  merge sorting   DT-15
  permutations in lex order   DT-17
  Towers of Hanoi   DT-19
Loop   GT-4, GT-11
  directed   GT-15

Machine independence   GT-37
Marginal distribution   Fn-28
Matrix
  permutation   Fn-11
Merge sorting   DT-15, GT-45
Merging sorted lists   DT-15
Monotone function   Fn-17
Multinomial coefficient   CL-20
Multiset   CL-3
  and monotone function   Fn-17

Nondecreasing function or list   Fn-17
Nonincreasing function or list   Fn-17
Normal distribution   Fn-36
Normal form
  conjunctive   DT-36
  disjunctive   DT-36
NP-complete problem   GT-44
NP-easy problem   GT-44
NP-hard problem   GT-44

Index

Numbers
    Bell CL-27
    binomial coefficients CL-15
    Fibonacci DT-48
    Stirling (set partitions) CL-25, Fn-15

Odds CL-32
One-line notation Fn-2
One-to-one function (injection) Fn-3
Onto function (surjection) Fn-3
Order
    direct insertion for permutations DT-6
    lexicographic (lex) CL-4
Order relation GT-24
Oriented simple graph GT-24

Parallel edges GT-11
Parent vertex DT-2, GT-27
Partial function DT-3
Partition
    set CL-25, Fn-14
    set (ordered) CL-20
    set and restricted growth function Fn-20
Path in a (directed) graph GT-16
Permutation CL-3, Fn-3, Fn-7
    cycle Fn-9
    cycle form Fn-9
    cycle length Fn-9
    derangement Fn-12
    direct insertion order DT-6
    identity Fn-7
    involution Fn-10
    is a bijection Fn-3
    matrix Fn-11
    powers of Fn-7
    random generation Fn-33
Poisson distribution Fn-35
Polynomial multiplication GT-47

Polynomial time algorithm (tractable) GT-43
Postorder sequence of vertices DT-8
Preorder sequence of vertices DT-8
Prime factorization DT-42
Prim's algorithm for minimum weight spanning tree GT-31
Probabilistic decision tree DT-30
Probability
    conditional DT-27
    conditional and decision trees DT-30
    function CL-28
    probability space CL-28
Probability distribution function
    see Distribution
Probability function CL-28, Fn-21
    see also Distribution
Probability space CL-28, Fn-21
    see also Distribution

Random generation of permutations Fn-33
Random graphs GT-7
Random variable Fn-22
    binomial Fn-34
    correlation of two Fn-25
    covariance of two Fn-25
    independent pair Fn-29
    standard deviation of Fn-25
    variance of Fn-25
Range of a function Fn-2
Rank (of a leaf) DT-4
Rate of growth
    Big oh notation GT-38
    comparing GT-43
    exponential GT-43
    little oh notation GT-40
    polynomial GT-40, GT-43
    Theta notation GT-38
Rearranging words CL-20
Recurrence
    see Recursion

Index

Recursion DT-43
  see also Recursive procedure
  binomial coefficients CL-23
  Fibonacci DT-48
  guessing solutions DT-45
  inductive proofs and DT-40
  set partitions (Bell
    numbers) CL-27
  set partitions (Stirling
    numbers) CL-25
  sum of first n integers DT-42

Recursive equation
  see Recursion

Recursive procedure
  see also Recursion
  0-1 sequences DT-15
  Gray code for subsets DT-25
  merge sorting DT-15
  permutations in lex
    order DT-17
  Towers of Hanoi DT-19

Reflexive relation GT-5

Relation Fn-4
  see perhaps Binary relation

Relative error CL-10

Restricted growth function and set
  partitions Fn-20

Root DT-2

Rooted graph GT-27

Rooted tree
  child DT-2, GT-27
  down degree of a vertex DT-2
  height of a vertex DT-3
  internal vertex DT-2, GT-27
  leaf DT-2, GT-27
  parent DT-2, GT-27
  path to a vertex DT-3
  siblings GT-27

RP-tree (rooted plane tree)
  see Decision tree

Rule
  absorption CL-15
  associative CL-15
  commutative CL-15
  DeMorgan's CL-15
  distributive CL-15
  double negation CL-15
  idempotent CL-15

Rule of Product CL-3

Rule of Sum CL-5

Sample space CL-28, Fn-21

SAT problem DT-36

Satisfiability problem DT-36

Sequence CL-2

Series
  geometric DT-50

Set CL-2, CL-14
  algebraic rules CL-15
  and monotone function Fn-17
  cardinality CL-3
  cardinality of CL-14
  Cartesian product CL-14
  complement CL-14
  complement of Fn-1
  difference CL-14
  intersection CL-14
  intersection of two Fn-1
  partition, *see* Set partition
  subset CL-14
  subsets of size k CL-15
  symmetric difference CL-14
  symmetric difference of
    two Fn-1
  union CL-14
  union of two Fn-1
  with repetition (multiset) CL-3

Set partition CL-25, Fn-14
  ordered CL-20
  recursion (Bell numbers) CL-27
  recursion (Stirling
    numbers) CL-25
  restricted growth function Fn-20

Simple event CL-29

Index

Simple graph GT-2
Simplest (base) cases for
    induction DT-42
Sorting (merge sort) DT-15, GT-45
Space
    probability CL-28
Spanning tree GT-30
    lineal (= depth first) GT-34
    minimum weight GT-31
Stacks and recursion DT-21
Standard deviation Fn-25
Stirling numbers (set
    partitions) CL-25
    image of a function Fn-15
Stirling's approximation for
    $n!$ CL-10
Strictly decreasing function or
    list Fn-17
Strictly increasing (or decreasing)
    function or list Fn-17
Strictly increasing function or
    list Fn-17
String
    see List
Subgraph GT-17
    cycle GT-18
    induced by edges or
    vertices GT-17
Subset of a set CL-14
Surjection Fn-3
Symmetric difference of sets Fn-1
Symmetric relation GT-5

Tchebycheff's inequality Fn-27

Theorem
    Bayes' DT-28, DT-32
    binomial coefficients CL-16
    binomial theorem CL-18
    bipartite and cycle
        lengths GT-34
    Central Limit Fn-38
    conditional probability DT-28
    correlation bounds Fn-26
    covariance when independent Fn-32
    cycles and multiple
        paths GT-18
    equivalence relations GT-5
    expectation is linear Fn-24
    expectation of a product Fn-32
    induction DT-41
    lists with repetition CL-3
    lists without repetition CL-9
    minimum weight spanning
        tree GT-31
    monotone functions and
        (multi)sets Fn-18
    permutations of set to fixed
        power Fn-10
    Prim's algorithm GT-31
    properties of Θ and O GT-38
    Rule of Product CL-3
    Rule of Sum CL-5
    Stirling's formula CL-10
    systematic tree traversal DT-9
    Tchebycheff's inequality Fn-27
    variance of sum Fn-32
    walk, trail and path GT-16
Towers of Hanoi DT-18
    four pole version DT-26
Tractable algorithm GT-43
Trail in a (directed) graph GT-16
Transitive relation GT-5
Traveling salesman problem GT-44
Traversal
    decision tree DT-8, GT-28

Index

Tree
    see also specific topic
    binary GT-36
    decision, see Decision tree
    height GT-35
    ordered tree, see Decision tree
    rooted, see Rooted tree
    RP-tree (rooted plane tree), see
        Decision tree
    spanning GT-30
    spanning, lineal (= depth
        first) GT-34
    spanning, minimum
        weight GT-31

Two-line notation Fn-5

Uniformly at random CL-28

Union of sets Fn-1

Variance Fn-25

Venn diagram CL-31

Vertex DT-2
    adjacent pair GT-3
    child DT-2, GT-27
    degree of DT-2, GT-4
    down degree of DT-2
    height of DT-3
    internal DT-2, GT-27
    isolated GT-11
    leaf DT-2, GT-27
    parent DT-2, GT-27

Vertex sequence GT-16
    breadth first DT-8, GT-29
    depth first DT-8, GT-29

Walk in a graph GT-16

Weakly decreasing function or
    list Fn-17

Weakly increasing function or
    list Fn-17

Words CL-11, CL-20